STATISTICS FOR MINING ENGINEERING

Statistics for Mining Engineering

Jacek M. Czaplicki

Mining Mechanization Institute, Silesian University of Technology, Gliwice, Poland

CRC Press
Taylor & Francis Group
Boca Raton London New York Leiden

CRC Press is an imprint of the
Taylor & Francis Group, an **informa** business

A BALKEMA BOOK

First issued in paperback 2017

CRC Press/Balkema is an imprint of the Taylor & Francis Group, an informa business, London, UK

© 2014 Taylor & Francis Group, London, UK

Typeset by Vikatan Publishing Solutions (P) Ltd., Chennai, India

Published by: CRC Press/Balkema
            P.O. Box 11320, 2301 EH Leiden, The Netherlands
            e-mail: Pub.NL@taylorandfrancis.com
            www.crcpress.com—www.taylorandfrancis.com

Library of Congress Cataloging-in-Publication Data

Czaplicki, Jacek M.
    Statistics for mining engineering / Jacek M. Czaplicki, Mining Mechanization Institute, Silesian University of Technology, Giliwice, Poland.
        p. cm.
    Includes bibliographical references and index.
    ISBN 978-1-138-00113-8 (hardback) -- ISBN 978-1-315-81503-9 (eBook PDF)
    1. Mining engineering--Statistical methods. I. Title.

TN272.7.C93 2014
519.5–dc23

                                                            2013043750

ISBN 13: 978-1-138-07558-0 (pbk)
ISBN 13: 978-1-138-00113-8 (hbk)

Contents

Preface and acknowledgements

The variety of technical devices that is available on the market is quite rich. The items offered have, in the majority of cases, the broad possibility of realizing their targets. Very often the devices that are selected are set up together, creating systems to execute predetermined functions. In mining engineering, mechanized systems are formed in order to excavate rocks, to haul broken material, to dump waste, to store excavated ore or to carry out ore treatment by mechanical, chemical or thermal means. The technical objects being used nowadays are complicated constructions. In order to apply them in an appropriate manner, we should follow the recommendations formulated by their producers and we should frequently observe the realization of their operation (exploitation)[1] process, which is understood here as the process of changes in their properties during the use of the object. Running changes are interesting first of all for the constructor and producer of the object. They are also interesting for the user of the equipment, who is concerned in its application to attain the best performance of the item purchased over the long term. Information gathered during the operation of the object creates statistical data.

Actually, the more complicated machines used in mining (and pieces of equipment from other engineering areas) are furnished with special systems which diagnose a continuously running machine, collect information on its exploitation events and send it to the main producer's computer. In many cases the information that is transmitted is very rich comprising a few tens of units of information. Again, this trio: the constructor, producer and user are interested in the accumulated material. And once again, this information has statistical character.

The information that is generated through the realization of the operation of a machine has a twofold meaning. It provides information on the technical parameters that are of interest at a given moment of time directly, e.g. power consumption, force applied during retardation, mass accumulated in the shovel etc. But very often there is concern about the average characteristics of the observed data, i.e. there is concern about the statistical characteristics of collected material.

It is easy to enumerate at least several further areas of mining engineering interest in which the important information that is gathered has a statistical character; investigations made during research on a stand in a lab generating records, collected data from an instrument with a new device installed and operating in real working conditions, sensors continuously tracing excessive shocks in a rock mass near mining production areas, information taken from an ore processing plant and connected with a given stage of ore treatment etc. All of these examples show areas that generate information of a statistical character. Information that should be analysed carefully; its synthesis should be conducted and statistical inference should be carried out.

The main goal of this book is to acquaint readers with a **basic knowledge** of mathematical statistics; knowledge that should be known by contemporary mining engineers. For this reason this book is addressed—first of all—to students of mining faculties and schools. The majority of the cases considered here are taken from the field of mine mechanization. However, there

---

[1]These two terms will be used here interchangeably, i.e. they are synonyms. Grounds for such an approach can be found in Chapter 1, Czaplicki 2010.

are also practical examples taken from different areas such as ore treatment, blasting, rock mechanics and some econometric problems associated with mining engineering.

Some parts of this book that deal with the reliability and exploitation problems of a variety of equipment will be interesting for reliability engineers. Similarly, some examples may be of interest for earthmoving engineers. Because of the universal character of the models and methods presented here, some parts may also be attractive for civil engineers, engineers involved in production as well as for engineers working at industrial institutes, especially mining ones. Students learning mathematical statistics and searching for practical applications of it will find some chapters interesting as well. Additionally, I hope that my academic colleagues working at mining universities will also find this book interesting, and perhaps useful in their educational work.

It is presumed that readers of the book have knowledge of the basics of probability theory. Nevertheless, in order to ensure a better understanding of the text, a short repetition from this field is presented in Chapter 1. It is also recommended that readers have some idea of the theory of reliability and the theory of exploitation. There are actually many books related to both probability and reliability theories and therefore no specific book is suggested here. There is a different situation related to the theory of exploitation (terotechnology, to a certain extent). One suggested book is "Mining equipment and systems. Theory and practice of exploitation and reliability" (2010) and also written by the author of this book. In order to make the considerations conducted here more communicative, several basic probabilistic terms are defined. The author has made a special effort to present definitions in such a form so as to be immediately understandable for even those readers with less knowledge of probability theory. Where terms of mathematical statistics are concerned, a special chapter at the end has been added where the majority of the essential terms are defined.

In the entire text terms and notations taken from the field of engineering that interlace with statistical terminology are commonly applied. In order to make the text more understandable examples are given; in many cases taken from author's practice. Data have been collected from different mines, different equipment producers and from several different countries. However, in several cases, data and results are taken from books and dissertations of colleagues; and the author is very grateful for their acceptance of the use of their great works in this book.

The construction of the examples is similar: the goal of the investigation is formulated, an examination is conducted and there is a need to work with the outcomes to get proper interpretation of the properties of the reality observed. It is necessary to apply the appropriate mathematical tools from mathematical statistics in order to obtain suitable results from the investigation. The next step is very sensitive: how to properly interpret the outcomes? This means that the information acquired must be translated into engineering language.

Often, during statistical inference, attention is paid to the nuances of the methods applied, to subtleties contained in the observations and to sensitive points of interference. This enriches the information that is the result of the investigation.

I would like to express my very warm thanks to Prof. Leon Dziembała, who passed away just recently, for support, Ms Michele Simmons for proofreading and Janjaap Blom, Senior Publisher CRC Press / Balkema—Taylor & Francis Group and his open and extremely professional team for their reliable and efficient cooperation.

Jacek M. Czaplicki
Mining Mechanization Institute
Silesian University of Technology, Gliwice, Poland

About the author

Jacek M. Czaplicki has been an academic lecturer for forty years and is continuously associated with his home University. He did, however, leave his school for a couple of years' lecturing in African universities.

He worked for three years at the School of Mines of the Kwara State College of Technology, Ilorin, Nigeria on a UNESCO project. A few years later, he was appointed to Zambia Consolidated Copper Mines Ltd for three years and worked as a lecturer at the School of Mines of the University of Zambia, Lusaka as part of a World Bank project.

Jacek Czaplicki received a Master of Science in Mine Mechanization from the Silesian University of Technology, Gliwice, Poland. He also obtained a Doctorate degree in Technical Sciences. Later, he submitted a dissertation and passed all of the requirements and was awarded a D.Sc. degree in Mining and Geological Engineering with a specialization in Mine Machinery at the same home University. Currently, he is a Professor (emeritus) of Mining Engineering.

He has published about a hundred and forty papers and more than ten books in Poland and abroad. His specialisation comprises mine transport, the reliability and computation of mine machinery systems and the reliability of hoist head ropes. In his research, he applies methods and models from mathematical statistics. He is an internationally recognised specialist in mine mechanisation.

List of major notations

$E(X)$ – expected value, mean value, mathematical hope of the random variable X

$F(n, m)$ – F Snedecor distribution with $F(n, m)$ degrees of freedom

H_0, H_1 – statistical hypotheses: basic and alternative

n – sample size

$N(m, \sigma)$ – normal distribution with parameters m and σ

$N(0, 1)$ – standardised normal distribution with parameters $m = 0$ and $\sigma = 1$

r_S, r'_S – Spearman correlation coefficient

R – linear Pearson correlation coefficient

R – non-linear correlation coefficient

$t(n)$ – Student's t probability distribution with n degrees of freedom

u – residual

u_α – quantile of order α standardised normal distribution $N(0, 1)$

z – Fisher's distribution

α – level of significance

$\phi_N(x)$ – probability density function of standardised normal distribution $N(0, 1)$

$\Phi_N(x)$ – probability cumulative function of standardised normal distribution $N(0, 1)$

$\chi^2(n)$ – χ^2 (chi-squared) distribution with n degrees of freedom

$\sigma^2(X)$ – variance of the random variable X

ρ – correlation coefficient in general population

$:$ – has distribution

Each example considered has following signs:

- ■ – beginning of example
- ◄ – end of example
- □ – end of a phase of consideration in the example; considerations will be continued

Random variables are usually marked in bold. Sets are usually marked by Frankenstein letters.

CHAPTER 1

Fundamentals

1.1 GOAL AND TASK OF STATISTICS

Originally, i.e. until the first half of the 19th century, the term statistics—a word taken from the Latin language: *status*—meant a set of data or numbers describing the state of a political body. The first notations in this regard were used in ancient China and Egypt and later in the Roman Empire. Such data were very useful and this was confirmed once again in the Middle Ages. Later on, the scope of the term was enlarged to include all sets of data. In the 20th century, methods for the analysis of the collected information were incorporated.

There is a difference in meaning in the English language as to whether the word *statistics* is used as a singular or plural noun.[1]

Statistics means:

a. A **discipline of science**, the branch of mathematics that deals with quantitative methods to investigate mass phenomena
b. A **set of data** that is a picture of the running changes of some phenomena (processes) in a certain space.

Statistics relates to status. Statistic means a characteristic describing the determined properties of a set of numbers. This characteristic is in fact an estimator (**function**) of an unknown parameter of the general population being investigated through taking a sample.

Statistics, as a science dealing with the methods used to gain, to present and to analyse data, has the **main goal** of obtaining generalised useful information about some phenomenon or property.

Such a phenomenon must have a mass character, i.e. it should be connected with a large number of cases. In such a situation, some regularity can be traced. In every case, this regularity is the result of the action of some causes; causes that create a set. Regularities of such a nature are called statistical regularities, i.e. they cannot usually be found by observing only one, singular event.

The reasons that generate statistical regularity can be divided into the:

• main ones or
• random ones.

The main reasons create a systematic component whereas the random reasons create a stochastic component.

The mass character of a phenomenon does not give a ground to apply statistical methods. These methods can only be used when the set is composed of similar elements that have non-identical properties.

The first stage of a statistical investigation consists of the construction of an investigation plan. It is composed of the goal and subject of the examination planned. The elements of the statistical population have to be identified together with their properties, which are

---

[1] There is no such difference in some other languages.

1

slightly different for each element. These are called statistical properties. These features can be:

- measurable (such as: mass, power, the number of hoists per day, the time of repair)
- immeasurable (e.g. the machine type, the type of conveyance applied, the existence or non-existence of a certain furnishing item in a given system).

Measurable properties can be:

- discrete (the number of cycles of a machine per unit of time, the number of device failures per tonne of mass transported etc.)
- continuous (for example: the time of renewal, the age of a piece of equipment)
- quasi-measurable (making order).

Quasi-measurable properties determine the intensity of the investigated feature in the population. Examples of quasi-measurable properties can be incomes or wages, which to a certain limit have continuous character, but later on become discrete in nature.

Each statistical unit can be described by one feature or many features, depending on the investigation goal formulated. In this context, we have one-dimensional populations or multi-dimensional ones.

Another division of statistical properties distinguishes:

- constant features
- variable features.

Constant features are described as answering the questions: What? When? Where? These features are common for all elements of the population of interest. They are not the subject of the investigation; they decide whether these elements belong to the population. The features that are the subject of investigation are variable ones.

When the subject of an investigation is fixed and the method of gathering data is decided, the next step is data collection and later—their control and ordering in a convenient way, grouping together, segregating etc. Such material is the input for the analysis of information accumulated.

To summarise—every statistical investigation consists of the following phases:

1. preparation
2. observation (data collection)
3. work with the gathered information
4. analysis
5. inference.

The main **task of mathematical statistics** is the analysis and interpretation of the results that arise from the application of appropriate methods.

The population whose elements are observed is called a sample, i.e. it is a subpopulation (subset) of the general population. Commonly, a sample is understood as a small part of anything or one of a number, which is intended to show the quality, style or nature of the whole; the specimen. However, the set observed is not always a trail in the statistical sense.

The decision to take only a sample has different grounds, e.g.:

- the general population is infinite in number (or almost infinite) and there is no possibility of testing all of the elements
- the investigation has a destructive character; to test all of the elements would to destroy the whole population; it makes no sense
- the cost of the testing of a large number of elements is very high
- we are interested in only a rough estimation of a certain property of the population; it makes no sense to test many elements.

The conclusions on the general population—formulated on a sample taken—will be correct if the sample is similar to the general population. This similarity means that the **sample is representative**. Errors due to the observation of a certain part of the population only are subject to the probability calculus law. It allows the value of an inaccuracy that is made to be estimated. However, one condition must be fulfilled here—the **sample has to be random**. This means that which elements will be in a sample depends on a purely chance event. From a statistical point of view, this is an element selection in which two conditions are fulfilled:

1. every elementary element of the population has an equal chance of being selected; a chance that has a positive probability
2. the possibility exists of estimating the probability of every subpopulation as a sample.

Mathematical statistics deals with the construction of rules to determine the properties of the general population based on the sample taken.

Taking into account the type of topic considered, the tasks of mathematical statistics can described as:

- the estimation of unknown parameters of the investigated feature of a statistical unit; this feature is treated as a random variable
- the verification of statistical hypotheses
- the identification of random processes
- decision-making in the case of uncertainty.

For proper understanding and for the appropriate usage of methods of statistical inference later on, it is necessary to have elementary information from probability calculus and the theory of random variables. Let us make a short review in this regard.

1.2   BASIC TERMS OF PROBABILITY THEORY

1.2.1   *Random events and some definitions of probability*

The foundation of mathematical statistics is the theory of probability which explains the 'mechanisms' that disclose statistical properties—the regularities occurring in mass phenomena.

In mining engineering we deal with results of some magnitude measurements or observations based on some experiments. Usually these experiments have a random character. A set that contains the effects of these experiments is called a **sample space** and every subset of it is called a **random event**. If a particular subset comprises indivisible elements we call it a set of **elementary events**. A set of subsets of elementary events is called a **Borel field of sets** whose elements are random events. A Borel field consists of:

- Elementary events
- Sum, product and difference of events
- Sure event
- Impossible event.

Due to the fact that events can be treated as elements of a set—let us recall briefly a few basic terms from event calculus.

The **sum of events** $A_1, A_2, ..., A_k$ is such event A

$$A = \bigcup_{i=1}^{k} A_i,$$

(1.1)

which consists of all of the elementary events belonging to at least one of the sets $A_1, A_2, ..., A_k$.

The sum of events is generalised for an infinite sequence of events.

The product of events A_1, A_2, ..., A_k is such event A

$$A = \bigcap_{i=1}^{k} A_i,$$ (1.2)

whose elements are common (joint) for all events A_1, A_2, ..., A_k.

The product of events is generalised for an infinite sequence of events.[2]

The difference between two events A_1 and A_2 is the event A

$$A = A_1 - A_2$$ (1.3)

that consists of the elementary events belonging to A_1 but not to A_2.

It follows from the symmetrical character of these definitions of the sum and the product that the operations of addition and multiplication are commutative, i.e. we have

$$A_1 \cup A_2 = A_2 \cup A_1 \quad \text{and} \quad A_1 \cap A_2 = A_2 \cap A_1$$

An important and particular case arises when A_1 coincides with the whole space $\mathbb{C}$. The difference

$$A_1^* = \mathbb{C} - A_1$$ (1.4)

is the set of all of the elements of a space which do not belong to A_1 and this event is called a **complementary event** or simply a complement of A_1.

The operations of addition and multiplication may be brought into relation with one another by means of the idea of complementary sets. If there is a given finite or numerable sequence A_1, A_2, ... the following relations hold:

$$\left(A_1 \cup A_2 \cup ... \right)^* = A_1^* \cap A_2^* \cap ... \qquad \left(A_1 \cap A_2 \cap ... \right)^* = A_1^* \cup A_2^* \cup ...$$

A **sure event** is an event comprising the whole space of the elementary events, and for this reason for any event A the following relation holds:

$$A^* \cup A = \mathbb{C}$$ (1.5)

An **impossible event** is an event that cannot occur under given circumstances. Therefore, the following relation holds:

$$\mathbb{C} = \Phi^*$$ (1.6)

Two events having a joint part but with no elements are called **separate (disjoint) events**.

One of the cardinal terms of this theory is **probability**. It has been defined differently.

The development of the theory of probability can be traced to the second half of the 19th century. Its beginnings were like those of an illegitimate child. One famous gambling addict—Chevalier de Meré—wanted to increase his chances in gambling and asked a well-known

---

[2] Notice, that here (operating on sets or events) we have different notations when symbols of multiplication and summation are concerned. This is necessary to differentiate from acting on numbers (in algebra).

contemporary mathematician, Pascal, what to do to be successful in games of hazard. Pascal was positively astonished by the problem formulated. He gave no answer but he started to study the problem.

In addition to Pascal, Fermat, Bernoulli and Laplace also considered the problem of games of chance.

In 1812 Pierre Simon de Laplace formulated the so-called **classical definition of the probability** of the occurrence of a random event as a measure of the chance of the appearance of this event. He determined this measure as the ratio of the number of events that were favourable for the occurrence of this given event (events mutually excluding themselves) and the number of all possible events identically possible. For example, when tossing a dice, the probability of the occurrence of an even number is 3/6 provided that the dice is made properly (honestly), i.e. all of the walls of the dice have the same chance of occurrence.

This classic definition has several defects because it can only be applied when the sets of events considered are finite and the structure of these sets is known. Moreover, this definition is a tautological one. It defines the probability of a random event taking into account a set of identically possible events, which means that they have the same probability.[3]

Along with the classic definition, some other definitions were formulated afterwards. Finally, in 1933 Andrey Kolmogorov formulated the axiomatic definition and this definition has become the ground for the modern theory of probability. Originally, this definition read as follows:

'Let $\mathfrak{E}$ be a collection of elements which we shall call elementary events and $\mathfrak{F}$ be a set of subsets of $\mathfrak{E}$; the elements of set $\mathfrak{F}$ will be called random events.

Axioms:

1. $\mathfrak{F}$ is a field of sets.
2. $\mathfrak{F}$ contains the set $\mathfrak{E}$.
3. A non-negative real number $P(A)$ is assigned to each set A in $\mathfrak{F}$. This number is called the probability of event A.
4. $P(\mathfrak{E}) = 1$.
5. If A and B have no element in common, then

$$P(A \cup B) = P(A) + P(B).'$$

In the above set of axioms only a finite number of events were taken into account; however, in the same publication, Kolmogorov extended his reasoning to an infinite number of events.[4]

Based on the theory of sets, it can easily be concluded that event $\mathfrak{E}$ is a sure event and $P(\Phi) = 0$, if Φ denotes an impossible event.

The **primitive notion** (not definable) in the axiomatic approach to probability theory is the whole **space $\mathfrak{E}$ of elementary events**—this is the set of all of the possible elementary, indivisible results of experiments or observations.

The subsets of the space of elementary events that belong to the enumerable additive field of subsets are called **random events** or events—for short. All subsets of the set of elementary events are events if this set is finite or enumerable.[5]

The probability is a **non-negative real number** supported at a [0, 1] interval.

Some mathematicians, however, (Kopociński 1973, Grabski and Jaźwiński 2001 for instance), are of the opinion that probability is a **function** in which the arguments are random events and

---

[3] A more comprehensive consideration in this regard was presented, for instance, by Papoulis (1965).
[4] In modern mathematics: $\mathfrak{E}$ is a sample space, $\mathfrak{F}$ is an event space.
[5] More on the additive classes of sets can be found for instance in Cramer (1999, Chapter 1.6).

the values are real numbers and this function fulfils the above axioms. The probability is a number for a fixed event. This approach is more general than the original one.

It is worth adding here that where the definitions of probability are concerned, there are also some other theories dealing with probability, i.e. the theory of **subjective probability** and **Bayesian theory**. The first theory formalises the personal assessment of the chances of the occurrence of a given random event. In practical applications individuals assessing the chances of this appearance are selected purposely by presuming that their estimation will be significant, which is important because they have substantial knowledge in the case being considered. The second theory applies two terms – *a priori* **probability** for the description of knowledge about the investigated phenomenon before the investigation is done and *a posteriori* **probability** for the description of enriched knowledge because the results of investigation are known.

The **conditional probability** denoted by $P(A|B)$ is the probability of the occurrence of event A provided that event B has appeared (i.e. $P(B) > 0$). It is calculated as the quotient of the joint probability of A and B and the probability of B[6,7]

$$P(A \mid B) = \frac{P(A \cap B)}{P(B)} \tag{1.7}$$

The probability of the sum of two events A and B

$$P(A \cup B) = P(A) + P(B) - P(A \cap B) \tag{1.8}$$

We say that the random variable A is stochastically independent[8] of an event B if

$$P(A|B) = P(A) \tag{1.9}$$

This means that the occurrence of one does not affect the probability of the other. The concept of independence extends to any dealing with collections of more than two events or random variables.

Considering the above, we can easily see that for two **independent events** the following equation holds:

$$P(A \cap B) = P(A)\,P(B) \tag{1.10}$$

Remark. Events can be pairwise independent but simultaneously jointly dependent.

Consider a set $\{A_1, A_2, ..., A_k\}$ of pairwise disjoint events whose union is the entire space. For any event B of the same probability space, the following relation holds:

$$P(B) = \sum_{i=1}^{k} P(A_i)\,P(B \mid A_i) \tag{1.11}$$

This is the formula for the **total probability**.

---

[6] This is the so-called classical Kolmogorov definition. A different approach was presented by Bruno de Finetti who preferred to introduce conditional probability as an axiom of probability.

[7] Notice, that if for a certain event B, the corresponding $P(B) = 0$, the expression $P(A|B)$ is undefined. Nevertheless, it is possible to define a conditional probability with respect to a σ-algebra of such events (such as those arising from a continuous random variable).

[8] The following terms can be found in the literature: statistically independent, marginally independent or absolutely independent.

The summation can be interpreted as a weighted average and consequently the marginal probability. Probability $P(B)$ is sometimes called 'average probability' or 'overall probability'.[9]

This law usually has one common application where the events coincide with a discrete random variable taking each value in its range.

Consider a set $\{A_1, A_2, ..., A_k\}$ of pairwise disjoint events whose union is the entire space. If $P(A_i)$ are known and also the conditional probabilities $P(B|A_i)$ then the conditional probability

$$P(A_i \mid B) = \frac{P(B \mid A_i)P(A_i)}{\sum_{i=1}^{k} P(A_i)P(B \mid A_i)}$$ (1.12)

This is the so-called the **Bayes' Theorem**. Probability $P(A_i|B)$ is called *a posteriori* whereas probabilities $P(A_i)$ are called *a priori*.

1.2.2 *Random variables, distribution function and probability density function*

In our previous considerations, there was no specific meaning given to the event being observed. Actually, to every result of an experiment ξ a number will be ascribed which means a function is to be constructed $x(\xi)$. Notice, that the independent variable ξ will not be a number but an element of set 𝕰.

A real **random variable** X **is a function** supported on the space 𝕰 of random events if:

a. the set $\{X \leq x\}$ is an event for any real number x,
b. the following equations holds:

$$P\{X = \infty\} = P\{X = -\infty\} = 0$$ (1.13)

In other words, a measurable function assigning real numbers to every outcome of the experiment is called a random variable.

Random variables will be marked in bold.

A random variable is a discrete one if it is supported by a finite or enumerable set of numbers. Examples of probability distributions for discrete variables will be given in Chapter 1.2.5.

In order to characterise a random variable, it is necessary to determine a set of its possible values and the corresponding probabilities.

A function $F(x)$, which is defined as the probability of an event $\{X \leq x\}$, is called a **distribution** (distribution function, cumulative function) of the random variable X, i.e.

$$F_X(x) = P\{X \leq x\}$$ (1.14)

The distribution is a non-decreasing monotonic function, continuous on the left and—as a probability—supported by a [0, 1] set.

If a distribution $F_X(x)$ of random variable X can be defined as

$$F_X(x) = \int_{-\infty}^{x} f_X(u)\, du$$ (1.15)

then the random variable X is continuous, its distribution is continuous and the function $f_X(x)$ is called a **probability density function**. Function $f_X(x)$ can be treated as a density mass on the

---

[9] Pfeiffer (1978), Rumsey (2006).

x axis. If function $f_X(x)$ is finite, it has a simple interpretation of a linear mass density, i.e. a mass on $(x, x + \Delta x)$ interval equals $f_X(x)dx$. The probability that random variable X takes a value from a certain interval equals a mass located on this interval. The distribution $F_X(x)$ equals the mass over $(-\infty, x)$ interval if the random variable is determined over the whole *x* axis.

Notice that

$$f_X(x) \geq 0 \quad \text{and} \quad \int_{-\infty}^{\infty} f_X(u)\, du = 1 \tag{1.16}$$

if the variable is determined over the whole *x* axis.

There are two types of random variables being commonly applied, namely:

- continuous random variables
- discrete random variables.

Having defined the term of random variable, we can return to the Bayes' Theorem considering random variables rather than random events. This problem is significant for instance, for strength of materials.

If the probability distribution of the load or strength of a given object is known and we get new information in this regard then—based on an equivalent of formula (1.12)—we can find a new, more likely probability distribution of the load (or strength) of the object. Melchers (1999) illustrates the relations between these distributions as is shown in Figure 1.1.

Notice that if additional information is of a low likelihood, i.e. its corresponding distribution is of a high spreading, then this information is given no significant input. In Fig. 1.1 the information is substantial and our knowledge is increased; the mean value shifted to the right (the first significant change) and the dispersion decreased (the second significant change). Our knowledge about the random variable of interest is more important, and thus more reliable than before the investigation.

It is obvious that for discrete random variables the integrals in patterns (1.15) and (1.16) must be replaced by a summation symbol and the formula for the probability function is different because it is now of a discrete type. Following forthcoming changes—if the considered random variable is continuous we have **probability density function**. If a random variable is discrete, we have a **probability mass function**.

The most comprehensive information on a random variable is given by its probability mass/density function.

1.2.3 *Descriptive parameters of a random variable*

Some constants connected with the probability mass/density function in a univocal way play very important roles in probability theory, in reliability theory and some other disciplines of science. These are the parameters of a random variable. Here are the most essential ones.

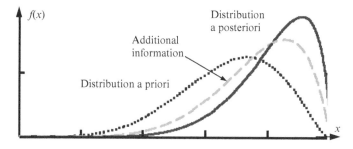

Figure 1.1.   Relation between information *a priori*, additional information and information *a posteriori*; information expressed by probability density functions.

The **expected value** of a random variable X is a number ascribed to the integral[10]

$$E(X) = \int_{\mathscr{X}} x f(x)\, dx \qquad (1.17)$$

where $\mathscr{X}$ is the domain of determinacy of X.

Several other terms are used to describe this parameter, which include mathematical expectation, mean and mathematical hope. The expected value of a random variable is, in fact, the weighted average of all possible values that this variable can take on. However, from a rigorous theoretical point of view, the expected value is the integral of the random variable with respect to its probability measure.

For a discrete random variable that takes the values x_i with probabilities p_i we have

$$E(X) = \sum_i x_i p_i \qquad (1.18)$$

For engineers, it is important to say that the expected value is the centre of gravity of its probability density.

Some essential properties can be formulated for this mean directly from the definition, namely:

- The expected value of a constant is simply the value of this constant

$$E(a) = a \qquad (1.19)$$

- The following relationship holds

$$E[(aX)^k] = a^k\, E(X^k) \qquad (1.20)$$

- The expected value of the linear function of a random variable equals the linear function of the expected value of this random variable, i.e.

$$E(aX + b) = aE(X) + b \qquad (1.21)$$

- For a random variable defined as:

$$Y = X - E(X) \quad \text{we have} \quad E(Y) = 0 \qquad (1.22)$$

In the theory of probability and in many of its diverse applications, the expected value is the most important parameter of a random variable. In engineering practice, the majority of the so-called nominal parameters of technical objects are just mean values.

The expected value may be intuitively understood by the law of large numbers: the expected value, when it exists, is almost surely the limit of the sample mean as the sample size increases to infinity. Nevertheless, the value cannot be expected in the everyday meaning—the expected value itself may be unlikely or even impossible. The expected value when throwing a dice is 3.5.

By knowing the properties of integrals, we can easily come to the conclusion that not all random variables have a finite expected value, since the integral may not converge absolutely.

---

[10] This definition differs from the definition usually presented in statistical books. Most commonly, there is a statement that an expected value is an integral or a sum. However, we obtain a number as the result of integration (or summation when a random variable is a discrete one) because the integral is a definite one (Riemann's integral). By the way, do not try to say to a mining engineer that the average number of—for instance—shearer knives worn at the mine is an integral.

And what is more, some random variables have no expected values. Usually, the Cauchy's distribution is given as an example. Its example probability density function is defined by the formula:

$$f(x) = \frac{1}{\pi} \frac{t}{t^2 + x^2} \qquad -\infty < x < \infty$$

as shown in Figure 1.2.

Moments
If for a given integer number k ($k > 0$) the function x^k is integrable with regard to the function $F(x)$, then the mean value

$$m_k = E(X^k) = \int_{\mathfrak{X}} x^k dF(x) dx \qquad (1.23)$$

where $\mathfrak{X}$ is the domain of the determinacy of X, is called a **raw (crude) moment** of the order k of the random variable X.

Obviously,

$$m_0 = 1 \quad m_1 = E(X) = m.$$

When the term 'moment' is mentioned, an engineer usually associates it with the corresponding idea which plays a major role in mechanics.[11] Formula (1.23) is defined as the moment of the probability density distribution.

If a denotes a certain constant then the formula:

$$E\{(X-a)^k\} = \int_{-\infty}^{\infty} (X-a)^k dF(x) \qquad (1.24)$$

is called a **moment of k-th order about a value** of the random variable X.

For $a = 0$ we have raw moments.
For $a = m$ we have **central moments**.

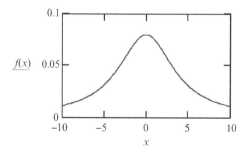

Figure 1.2.   Probability density function for Cauchy's distribution.

---

[11]Generally, in physics, the term moment can refer to many different concepts such as the moment of force, the magnetic moment, the electric dipole moment etc.

Of great particular usage in the engineering world is the second central moment called **variance** that is determined by the pattern:

$$E\{(X-m)^2\} = \sigma^2(X) = E(X^2) - [E(X)]^2 \tag{1.25}$$

The positive square root of variance is called the **standard deviation**.

Standard deviation is a measure[12] of the average dispersion of the values of the random variable around its expected value. A low standard deviation indicates that the data points tend to be very close to the mean, whereas a high standard deviation indicates that the data points are spread out over a large range of values.

Considering the definition of variance, some important properties can be formulated. The standard deviation is invariant under changes in location and scales directly with the scale of the random variable. Thus, for a constant a and random variables X and Y we have:

- The variance of constant a is zero

$$\sigma^2(a) = 0 \tag{1.26}$$

- Adding a constant to the random variable does not change the variance:

$$\sigma^2(X+a) = \sigma^2(X) \tag{1.27}$$

- The following equation holds:

$$\sigma^2(aX) = a^2\sigma^2(X) \tag{1.28}$$

- For a linear function of a random variable we have:

$$\sigma^2(aX+b) = a^2\sigma^2(X) \tag{1.29}$$

- For a random variable Y defined as:

$$Y = X/\sigma(X) \quad \text{we have} \quad \sigma^2(Y) = 1 \tag{1.30}$$

- For any constant $a \neq E(X)$ the following inequality holds:

$$\sigma^2(X) < E(X-a)^2 \tag{1.31}$$

- If the investigated population is divided into k groups, then the variance of the whole population is the sum of two components:

  - the arithmetic mean of the variance within groups
  - the variance of the means of the groups, i.e. the variance between the groups.

The role of standard deviation as a parameter describing the distribution of random variables is proven by the Chebyshev's inequality (also spelled the Tchebysheff's inequality). This will be given in Chapter 3, formula (3.14).

Using formulas (1.22) and (1.30), we can make a **standardisation of a random variable**.

---

[12] In this book the term 'measure' appears very often. Let us recall its definition. If $\mathfrak{M}$ denotes any σ-algebra of any space X and if μ is a real function of the space X as determined in $\mathfrak{M}$ then every real function μ is called a measure in the σ-algebra. Moreover, every subset belonging to $\mathfrak{M}$ is called a measurable with regard to measure μ. If μ takes values not greater than 1, this measure is called a probabilistic one.

If a random variable of the expected value $E(X)$ is given and the corresponding standard deviation $\sigma(X)$, then the random variable:

$$U = \frac{X - E(X)}{\sigma(X)} \tag{1.32}$$

is called the standardised random variable. Its expected value is zero and its standard deviation is 1.

Generally, the standardisation of a random variable is a transformation aimed at bringing different distributions of random variables of different means and degrees of differentiation to comparability. The difference between a certain probability density function before and after standardisation is shown in Figure 1.3.

In some cases, the standardisation of random variables is applied in the structural reliability analyses conducted in connection with the uncertainties related to the resistance and load of technical objects. Both resistance and load are generally treated as random variables that have known distribution functions, although in some practical cases these functions are unidentified. However, there is still a possibility of carrying out a reliability analysis by applying the third-order polynomial normal transformation technique engaging the first four central statistical moments and the explicit fourth-moment standardisation function (Yan-Gang Zhao and Zhao-Hui Lu 2007).

The root of equation

$$F(x_\alpha) = \alpha \quad 0 < \alpha < 1 \tag{1.33}$$

is called a **quantile of the order** α in the distribution $F(x)$. Figure 1.4.

There are many such roots which used to be applied in analyses of data in descriptive statistics. The most important and the most frequently used is the quantile of the order ½ which is called **median**, i.e.

$$Me(X) = x_{1/2} \quad \text{where} \quad F(x_{1/2}) = \frac{1}{2} \quad \text{which means that} \quad \int_{-\infty}^{x_{1/2}} f(x)dx = \frac{1}{2} \tag{1.34}$$

Practically, this is a very interesting statistical measure because of its property. The probability that a random variable takes a value not greater than the median equals the probability that the random variable takes a value not lower than the median and this probability is ½. A median divides the mass/density of probability in half. It is only defined on one-dimensional

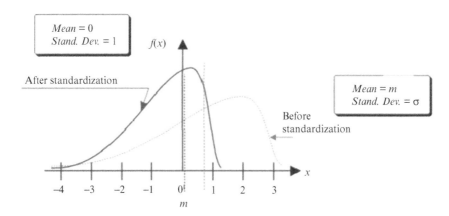

Figure 1.3.　Probability density functions before and after standardisation.

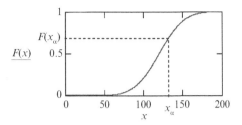

Figure 1.4. Quantile of the order α in the distribution $F(x)$.

data[13] and is independent of any distance metric. In a sample of data or a finite population, there may be no member of the sample whose value is identical to the median (in the case of an even sample size), and, if there is such a member, there may be more than one so that the median may not uniquely identify a sample member.

The medians of certain types of distributions can be easily calculated from their parameters, e.g. the median of a Cauchy's distribution with the location parameter x_0 and the scale parameter y is x_0, the location parameter. But for some distributions median formulas are complicated or there is no single formula that can be used to find the median (for example, Binomial distribution).

The **mode** of a random variable X ($Mo(X)$) is such that its value corresponds with:

– the greatest probability of the occurrence when the random variable is a discrete one
– the local maximum of the probability density function when the random variable is a continuous one.

Like the mean and median, the mode is a way of expressing, in a single number, important information about a random variable or a population. The mode is not necessarily unique, since the same maximum frequency may be attained at different values. The most extreme case appears in uniform distribution, where all values occur equally frequently.

Information on the mode (sometimes called a dominant) of a given random variable is important when the prognosis of realisation of this random variable should be done and this prognosis will be done only once (not repeated). The proposed value should be simply the mode because it has the greatest chances of occurring. This prognosis can be different if additional information is at hand.

Of particular value are order statistics and their functions, which are used to estimate order parameters.

Let X_1, X_2, ..., X_n be an n-dimensional random vector and let x_1, x_2, ..., x_n be its realisation. The **order statistic** $X_{k,n}$ is a function of the random variables X_1, X_2, ..., X_n taking the k-th largest value in each sequence of values x_1, x_2, ..., x_n. Order statistics are important, for instance, in durability studies.

1.2.4 *Relationships between parameters of random variables*

When the probability distribution of a random variable is a symmetric one, then its expected value, its mode and its median take the same value[14], e.g. for the Gaussian distribution. In a case where a given distribution is asymmetric, these basic parameters of the random variable are different in value. A moderately asymmetric probability density function having a positive skew and positions of the mean, median and mode are shown in Figure 1.5.

---

[13] A geometric median, on the other hand, is defined in any number of dimensions.
[14] Some exceptions such as the Cauchy's distribution or the uniform one, for instance, are neglected here.

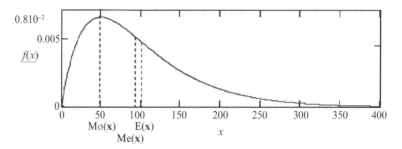

Figure 1.5.   A moderately asymmetric probability density function.

A few terms have been used here that need to be defined.

Skewness is a property of the probability distribution of a random variable. Its particular cases are the symmetry and asymmetry of the shape of the probability mass/density function. There are several measures of skewness in mathematical statistics but two are applied most frequently. The first one is defined as:

$$\gamma_1 = E\left[\left(\frac{X - E(X)}{\sigma(X)}\right)^3\right] = \frac{E[(X - E(X))^3]}{\{E[(X - E(X))^2]\}^{3/2}}$$ (1.35)

Someone suggested using the following ratio as a measure of skewness:

$$\varpi = \frac{E(X) - Mo(X)}{\sigma(X)}$$ (1.36)

(Weisstein).

The skewness value can be positive, negative or zero. Qualitatively, a negative skew indicates that the *tail* on the left side of the probability density function is *longer* than the one on the right side and the bulk of the values (possibly including the median) lie to the right of the mean. A positive skew indicates that the *tail* on the right side is *longer* than the one on the left side and the bulk of the values lie to the left of the mean. A zero value indicates that the values are relatively uniformly distributed on both sides of the mean, typically (but not necessarily) implying a symmetric distribution.[15]

In availability investigations randomisation[16] of the steady state availability is sometimes done and it is presumed that beta distribution is a good model to describe its distribution for a group of identical items. Obviously, this presumption needs to be verified. This is similar when research concerns the utilisation of a given piece of equipment on a daily basis. And here again the beta probability distribution[17] is applied as a rule because it is supported on an [0, 1] interval. (Utilisation can also be expressed in percentages). Depending on the values taken by parameters of this distribution, the shapes of the probability density

---

[15] These considerations are correct provided that the probability distribution is not a multimodal one. Such a distribution may occur if the population being investigated is not homogeneous. There are also some further probability distributions for which the above regularities do not hold but these distributions are considered mainly in theory (e.g. heavy-tailed distributions).

[16] **Randomisation** relies on making something random. Randomisation can concern different things (e.g. the randomisation of an experiment) but here in our discussion a constant parameter is treated as a random one.

[17] This distribution is defined by formula (1.61).

function are different. However, because the availability of actually constructed machines is high as a rule and their utilisation also, thus the probability distribution that is applied has a negative skew and the positions of the mode, median and mean is opposite to the one in Figure 1.5. An example of the probability distribution of the utilisation of a hoist is shown in Figure 1.6.

The last term that will be defined in this chapter is a **class of probability distributions**.

A class of probability distribution creates a family of distributions that have the same cumulative function (or mass/density function) but whose structural parameters can be different.

The parameters of probability density functions can be divided into three groups:

- location parameters
- shape parameters
- scale parameters.

If there is a change in the value of the location parameter, the probability density function is displaced along the x axis (Figure 1.7).

If there is a change in the value of the shape parameter, the probability density function changes its profile (Figure 1.8).

A change in the value of the scale parameter makes the probability density function fatter or causes it to be raised up (Figure 1.9).

A function of a random variable is a random variable as well.

Let us review some of the most frequently applied probability distributions in mining engineering.

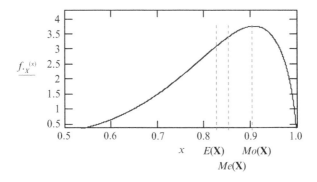

Figure 1.6.   Probability density function of the utilisation of a hoist; $Mo(X) > Me(X) > E(X)$.

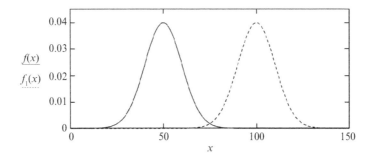

Figure 1.7.   Two probability density functions with different values of the location parameter only.

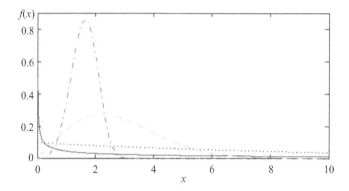

Figure 1.8.   Probability density functions with different values of the shape parameter only.

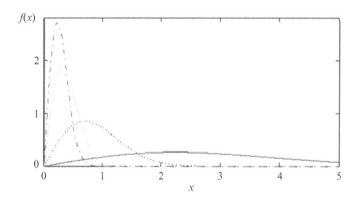

Figure 1.9.   Probability density functions with different values of the scale parameter only.

1.2.5   *Probability distributions of discrete random variables*

Bernoulli distributions

A random variable X that takes an x_1 value with the probability p, $0 < p < 1$ and an x_2 value with the probability $1 - p$ is called a **double-point Bernoulli** random variable. Its probability **distribution** is given by the formula:

$$F(x) = \begin{cases} 0 & x \leq x_1 \\ p & x_1 < x \leq x_2 \\ 1 & x > x_2 \end{cases} \qquad (1.37)$$

Usually an alternative formula is given stating:

$$P(X = x_1) = p \quad \text{and} \quad P(X = x_2) = 1 - p \qquad (1.37a)$$

An example of the probability mass function and corresponding cumulative function for the double-point Bernoulli distribution are shown in Figure 1.10.

In probability theory and statistics, the **Bernoulli distribution**, named after the Swiss scientist Jacob Bernoulli, is a discrete probability distribution, which takes value 1 with success probability p and value 0 with probability $1 - p$. Thus

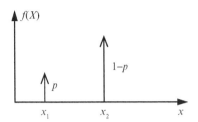 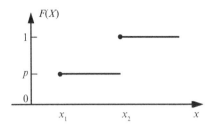

Figure 1.10. The probability mass function *f(x)* and the corresponding cumulative function *F(x)* for a double-point Bernoulli distribution.

$$P(X = 1) = 1 - P(X = 0) = p$$

This is the particular case of the double-point Bernoulli distribution.
The expected value and the variance for both distributions are the same:

$$E(X) = p \quad \sigma^2(X) = p(1 - p) \tag{1.38}$$

If a point of interest is the certain state in which a given technical object can be in given moment and there are only two possibilities (e.g. two alternative states: work and repair), this means that this point of interest is the random variable of the double-point Bernoulli distribution. We may ascribe the notation 1 to the work state and the notation 0 for the repair state.

Binomial distribution
Let X_1, X_2, ..., X_n be independent random variables of the identical double-point Bernoulli probability distribution with the parameter p. Such a defined sequence of random variables is called a sequence of independent Bernoulli experiments (trials). It is presumed that if the random variable X_i takes the value 1 then it is classified as a success. If in the i-th Bernoulli trial the random variable takes the value 0, it is classified as a failure.
The probability that in n independent Bernoulli trials success occurs exactly k times is given by the formula:

$$P(X = k) = \binom{n}{k} p^k (1 - p)^{n-k} \tag{1.39}$$

where $\binom{n}{k} = \frac{n!}{k!(n-k)!}$ is the binomial coefficient.

Formula (1.39) defines the Bernoulli probability function.
An example of the Bernoulli probability density function is shown in Figure 1.11.
The main parameters—the expected value and the variance—of the random variable are determined as follows:

$$E(X) = np \quad \sigma^2(X) = np(1 - p) \tag{1.40}$$

Presume that there is a given mechanised system applied in an underground coal mine and that this system consists of a shearer, two scraper chain conveyors and a certain number of belt conveyors that deliver broken coal from the wall to the shaft. The system is observed over n production shifts and the point of interest is the work state of the system in a given moment of time in a shift. The probability that the system will be k times ($k \le n$) in the work state in a given moment is determined by formula (1.39).

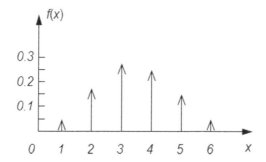

Figure 1.11.   The Bernoulli probability density function.

Geometric distribution
This distribution is defined in a twofold manner:

a. as the probability distribution of the number X of independent Bernoulli trials needed to get one success; this distribution is supported on the set $\{1, 2, ...\}$
b. as the probability distribution of the number $Y = X - 1$ of failures before the first success; this distribution is supported on the set $\{0, 1, 2, ...\}$

If the probability of success in every trial is p, then the probability that in k-*th* trial will be success is determined by the pattern:

$$P(X = k) = (1 - p)^{k-1}p \quad \text{for } k = 1, 2, ... \tag{1.41a}$$

Similarly, if the probability of success in every trial is p, then the probability that there will be k-th failures preceding the first success is determined by the pattern:

$$P(Y = k) = (1 - p)^k p \quad \text{for } k = 0, 1, 2, ... \tag{1.41b}$$

Often, the name *shifted* geometric distribution is adopted for the distribution of the number X. This 'displacement' is visible looking at the mode because for X the random variable is 1 whereas for the second random variable it equals 0.

The expected value and the variance in these probability distributions are the following:

$$E(X) = 1/p \quad E(Y) = (1 - p)/p \quad \sigma^2(X) = \sigma^2(Y) = (1 - p)/p^2 \tag{1.42}$$

The geometric probability distribution is the discrete equivalent of the exponential distribution[18] and it has no memory as does its exponential counterpart.[19]

An example of the probability mass distribution for geometric distribution is given in Figure 1.12.

Considering the number of hoist winds up to the moment of the over-winding at an extreme level, it is easy to see that this number is a random variable that is geometrically distributed in which p denotes the probability of over-winding in a single hoist work cycle.

---

[18] This distribution is given by formula (1.52).
[19] The idea of a lack of memory is extended into the area of stochastic processes, e.g. the Poisson process in which the time from the occurrence of one event to the next one is an exponential one that is characterised only by lack of memory. This lack of memory means that the probability of the appearance of k events in the interval $(T, T + t)$ does not depend on how many events and in which manner they have occurred up to the moment T (Gnyedenko and Kovalenko 1966).

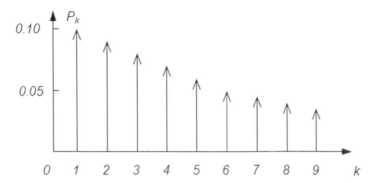

Figure 1.12. An example of the probability mass distribution for a geometric distribution.

Poisson distribution
This distribution was first introduced by Siméon Denis Poisson and published together with his probability theory (1837).

A practical application of this distribution was made by Władysław Bortkiewicz in 1898 when he was given the task of investigating the number of soldiers in the Prussian army who were accidentally killed by horse kicks. This investigation led to the Poisson distribution and according to some researchers to the field of reliability engineering.[20]

If the random variable X takes values at points $k = 0, 1, 2, \ldots$ with the probability

$$P(X = k) = \frac{\lambda^k}{k!} e^{-\lambda} \quad \lambda > 0 \tag{1.43}$$

then we say that this random variable has a Poisson distribution (Figure 1.13).

The expected value and the variance are identical and

$$E(X) = \lambda \quad \sigma^2(X) = \lambda \tag{1.44}$$

If we are interested in the probability distribution of the number of failures of a given type in a given time interval and the events that occur create a memoryless stream of arrivals, then the probability distribution of this random variable is given by formula (1.43).

By observing the process of the operation of belt conveyors in mining, we can easily determine that the mean time between two neighbouring failures is long compared to the mean time of repair. Thus, it is usually presumed that the operation process from a reliability point of view is described by the **Poisson process** of which the probability of the occurrence of k failures in time t is given by formula (1.43) where the parameter is λt (t – time).

The Poisson distribution can also be applied to systems with a large number of possible events, each of which is rare and the process of occurring events can usually be described by the Poisson process. The Poisson distribution is sometimes called a 'Poissonian'.

1.2.6 *Probability distributions of continuous random variables*

Uniform distribution
The uniform distribution, sometimes also known as the **rectangular distribution**, is a distribution that has a constant probability.

---

[20] *Vide*: http://en.wikipedia.org/wiki/Poisson_distribution#cite_note-4.

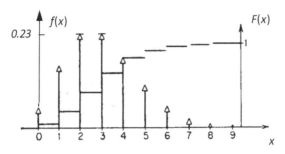

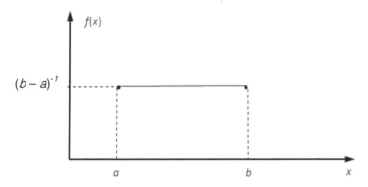

Figure 1.13. An example of the probability mass function and the cumulative function for the Poisson distribution.

Figure 1.14. The probability density function of the uniform distribution.

A continuous random variable X which has the probability density function given by the pattern:

$$f(x) = \begin{cases} (b-a)^{-1} & \text{for } x \in [a, b] \\ 0 & \text{otherwise} \end{cases} \quad \text{(Figure 1.14)}. \tag{1.45}$$

A frequently used notation is X: $U(a, b)$.
The expected value and the mean are the same:

$$E(X) = Me(X) = \tfrac{1}{2}(a+b) \tag{1.46}$$

Depending on definition of mode some researchers are of the opinion that the mode is any value in the interval $[a, b]$; some that the mode does not exist.
The variance is determined by the formula

$$\sigma^2(X) = (b-a)^2/12 \tag{1.47}$$

The uniform distribution is applied to estimate the parameters for probability distributions, making use of additional information (*a priori*) on estimating the parameters of a random variable. If all of the values that are taken into account are equally probable, the uniform distribution is used.

This additional information is described by the probability distribution that is chosen and then the Bayesian approach is applied. The Bayesian inference can still be used when one or even two limited values *a* and *b* are $-\infty$ and ∞. It was proved (see de Groot 1970, for example) that even presuming an improper distribution, there is still a possibility of conducting the Bayesian procedure and the distribution *a posteriori* will be the proper one.

The uniform distribution has several interesting properties, among other things:

- If the random variable *X* has the a uniform distribution supported on the (0, *b*) interval, then the random variable $Z = \ln(b/X)$ has the exponential distribution of the expected value equals 1.
- If random variables $X_1, X_2, \ldots, X_n$ are stochastically independent and have the same uniform distribution supported on the (0, 1) interval and if *n* tends to infinity, then the probability distribution of their sum tends to the normal distribution $N\left(\frac{n}{2}, \sqrt{\frac{n}{12}}\right)$ and the distribution of their mean tends to the normal distribution $N\left(\frac{1}{2}, \sqrt{\frac{1}{12n}}\right)$.

Gamma distribution

If the probability density function of the random variable *X* is given by the formula

$$f(x) = \frac{v^\xi}{\Gamma(\xi)} x^{\xi-1} e^{-vx} \quad x \geq 0, v > 0, \xi > 0 \tag{1.48}$$

then we say that this random variable has a gamma distribution with the shape parameter ξ and the scale parameter v.

A frequently used notation is *X*: *Ga*(ξ, v).

The expected value and the variance are as follows:

$$E(X) = \xi/v \quad \sigma^2(X) = \xi/v^2 \tag{1.49}$$

The **gamma function**, being a component of the function (1.48), is determined by the pattern:

$$\Gamma(\xi) = \int_0^\infty x^{\xi-1} e^{-x} dx \tag{1.50}$$

The gamma function is an extension of the factorial function and it has the following property:

$$\Gamma(\xi) = (\xi - 1)\,\Gamma(\xi - 1) \tag{1.50a}$$

For ξ being an integral number

$$\Gamma(\xi) = (\xi - 1)! \tag{1.50b}$$

For large values of ξ, the gamma function can be calculated using Stirling's formula and then

$$\Gamma(\xi+1) \approx \left(\frac{\xi}{e}\right)^\xi \sqrt{2\pi\xi} \left(1 + \frac{1}{12\xi}\right) \tag{1.51}$$

If a scale parameter is a multiple of ½, then the distribution is named the chi-squared or χ^2-distribution[21]. This distribution is extremely significant in mathematical statistics.

---

[21] See formula (1.109).

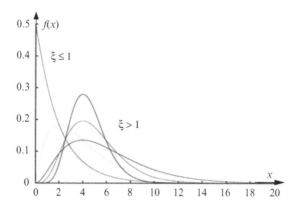

Figure 1.15.　Gamma probability density functions.

If the scale parameter is not larger than 1, the distribution is extremely asymmetric. Different shapes of the gamma probability distribution are shown in Figure 1.15.

The gamma distribution is applied to describe the load and strength of some machine elements (Warszyński 1988). It is also employed for the modelling of the lifetime of items that are subject to cumulating agents. It is suitable for modelling the processes of the wearing of the elements of machines and devices (Gertzbah and Kordonsky 1966).

In mining, winders have probability distributions of times of work and repair states that can be satisfactorily described by the gamma distribution. The repair time of dumpers used in open pit mines as well as of power shovels can be described by this distribution. The gamma distribution is frequently applied in the mathematical description of the processes of the wearing of tyres.

This probability distribution was also applied in the study of the processes of the separation of grains in pulsators. It was assumed that the length of a single jump of a particle during water pulsation is a random variable of the gamma distribution (Smirnow 1979). The application of a generalised gamma distribution for the approximation of the composition functions for grains was presented by Nipl (1979). Some particular cases of the gamma distribution have been used in the description of certain material characteristics during ore enrichment processes (*vide*: Tumidajski and Saramak 2009 p. 63).

If the shape parameter is 1 in the gamma distribution, then the probability density function is given by the formula

$$f(x) = v e^{-vx} \qquad v > 0 \qquad\qquad (1.52)$$

and this **distribution** is called an **exponential** distribution (Figure 1.16).

This probability distribution has many applications in engineering practice because it is easy to use. Many electric and electronic parts have a lifetime probability distribution of an exponential character. When the first models of the processes of the changes of states for technical objects occurred in literature, they were almost exclusively composed of exponential random variables. The processes of changes of states in which the times of the states are mutually independent and described by exponential distributions are called Markov processes. Their parameters and characteristics are shown in an explicit form, which is easy to use. For this reason, exponential distribution was widely applied during the early days of reliability development as well as in the theory of the exploitation of premature growth. This distribution was abused in those days, but excuses can be found in the poor development of advanced modelling by applying other, more complicated distributions. In those days,

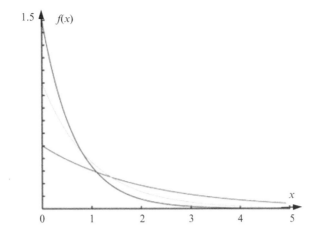

Figure 1.16.   Probability density functions of an exponential distribution.

engineers working with the data gathered applied this distribution even when the goodness of the description was evidently poor (e.g. Barbaro and Rosenshine 1986 analysing the operation[22] of a shovel-truck system). Unfortunately, engineers who are engaged in research have fallen into the habit of applying exponential distribution and they still use it, even in cases where it should not be applied. One can get the impression that engineers have not noticed that the actual models are much more advanced using different probability distributions. Some modern models allow any probability distribution to be employed, i.e. such a distribution can be exponential as well as another different model (e.g. Sivazlian and Wang 1989). However, it should be clearly stated that these modern models are more advanced and more difficult to use in analyses and application.

Nevertheless, many pieces of equipment, also in mining (e.g. conveyors of different types, excavating machines, stackers), have probability distributions of work times that can be satisfactorily described by exponential distribution. Similarly, many machines used in earthmoving engineering have probability distributions of a work time exponential.

The probability distribution of work time is also described by exponential distribution in many modern sophisticated electronics systems. In some cases, the repair time has a probability distribution of exponential character as well.

Erlang distribution
The sum of k independent random variables distributed exponentially has a gamma probability distribution with the parameter $\xi = k$. This distribution is called the Erlang distribution of the order k. Its cumulative function is given by the formula:

$$F(x) = 1 - \sum_{i=1}^{k-1} \frac{(\lambda x)^i}{i!} e^{-\lambda x} \quad x \geq 0, \lambda > 0 \tag{1.53}$$

It is easy to notice that the gamma distribution of the scale parameter taking a natural number is in fact the Erlang distribution.

---

[22]The terms 'exploitation' and 'operation' are used interchangeably. Recall, according to the classical definition used in the theory of exploitation that the exploitation process is the process of changes in the properties of an object. This process does not concern either the design/construction or production phases.

The Erlang distribution was developed by A.K. Erlang, a Danish mathematician, statistician and engineer who invented and developed the fields of traffic engineering and queuing theory. In the early years of the 20th century, he examined the number of telephone calls which could be made at the same time to the operators of switching stations. This work on telephone traffic engineering has been expanded to consider waiting times in queuing systems in general. The distribution is now also used in the fields of stochastic processes, telecommunication and biomathematics.

When the scale parameter of the Erlang distribution is 2, then the distribution simplifies to the chi-squared distribution with $2k$ degrees of freedom. It can therefore be regarded as a generalised chi-squared distribution.

The term 'degrees of freedom' was used here and therefore needs explanation.

In statistics, the number of degrees of freedom is the number of values in the final calculation of a statistic that are free to vary. The **degrees of freedom** of an estimate of a statistical parameter is equal to the number of independent outcomes that go into the estimate minus the number of parameters used as the intermediate steps in the estimation of the parameter itself.

The Erlang distribution is extremely significant in the queuing theory that is called the mass servicing theory in Central Europe. More than half a century ago the so-called Erlangian systems, which can be analysed based on the decomposition of the Erlang random variable into the sum of independent random variables exponentially distributed, came into existence. An example of an analysis of such a machinery system—the system applied in surface mining— was presented in Czaplicki's book (2004 and 2010).

Weibull distribution

If the probability density function of the random variable X is determined by the formula:

$$f(x) = \alpha\lambda\, x^{\alpha-1} e^{-\lambda x^{\alpha}} \quad x \geq 0, \alpha > 0, \lambda > 0 \tag{1.54}$$

then we say that this random variable has the Weibull probability distribution.

The notation that is sometimes used for this random variable is $X\!: W(\alpha, \lambda)$.

This function has two structural parameters. The parameter α is the shape parameter whereas the parameter λ is the scale parameter.

The picture showing the probability density functions for the Weibull distribution is quite similar to the gamma distribution. Both distributions are particular cases of the generalised gamma distribution.

The expected value and the variance for the Weibull distribution are determined as follows:

$$E(X) = \Gamma(1+\alpha^{-1})\lambda^{\frac{-1}{\alpha}} \quad \sigma^2(X) = \left[\Gamma\!\left(1+\frac{2}{\alpha}\right) - \Gamma^2(1+\alpha^{-1})\right]\!/\lambda^{\frac{2}{\alpha}} \tag{1.55}$$

The name of this distribution came from Waloddi Weibull, who described and applied this distribution to the strength of materials (Weibull 1939), although some researchers used it earlier. It is said that it was identified by Fréchet who published it in a Polish mathematical journal (1927) and was first applied by Rosin and Rammler (1933) to describe the size distribution of particles. However, a comprehensive mathematical analysis was done by Gnyedenko (1941). This distribution was generalised by Weibull, who added the parameter of the displacement of the origin of the coordinate system, in 1956.

The probability distributions of the lifetime of many mechanical parts can be described by the Weibull distribution.

It was proved (Gercbach and Kordonsky 1968) that if the elements of a system have work times of a gamma distribution of slightly different values of the parameters, then the probability distribution of the work time of the system is the Weibull distribution when the number of elements is large. Internal combustion engines have several cylinders, gas turbines have many vanes and many electronic systems have many identical parts. All systems having the same elements are characterised by the fact that the work time probability distribution tends to be the Weibull distribution if the number of these elements is increased.

In some cases, the Weibull distribution is obtained from a theoretical analysis that is associated with an empirical situation. Consider a system consisting of a certain number of elements connected in a sequence with each element consisting of a pair of items working in a parallel way. A failure of the system occurs when both elements in any pair fail. If the reliability of an element is described by the exponential distribution then it was proved (Gnyedenko et al. 1969, Chapter 2.2) that when two simple mathematical conditions are fulfilled, the reliability (survival) function has a limit and this boundary is the Weibull distribution.

An analysis of the process of operation of powered support of operating in underground coal mines is presented in Czaplicki's book of 2010 (Chapter 7.6) and, based on empirical data, the probability distributions of the times of states are described by the Weibull family of functions. An analysis of the system of power shovel-crusher-conveyors is also presented in the same chapter. The author took into account that the empirical distributions of the random variables of work and repair times are often different to the exponential and for this reason he applied the Weibull function. He stated that the process of changes of states is a semi-Markov one. This probability function was used to describe the distribution of the instantaneous productivity of bucket wheel excavators (Jurdziak 2006, Dworczyńska et al. 2012).

The Weibull distribution is extremely significant in the reliability of objects working until the first failure occurrence. The hazard function[23] (a function of the conditional intensity of failure) of an item for which the lifetime is a random variable described by the Weibull distribution has a variety of different shapes (Figure 1.17) and because of this property, the Weibull distribution can be applied in many empirical cases[24].

For $\alpha = 1$ the Weibull distribution becomes an exponential one.

Normal (Gaussian) distribution
If the probability density function of the random variable X is determined by the pattern:

$$f(x) = \frac{1}{\sigma\sqrt{2\pi}}\exp\left[-\frac{(x-m)^2}{2\sigma^2}\right] \qquad x \in (-\infty,\infty), m \in (-\infty,\infty), \sigma > 0 \qquad (1.56)$$

then we say that the random variable has a normal distribution.

A frequently used notation is $X: N(m, \sigma)$.

This function has two structural parameters: m and σ. The first one is the location parameter and the second is the scale parameter.

---

[23] A **hazard function** is associated with items working until the occurrence of the first failure. It is the probability that an item will fail in the near future (a brief time interval) provided that the item has been functioning well until now. Mathematically, it is the conditional intensity of the failure of an item and is determined by the formula: $\lambda(t) = f(t)/R(t)$, where $R(t)$ is the reliability function.

[24] The Weibull distribution has had a tremendous career amongst practitioners, especially engineers. In 1977 Weibull himself collected 1019 references (articles) and 36 titles of books in which his model is mentioned in a technical report of Förvarets Teletekniska Laboratorium (in Stockholm, Sweden) (all these titles are only in English). Due to the enthusiasm related to this distribution, Giorski (1968) draws the attention of what he calls 'Weibull euphoria' arguing that the model is very useful but it is not 'universal' (one year later, Ravenis (1969) proclaimed that Weibull's model is a *potentially universal p.d.f. for scientists and engineers*…).

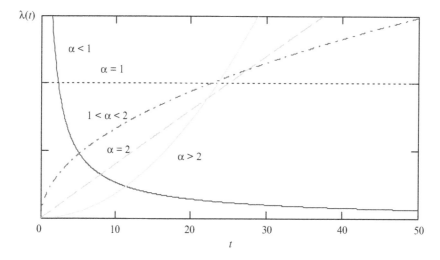

Figure 1.17.   Hazard function for the Weibull distribution of different parameters.

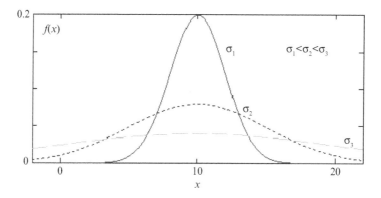

Figure 1.18.   The probability density function of a normal distribution.

The expected value and the variance are determined as follows:

$$E(X) = m \quad \sigma^2(X) = \sigma^2 \tag{1.57}$$

The visible different normal density functions that have different structural parameters are shown in Figure 1.18.

The Gaussian distribution is considered to be the most prominent probability distribution in the field of statistics. There are some significant reasons for this. Firstly, the normal distribution arises from the central limit theorem,[25] which states that under mild conditions, the mean of a large number of random variables independently drawn from the same distribution is distributed approximately normally, irrespective of the form of the original distribution. Secondly, the normal distribution is very tractable analytically, that is, a large number

---

[25]This Theorem may be expressed in the following way: Whatever the distributions of the independent random variables are—subject to certain very general conditions—their sum is asymptotically normal, where the summarised expected value is the sum of the expected values of the random variables that are the components of the sum and the variance is the sum of all component variances.

of results involving this distribution can be derived in an explicit form. Moreover, the normal distribution is often used as the first approximation to describe the random variables that gather around a single mean value.

In mining, the probability distribution of a stream of rock won in a time unit that is generated by winning machines during the excavation process can be satisfactorily described by the normal distribution. As a rule, the times of the work cycle phases for trucks, i.e. load-haul-dump-return can be described by the Gaussian distribution. The total mass of broken rock delivered to a shaft during a shift in an underground mine can usually be modelled by the Gaussian distribution. The times of loading and unloading for many transporting machines can be depicted by a bell-shaped distribution, i.e. can be described by the normal distribution. The distribution of random measuring errors of different physical magnitudes is also normal most of the time.

The particular case of the normal distribution is the standardised normal distribution for which the expected value is zero and the standard deviation is one. Keeping in mind formula (1.32), we can conclude that the relationship between the random variable X of the normal distribution and a random variable U that has a standardised normal distribution is as follows:

$$U = (X - m) / \sigma_x$$

About 68% of the values drawn from the normal distribution are within one standard deviation σ away from the mean, about 95% of the values lie within two standard deviations and about 99.7% are within three standard deviations. This fact is known as the 68-95-99.7 *rule* or the *empirical rule* or the *3-sigma rule*.

This probability distribution has very broad applications in mathematical statistics and it is frequently used in research in engineering practice.

Log-normal distribution

A random variable that includes only positive real numbers has a log-normal distribution if its logarithm has a normal distribution. Depending on which logarithm is taken into account, we have slightly different shapes of the probability density functions and slightly different formulas to determine the parameters of these random variables. The decision as to which logarithm is taken into account depends as a rule on further parts of the research procedure. The logarithm that is chosen for the investigation is the one which will be more convenient for further analysis.

The probability density function of this random variable is given by the formula:

$$f(y) = \frac{1}{y\sigma\sqrt{2\pi}} \exp\left[-\frac{(\ln y - \mu)^2}{2\sigma^2} \right] \quad \text{for } y > 0, \mu > 0, \sigma > 0 \ \ (\text{Figure}\,1.19). \qquad (1.58)$$

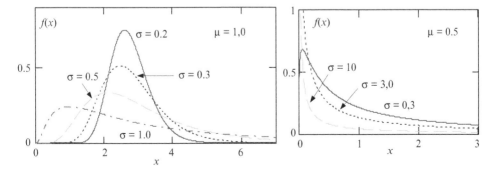

Figure 1.19.   The probability density functions of the log-normal distribution.

A variable might be modelled as log-normal if it can be thought of as the multiplicative product of many independent random variables each of which is positive. This is justified by considering the central limit theorem in the log-domain. For example, in finance, the variable can represent the compound return from a sequence of many trades (each expressed as its return + 1) or a long-term discount factor can be derived from the product of short-term discount factors. In wireless communication, the attenuation caused by shadowing or slow fading from random objects is often assumed to be log-normally distributed.

The expected value and the variance of the random variable X is given by the formulas:

$$E(X) = e^{m+(\sigma^2/2)}$$
$$\sigma^2(X) = (e^{\sigma^2} + 2)\, e^{2m+\sigma^2} \qquad (1.59)$$

The median and mode are as follows:

$$Me(X) = e^m$$
$$Mo(X) = e^{m-\sigma^2} \qquad (1.60)$$

In hydrology, the log-normal distribution is used to analyse the extreme values of such variables as the monthly and annual maximum of daily rainfall and river discharge volumes. It was proposed that coefficients of friction and wear may be treated as having a log-normal distribution (Steele 2008). This type of distribution is applied to describe fatigue strength and strength against long-lasting stresses (Gercbah and Kordonsky 1966). Also, durability measured in the number of work cycles of technical objects is often described by a log-normal distribution. Some trials were performed to describe the times of truck work cycles using this distribution (Hufford et al. 1981, Griffin 1989).

A theoretical approach to the determination of the distribution of grain size during the shredding process was analysed by Kolmogorov (1941) and Epstein (1947). They proved that under certain conditions related to long-lasting repeated crushing, the grain size distribution can be satisfactorily described by a log-normal distribution. A log-normal distribution cumulative function was applied in ore dressing to describe the grain composition function of the variance being the function of some empirical constants (Andreyev et al. 1959).

Beta distribution
If the probability density function of the random variable X is determined by the pattern

$$f(x) = B^{-1}(c, d)\, x^{c-1}\, (1-x)^{d-1} \quad 0 \le x \le 1,\, c > 0,\, d > 0 \qquad (1.61)$$

where:

$$B(c, d) = \int_0^1 x^{c-1}(1-x)^{d-1}\,dx = \frac{\Gamma(c)\Gamma(d)}{\Gamma(c+d)} \quad \text{is the \textbf{beta function}} \qquad (1.62)$$

then we say that the random variable has a beta distribution.

Plots of the probability density functions of the beta distribution of different parameters are shown in Figure 1.20.

A frequently used notation is $X: Be(c, d)$.

The expected value and the variance in this probability distribution are

$$E(X) = \frac{c}{c+d} \quad \sigma^2(X) = \frac{cd}{(c+d)^2(c+d+1)} \qquad (1.63)$$

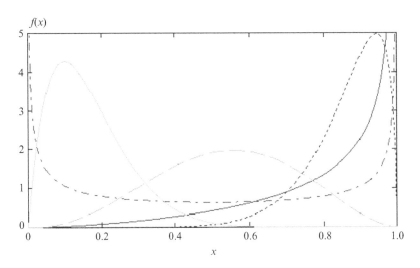

Figure 1.20.   The probability density functions of a beta distribution.

The probability distribution of steady-state availability—treated as a random variable—for identical pieces of equipment are described by the beta distribution. Similarly, as a rule the probability distribution of the utilisation of equipment is described by this type of probability distribution.

There are several further probability distributions that are applied in mining engineering (e.g. the Pareto distribution, logistic distribution, M1 and M2 distributions that are based on the hyperbolic cosine and used in the pulverisation of coal (Mianowski 1988, Tumidajski 1992)). However, their applications are not very frequent and for this reason they are not considered in this text.

Let X be a random variable and $\mathfrak{B}$ be the Borel set[26] on the x axis that $0 < P(X \in \mathfrak{B}) < 1$. The conditional distribution determined for a real number x by the expression $P(X \leq x, X \in \mathfrak{B})$ is called the **truncated distribution** of the random variable X.
Example. If the probability density function is determined as

$$f_1(u) = \begin{cases} \dfrac{f(x)}{F(b) - F(a)} & \text{for } a \leq u \leq b \\ 0 & \text{elsewhere} \end{cases} \tag{1.64}$$

where function $f(x)$ is determined by formula (1.56) and $F(x) = \int_{-\infty}^{x} f(v)dv$ (the integrand is the density function before truncation), then we say that the random variable U has a **truncated normal distribution** supported on interval $[a, b]$.
The expected value and the variance of the random variable above can be defined as follows:

$$E(U) = \mu = m + \sigma \frac{I_1}{I_0} \tag{1.65}$$

---

[26]A **Borel set** is any set in a topological space that can be formed from open sets (or, equivalently, from closed sets) through the operations of a countable union, countable intersection and relative complement.

$$\sigma^2(U) = (I_2\, I_0^2 - I_1^2)\, (\sigma/I_0)^2 \tag{1.66}$$

where:

$$I_0 = \Phi_N\left(\frac{b-\mu}{\sigma}\right) - \Phi_N\left(\frac{-\mu}{\sigma}\right) \tag{1.67}$$

$$I_1 = \phi_N\left(\frac{\mu}{\sigma}\right) - \phi_N\left(\frac{b-\mu}{\sigma}\right) \tag{1.68}$$

$$I_2 = I_0 - \frac{\mu}{\sigma}\phi_N\left(\frac{\mu}{\sigma}\right) - \frac{b-\mu}{\sigma}\phi_N\left(\frac{b-\mu}{\sigma}\right) \tag{1.69}$$

$\Phi_N(y)$—the distribution function of a normal standardised random variable $N(0,\ 1)$; see Table 9.1 at the end of this book,

$\phi_N(y)$—the probability density function of the normal standardised random variable $N(0,\ 1)$.

The distribution of a stream of excavated rock which flows over a belt conveyor system in a unit of time can be described by the truncated normal distribution. The left side boundary is zero ($x \geq 0$) whereas the right side boundary is defined by the inequality: $x \leq x_{max}$, where x_{max} is the maximum conveyor output (the maximum mass of excavated rock that can physically be located on the conveyor, in unit of time).

1.2.7   *Two-dimensional random variable and its moments*

The considerations examined up to now have concerned a one-dimensional random variable. However, in some empirical cases, we may have two or more random variables. It is necessary to consider the statistical measures that allow these variables to be described. Let us confine our attention to a two-dimensional case.

The probability distribution $F(x, y)$ of a two-dimensional random variable is defined as:

$$F(x, y) = P(X \leq x, Y \leq y) \int\limits_{-\infty}^{x} \int\limits_{-\infty}^{y} f(u, v)\,du\,dv. \tag{1.70}$$

if the random variable is a continuous one.

We frequently interpret the probability distribution by means of the distribution of a unit of mass over the (x, y) plane. By projecting the mass in a two-dimensional distribution on one of the coordinate axes, we obtain the marginal distribution of the corresponding variable. Thus, we have:

$$F_X(x) = P(X \leq x) \quad \text{and} \quad F_Y(y) = P(Y \leq y) \tag{1.71}$$

It is easy to enlarge these considerations to include discrete random variables.

If a two-dimensional random variable $(X,\ Y)$ is given, then the raw statistical moment of the order $k + l$; $k, l = 0, 1, \ldots$ of this variable is given by the pattern:

$$m_{kl} = E(X^k Y^l) = \begin{cases} \displaystyle\sum_i \sum_j x_i^k y_j^l p_{ij} & \text{for both discrete random variables } X \text{ and } Y \\[2ex] \displaystyle\int\limits_{-\infty}^{\infty} \int\limits_{-\infty}^{\infty} x^k y^l f(x, y)\,dx\,dy & \text{for both continuous random variables } X \text{ and } Y \end{cases}$$

$$\tag{1.72}$$

Therefore:

$$m_{10} = E(X) \quad m_{01} = E(Y) \quad m_{20} = E(X^2) \quad m_{02} = E(Y^2) \quad m_{11} = E(XY) \tag{1.73}$$

The last moment above, m_{11} is called the second order product moment or more concisely, the **mixed moment**.

Similarly, we can define the central moments for a two-dimensional random variable (X, Y). The central moment of the order k, l is given by the formula:

$$v_{kl} = E\{[X - E(X)]^k [Y - E(Y)]^l\} \tag{1.74}$$

If $k = l = 1$, then

$$v_{11} = E\{[X - E(X)] [Y - E(Y)]\} = \mathrm{Cov}(X, Y) \tag{1.74a}$$

is called the **covariance** of (X, Y).

This statistical parameter has some interesting properties, namely:

$$\begin{cases} \mathrm{Cov}(X, Y) = \mathrm{Cov}(Y, X) \\ \mathrm{Cov}(X, a) = 0 \qquad\qquad a - \text{any constant} \\ \mathrm{Cov}(X, X) = \sigma^2(X) \\ \mathrm{Cov}(X, Y) = E(XY) - E(X) E(Y) \end{cases} \tag{1.75}$$

If the random variable X is independent of the random variable Y, then their covariance is zero. The converse theorem does not hold.

If the random variables X and Y have $\mathrm{Cov}(X, Y) = 0$, then we say that they are uncorrelated linearly.

Further covariance properties are:

$$-\sigma(X)\,\sigma(Y) \le \mathrm{Cov}(X, Y) \le \sigma(X)\,\sigma(Y) \tag{1.76}$$

$$\sigma^2(X \pm Y) = \sigma^2(X) + \sigma^2(Y) \pm 2\,\mathrm{Cov}(X, Y) \tag{1.77}$$

A generalisation of the concept of variance is the variance matrix or the **variance-covariance matrix**.[27] Such a matrix for the random vector $\left(X_1, X_2, \ldots, X_n \right)$ has the form:

$$\begin{bmatrix} \sigma_1^2 & \mathrm{cov}(X_1, X_2) & \ldots & \mathrm{cov}(X_1, X_n) \\ \mathrm{cov}(X_2, X_1) & \sigma_2^2 & \ldots & \mathrm{cov}(X_2, X_n) \\ \ldots & \ldots & \ldots & \ldots \\ \mathrm{cov}(X_n, X_1) & \mathrm{cov}(X_n, X_2) & \ldots & \sigma_n^2 \end{bmatrix} \tag{1.78}$$

This matrix is symmetric and its determinant is non-negative.

A very important parameter called the **linear correlation coefficient** (ρ) is related to the covariance. It is defined in the following way:

$$\rho = \mathrm{Cov}(X, Y)/\sigma(X)\,\sigma(Y) \tag{1.79}$$

---

[27] Some statisticians call this matrix the variance of the random vector because it is a natural generalisation to the higher dimensions of the one-dimensional variance (see for instance Feller 1957). Others call it the covariance matrix because it is the matrix of covariances between the scalar components of the random vector. The term cross-covariance is also sometimes used. Finally, very often the matrix is called the variance-covariance matrix since the diagonal terms are in fact variances.

This coefficient has several important properties, namely:

- it is supported on [−1, 1] interval;
- if two random variables are stochastically independent linearly, then their correlation coefficient is zero;
- the necessary and sufficient condition for $\rho^2 = 1$ is that the following relationship holds:

$$P(Y = aX + b) = 1 \quad \text{for } a > 0 \tag{1.80}$$

Because of the last relation, the coefficient is called a linear correlation coefficient.

A particular case of a two-dimensional random variable is the Gaussian random variable which is characterised by the probability density function:

$$f(x, y) = \frac{1}{2\pi\sigma_x\sigma_y} e^{\frac{-1}{2(1-\rho^2)}\left[\frac{(x-m_{10})^2}{\sigma_x^2} - 2\rho\frac{(x-m_{10})(y-m_{10})}{\sigma_x\sigma_y} + \frac{(y-m_{01})^2}{\sigma_y^2}\right]} \quad \text{(Figure 1.21)} \tag{1.81}$$

where x, y, m_{10}, m_{01} are supported on an infinite set of numbers, σ_x and σ_y takes only positive values and $\rho \in [−1, 1]$. The expected values m are determined by (1.73).

The expected value of a conditional two-dimensional probability distribution (the conditional expected value) of the random variable X provided that $Y = y$ is determined by the pattern:

$$E(X \mid Y = y_j) = \begin{cases} \sum_i x_i \dfrac{p_{ij}}{p_{.j}} & \text{for a discrete random variable} \\ \displaystyle\int_{-\infty}^{\infty} x \dfrac{f(x, y)}{f_2(y)}\,dx & \text{for a continuous random variable} \end{cases} \tag{1.82}$$

where $\frac{p_{ij}}{p_{.j}}$ and $\frac{f(x,y)}{f_2(y)}$ are the probability mass functions and the probability density function in the conditional distribution of the random variable X, respectively.

There are some two-dimensional random variables that are used in mining engineering problems but their applications are rare; perhaps they will be more useful in the future. Two of these distributions are as follows.

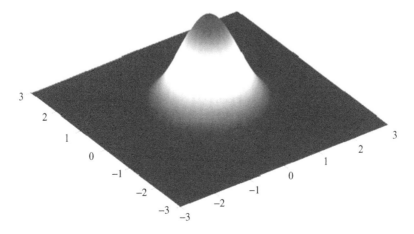

Figure 1.21.   The probability density function of a two-dimensional normal random variable.

a. The two-dimensional composition of a gamma distribution and a standardised normal distribution; this probability density function is given by the formula:

$$f(x, y) = \begin{cases} \alpha \left[y^{a-1} e^{-\left(\beta y + \frac{x^2}{2} \right)} + x^{b-1} e^{-\left(\gamma x + \frac{y^2}{2} \right)} \right] & \text{for } (x, y) \in A \\ 0 & \text{elsewhere} \end{cases} \quad (1.83)$$
(Figure 1.22)

where: $a > 0, b > 0, \alpha > 0, \gamma > 0, \quad A \in \{(x, y): x > 0, y > 0\}$

$$\alpha = \sqrt{\frac{2}{\pi}} \frac{\beta^a \gamma^b}{\gamma^b \Gamma(a) + \beta^a \Gamma(b)} \quad (1.83a)$$

The random variables X and Y that are components of the above function are stochastically dependent because:

$$f(x, y) \neq f(x)\, f(y)$$

It is interesting that the function (1.83) has a constant regression function of the first type equals

$$m = 4[1 + 2\varphi(0)] \cong 7.2$$

This function was applied in a study concerning reliability (Czaplicki 1981).

b. The Morgenstern distribution (Morgenstern 1956, Butkiewicz 1997, Butkiewicz and Hys 1977) for which the probability density function is given by the formula:

$$f(x, y) = f_1(x) f_2(y) \left[1 + \mu(1 - 2F_1(x))(1 - 2F_2(y)) \right] \quad (1.84)$$

where: $\mu \in [-1, 1]$

$$F_1(x) = 1 - \exp\left[-\left(\frac{x}{a_1} \right)^{b_1} \right]$$
$$F_2(y) = 1 - \exp\left[-\left(\frac{y}{a_2} \right)^{b_2} \right] \quad (1.84a)$$

This probability distribution is significant in investigations related to grain size, their magnetic susceptibility and perhaps some further physical-and-chemical properties of grains (Tumidajski and Saramak 2009).

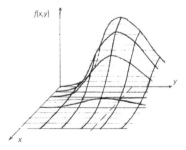

Figure 1.22.   An example of the probability density function (1.83).

1.2.8 *Stochastic process*

Generalisation of the concept of a random variable is a stochastic (random) process.

A **stochastic process** $X(t)$ is a set of random variables depending on the parameter t that belong to a certain set of real numbers. Very often it is presumed that this parameter is time. However, in engineering practice it is much better to analyse many stochastic processes that are dependent on different parameters[28] such as: the total mass of transported material, the work executed by a given machine, the number of work cycles of a piece of equipment, the productivity of a system etc.

When parameter t is fixed, the stochastic process is a random variable. When an elementary event is fixed, a random process is the function of time $x(t)$ that is called the **realisation of the stochastic process**. In order to avoid any misunderstandings that may arise when using the word value (the value of parameter and the value of process), the values that take the random variable $X(t)$ will be called **process states**.

During empirical investigations we observe the realisation of a certain stochastic process only. When the investigation of the process is repeated, we observe stochastic copies of it. A set of such realisations and conditions of the choice of the observation is the statistical information based on which a statistical inference is being made.

We say that a stochastic process $X(t)$ is completely described when for every set of moments $t_1, t_2, ..., t_n$ the joint distribution of the random variables $X(t_1), X(t_2), ..., X(t_n)$ is known. Knowing this distribution, we are able to find the basic characteristics of the process such as: the expected value, its variance, the limited probability distribution of the process states and so on.

An important property of a stochastic process is its stationarity.

Two types of stationarity are distinguished in theory.

We say that a stochastic process has **strict (strong) stationarity** if its joint probability distribution does not change when shifted in time or space. Accordingly, parameters such as the average value and variance, if they exist, also do not change over time or position.

If the first moment and covariance do not vary with respect to the time of a certain stochastic process, we say that such a process has a **wide-sense (weak-sense) stationarity**.

Two types of stochastic processes are widely applied in engineering analyses. These are processes that depend on the continuous parameter t and processes that are determined on the set of nonnegative integer numbers. The latter ones are called **stochastic chains**.

The stochastic processes with the broadest applications in engineering practice are the Markov processes.

A stochastic process $X(t)$ of the continuous parameter t, $t \geq 0$ (or a stochastic chain $X(t)$, $t = 0, 1, ...$) is called a **Markov process (Markov chain)** when for every system of moments $t_1 < t_2 < \cdots < t_n < t$ and any set of real numbers $x_0, x_1, ..., x_n, x$ the following equality holds;

$$P(X(t) < x \mid X(t_i) = x_i, i = 0, 1, ..., n) = P(X(t) < x \mid X(t_n) = x_n) \qquad (1.85)$$

In other words, if the state of the Markov process in moment t_n is known, then any additional information on the states of the process before t_n does not have any influence on the distribution of the process values in later moments. The Markov process can be thought of as 'memoryless'. It is the characteristic property that the process of the changes of the states of a continuous parameter of the Markov type has all of the states of the exponential distribution.

The idea of the Markov chain can be described as follows.

A set of states $\mathfrak{s} = \{\mathfrak{s}_1, \mathfrak{s}_2, ..., \mathfrak{s}_n\}$ is given. The process starts in one of these states and moves successively from one state to another. Each move is called a *step* or *jump*. If the process is

---

[28] Some researchers are of the opinion that time does not exist; it is only a very convenient measure that was invented and introduced by people. Notice that devices measuring time in fact show the progress of a certain physical process only.

currently in state $\mathbf{s}_i$, then it moves to state $\mathbf{s}_j$ at the next step with a probability denoted by p_{ij} and this probability does not depend on which state the process was in before the current state (memoryless property). The probabilities p_{ij} are called transition probabilities. The process can remain in state i with the probability p_{ii}. An initial probability distribution, defined on $\mathbf{s}$, specifies the starting state. In a continuous space, the probability distribution that has a memoryless property is an exponential distribution whereas in a discrete space, it has a geometric distribution. The process is called a 'chain' because not only are the states discrete but the moments when the jumps occur are also discrete ones.

The processes of operation (exploitation), which are understood as processes of changes of states of mechanised systems in mining and also in earthmoving engineering, can be described by the Markov processes in many cases (Sajkiewicz 1982, Czaplicki 2010a for instance). The process of the fluctuation of a mass in bins can be modelled by the Markov process in some cases (Benjamin and Cornell 1970) as can the process of road surface degradation (ibidem).

Nowadays, a growing interest in the application of semi-Markov processes can be observed in engineering practice (Brodi and Pogosjan 1973, Barlow and Proschan 1975, Bousfiha et al. 1996, Bousfiha and Limnios 1997, Limnios and Oprişan 2001, Grabski and Jaźwiński 2001). This trend also concerns the mining engineering field (Czaplicki 2010a and 2010b, Czaplicki and Kulczycka 2012).

A semi-Markov process is a random process $X(t)$ with a finite or countable set of states $\mathbf{s}$ that have stepwise trajectories with jumps at times $0 < \tau_1 < \tau_2 < \dots$ such that

- the values $X(\tau)$ at its jump times form a Markov chain with transition probabilities:

$$p_{ij} = P\{X(\tau_n) = j \mid X(\tau_{n-1}) = i\} \tag{1.86}$$

- the distributions of the jumps τ_n are described in terms of the distribution functions $F_{ij}(t)$ as follows:

$$P\{\tau_n - \tau_{n-1} \le t, \, X(\tau_n) = j \mid X(\tau_{n-1}) = i\} = p_{ij} \, F_{ij}(t) \tag{1.87}$$

This process is called a semi-Markov process (see Korolyuk and Turbin 1976, Limnios and Oprişan 2001, Harlamov 2008)[29].

It is of practical importance to emphasise that states in this type of process are mutually independent.

If $F_{ij}(t)$ are exponential then the semi-Markov process is a continuous-time Markov chain or simply a Markov process. If all of the distributions degenerate to a certain point, the result is a discrete-time Markov chain. In analytic terms, an investigation of semi-Markov processes can be reduced to a system of integral equations (Korolyuk and Turbin 1976).

1.3 BASIC TERMS OF STATISTICAL INFERENCE

An essential task of mathematical statistics is statistical inference. It can only be applied in relation to the outcomes of random experiments. Two cardinal parts of this field of science are the theory of estimation and the theory of the verification of statistical hypotheses.

---

[29] Semi-Markov processes were introduced by Levy (1954) and Smith (1955). Takács (1954, 1955) investigated similar processes. The foundations of the theory of semi-Markov processes were mainly laid by Pyke (1961a, b), Pyke and Schaufele (1964), and Korolyuk and Turbin (1976).

The theory of estimation deals with the assessment of unknown parameters of the general population that are based on a random experiment.

1.3.1 *Estimator*

A basic term of the theory of estimation is an **estimator**. This is understood as a clearly defined function of the outcomes of a random trial implied by an unknown parameter θ of the general population. If so, the **estimator is a random variable**.

Notice that if the point of interest is an unknown parameter θ of the general population and that a random sample of size n is taken to estimate its value, then by applying the estimator T_n we have the following situation:

$$T_n = f(X_1, X_2, ..., X_n \mid \theta) \tag{1.88}$$

because every element of the sample can be treated as a random variable.

A value calculated based on a random sample taken t_n of the estimator is called an **estimate** of the unknown parameter θ. This estimate is given by the equation:

$$t_n = f(x_1, x_2, ..., x_n) \tag{1.89}$$

It is a number, the **deterministic value**.

1.3.2 *Properties of estimators and methods of their construction*

Theoretically, an infinite number of estimators can be constructed. Some of them will give a worse assessment of the unknown parameter, some—better. Thus, we must have the possibility of making an evaluation of which estimator is better and which is worse. Or, to be more precise, which estimator has better properties that will usually yield better estimates because the estimator is a function. This gives us a *tool* with which to select a *good* estimator.

There are two important items connected with each estimator, namely:

- the error of estimation
- the properties that characterise the selected estimator.

It is obvious that one can be almost certain that when making an estimation an error will occur because a statistical inference is made based on a trial only. This error is called the **error of estimation**.

It can be defined as:

$$d = T_n - t_n \tag{1.90}$$

The right side of the equation determines a certain random variable because the function of a random variable is also a random variable. Therefore, the left side of the equation is also a random variable. Therefore, the estimation error is a random variable. It has its own probability distribution and we are able analyse its basic parameters.

To achieve a precise estimation, i.e. to ensure a small error of estimation, it is necessary to pay attention to both the correct sampling and the selection of an estimator with good statistical properties.

A *good* estimator should be, among the other things, characterised by following properties:

- unbiasedness
- consistency
- efficiency
- sufficiency.

It is said that an **estimator** T_n of parameter θ **is unbiased** if its expected value is the same as the value of the estimated parameter[30], i.e.

$$E(T_n) = \theta \tag{1.91}$$

Notice that the **expected value of a random variable is a deterministic value**[31].

An unbiased estimator allows the unknown parameter to be estimated without systematic errors.

The difference:

$$\Delta = E(T_n) - \theta \tag{1.92}$$

is called a **bias of the estimator** T_n. Obviously, the bias of an unbiased estimator is zero.

Let us consider the estimation of an unknown expected value m in a certain general population $\mathfrak{X}$. Let us use the arithmetic mean formula:

$$\bar{x} = \frac{1}{n} \sum_{i=1}^{n} x_i. \tag{1.93}$$

This estimator is an unbiased one since:

$$E(\bar{x}) = E\left(\frac{1}{n} \sum_{i=1}^{n} X_i \right) = \frac{1}{n} \sum_{i=1}^{n} E(X_i) = \frac{1}{n} nm = m$$

The second statistical parameter used most frequently, besides the expected value, is the standard deviation σ or its square—the variance. Presume now that for the estimation of the unknown variance σ^2 of the general population, the following estimator was applied:

$$s^2(X) = \frac{1}{n} \sum_{i=1}^{n} (x_i - \bar{x})^2 \tag{1.94}$$

Let us test the bias of the estimator above. Here we have:

$$E[s^2(X)] = E\left[\frac{1}{n} \sum_{i=1}^{n} (x_i - \bar{x})^2 \right] = E\left\{ \frac{1}{n} \sum_{i=1}^{n} [(x_i - m) - (\bar{x} - m)]^2 \right\}$$

$$= E\left[\frac{1}{n} \sum_{i=1}^{n} (x_i - m)^2 - 2(\bar{x} - m) \frac{1}{n} \sum_{i=1}^{n} (x_i - m) + (\bar{x} - m)^2 \right]$$

$$= \frac{1}{n} \sum_{i=1}^{n} E[(x_i - m)^2] - E[(\bar{x} - m)^2] = \frac{1}{n} \sum_{i=1}^{n} \sigma^2 - \frac{1}{n} \sigma^2 = \frac{n-1}{n} \sigma^2 \neq \sigma^2$$

Thus, the above estimator is a biased one. Its bias is:

$$E[s^2(X)] - \sigma^2 = \frac{n-1}{n} \sigma^2 - \sigma^2 = -\frac{1}{n} \sigma^2$$

---

[30] There are more general notions of bias and unbiasedness. Here 'bias' is *de facto* '*mean*-bias' in order to distinguish *mean*-bias from the other notions with the most noteworthy ones being '*median*-unbiased' estimators. See for example Rojo (2012).

[31] In some statistical investigations, a randomisation of the expected value is made, i.e. we treat that value as a random variable although this is a targeted exception from the rule.

Notice, that when the sample size taken is large, its bias is:

$$\lim_{n \to \infty} \left(-\frac{1}{n} \sigma^2 \right) = 0$$

Therefore, this **estimator** is **asymptotically unbiased** for this reason. It can be applied when the sample size is large; practically $n > 30$.

By analysing the bias of the investigated estimator, a conclusion can be formulated that it gives estimates of the variance that are too low.

It is easy to prove that an estimator:

$$\hat{s}^2(X) = \frac{1}{n-1} \sum_{i=1}^{n} (x_i - \bar{x})^2 \qquad (1.95)$$

of the unknown variance of the general population is an unbiased one. It can be applied in any sample size.

If a sample of N elements is given and it consists of k groups of size n_i; $i = 1, 2, ..., k$, the following equation holds:

$$s_x^2 = \bar{s}_i^2 + s^2(\bar{x}_i) = \frac{\sum_{i=1}^{k} s_i^2 n_i}{N} + \frac{\sum_{i=1}^{k} (\bar{x}_i - \bar{x})^2 n_i}{N} \qquad (1.96)$$

where: $\bar{s}_i^2$—the estimate of variance within the i-th group
$\bar{x}$—the arithmetic mean of the whole sample
$\bar{x}_i$—the arithmetic mean of the i-th group
$s^2(\bar{x}_i)$—the variance between groups.

By looking more carefully at pattern (1.96), a simple conclusion can be drawn—the more differentiated the population, the greater the value of its variance. The relationship (1.96) is called the '**variation identity**' (Sobczyk 1996 p. 46). This pattern is useful in the calculation of some combined machinery systems (Czaplicki 2010a p. 230) as well as in calculations connected with the homogeneity of shovel-truck systems (*ibidem* p. 266).

The second important property that must be investigated before the application of a given estimator is its **consistency**.

It is said that estimator T_n of the **parameter** θ is **consistent** if it converges in probability (stochastic convergence) to the true value of the parameter, i.e. the following equation holds:

$$\lim_{n \to \infty} P\{|T_n - \theta| < \varepsilon\} = 1 \quad \varepsilon > 0 \qquad (1.97)$$

Looking at this relationship, one may easily come to the conclusion that enlarging the sample size leads to a situation in which the estimates that are obtained will be closer and closer to the real value of the unknown parameter θ.

Suppose one has a sequence of observations $\{x_1, x_2, ...\}$ from a normal $N(\mu, \sigma)$ distribution. In order to estimate the unknown expected value μ, one uses the sample mean determined by formula (1.93). Now assume that every element of the sample is a random variable. If so, the estimator (1.93) becomes a random variable. Denote it by T_n. From the properties of the normal distribution, we know that T_n is itself normally distributed with the mean μ and the variance σ^2/n. Equivalently, the random variable $(T_n - \mu)/(\sigma/\sqrt{n})$ has a standard normal distribution. Therefore, the following relationship holds:

$$P(|T_n - \mu| \geq \varepsilon) = P[(|T_n - \mu| \sqrt{n} / \sigma \geq (\varepsilon \sqrt{n}/\sigma)] = 2 [1 - \Phi_N(\varepsilon \sqrt{n}/\sigma)] \to 0$$

as n tends to infinity for any fixed $\varepsilon > 0$. Thus, the estimator T_n of the sample mean is consistent for the population mean μ.

An estimator can be unbiased but not consistent. For example, for independent and identically distributed[32] random variables that are components of a sample $\{X_1, ..., X_n\}$, one can use $T_n = x_1$ as the estimator of the mean $E(X)$.

Alternatively, an estimator can be biased but consistent. For example, if the mean is estimated by the formula $[(1/n) + (\Sigma x_i/n)]$, this estimator is biased, but as sample size n tends to infinity, it approaches the correct value and so it is consistent.

A significant property of estimator is its **efficiency**.

A given estimator is the most efficient if it is unbiased and of the lowest variance among all of the possible the unbiased estimators constructed on samples.

A small value of the variance of a given estimator ensures a small dispersion of the estimates of the unknown value of the parameter that we can obtain from it.

When investigating the efficiency of a given estimator, we compare at least the variances between two estimators of the same parameter. The estimator with a lower variance is more efficient than the other.

Compare, for instance, an efficiency of the estimators of variances basing on the mean and the median. We have:

$$Ef = \frac{D^2(\bar{x})}{D^2(Me)} = \frac{\dfrac{\sigma^2}{n}}{\dfrac{\pi \sigma^2}{2n}} = \frac{2}{\pi} \cong 0.64$$

The final property of estimators to consider is sufficiency.

An **estimator** of an unknown parameter is **sufficient** if it contains all of the information comprised in a sample taken and there is no other estimator that gives more information on the parameter being estimated.

Consider, for example, two unbiased estimators of the expected value $E(X)$. One defined by formula (1.93) and the second determined by the pattern:

$$\tilde{x} = \frac{1}{2}(x_{min} + x_{max}),$$

where x_{min} and x_{max} are the lowest and the highest value in the sample taken, respectively.

The estimator $\tilde{x}$ is insufficient because it takes only two values from the sample. The arithmetic mean takes all of the sample elements into account.

Further considerations concerning the theory and practice of estimation will be conducted in Chapter 4 where a synthesis of the information obtained from a statistical investigation is analysed.

1.3.3 *Statistical hypotheses and their types*

The basic terms in the theory of verification of statistical hypotheses are: statistical hypothesis[33] and statistical test.

A **statistical hypothesis** is any conjecture (supposition) concerning the general population.

In practice, we almost always have some information on the population of interest, e.g. the investigated random variable is a continuous one and we know its physical limits, what values the random variable takes and so on. This information determines a certain **set of admissible**

---

[32] Sometimes the abbreviation 'i.i.d.' is used for the term 'independent and identically distributed'.
[33] A hypothesis (from Greek ὑποτιθέναι—*hypotithenai*, Latin *hypothesis* meaning both 'to put under' or 'to suppose') is a proposed explanation for a phenomenon; a statement requiring verification.

(possible) **hypotheses**. Denote this set by $\mathfrak{B}$. This set determines a set of probability distributions and we know that these distributions may characterise the population. These distributions can be different in both, in formulas that indicate a class of distribution and the values of the parameters can be different (differences in a given class).

Each formulated statistical hypothesis separates a certain subset $\mathfrak{b}$ from the set $\mathfrak{B}$. It can be written as:

$$H : F(x) \in \mathfrak{b} \quad \mathfrak{b} \in \mathfrak{B}$$

If the subset contains only one element (i.e. determines one distribution only), then such a hypothesis is called a simple hypothesis. Otherwise, the hypothesis is a composite (complex) one.

Let us divide the statistical hypotheses remaining in a given class of distribution.
A supposition that is formulated can concern:

- the parameter of the population
- the class of the population.

If a hypothesis is formulated in relation to a parameter of the random variable, we say that the **hypothesis** is a **parametric** one provided that the distribution is known. If a hypothesis is formulated and there is no information on the population, we say that the **hypothesis** is a **nonparametric** one.

Let us presume that the random variable of our interest is a discrete one (e.g. the number of failed machines, the number of occupied service stands, the number of spare parts) etc. Therefore, the set of admissible hypotheses comprises all of the possible distributions of discrete random variables that are nonnegative. If we have a ground to guess that the random variables of interest may be described by a binomial distribution, our hypothesis is both a parametric one and a complex one. A subset $\mathfrak{b}$ contains all of the binomial distributions with different values of parameters. If, in turn, a hypothesis was formulated that the binomial distribution has the parameter $p = 0.1$, then the hypothesis is simple.

The formulation of a statistical hypothesis is a very important part of statistical analysis; however, it sets the challenge of verifying this supposition.

1.3.4 *Statistical tests and critical region*

A **statistical test** is any rule of conduct used to predicate whether the verified statistical hypothesis should be rejected or whether there is no basis to do this. The statement that there is no ground to reject the hypothesis is not the same as stating that the hypothesis is a true one. It may happen that based on the result of a different sample taken for the verification of this supposition, an inference may be altered—in which case the hypothesis should be rejected.

The division of statistical hypotheses into parametric and nonparametric ones means that all tests in statistics are divided into **parametric tests** and **nonparametric tests**.

A statistical test is constructed according to some rules.

Firstly, a hypothesis is formulated that will be the subject of verification. This **hypothesis** is called a **null one** and is noted as:

$$H_0 : F(x) \in \omega_0 \quad \omega_0 \in \mathfrak{B}$$

In addition, an **alternative hypothesis** is also formulated which is different to the null one. Often, it is a denial statement compared to the verified hypothesis. It can be noted as:

$$H_1 : F(x) \in \omega_1 \quad \omega_1 \in \mathfrak{B}$$

and this hypothesis is accepted as the true one when the null hypothesis is rejected.

Notice that a sample $W_n = (x_1, x_2, ..., x_n)$ can be treated as a certain point in the n-dimensional space of trials. Denote a set of all possible results of trials by $\mathfrak{Z}$. A statistical test relies on the determination of such a region $\mathfrak{L}$ that if $W_n \in \mathfrak{L}$, then the verified hypothesis should be rejected. This region is a critical one. If $W_n \in \mathfrak{Z} - \mathfrak{L}$, then the verified null hypothesis can be accepted.

The region $\mathfrak{L}$ is the **area of the rejection of the verified hypothesis** and also the **critical region of the test. The area of the acceptance of the hypothesis** is obviously determined by: $\mathfrak{Z} - \mathfrak{L}$.

Because inference about the properties of the investigated population is conducted based on a sample, there is a real possibility that the deduction will produce an incorrect result. The information contained in the sample may be such that we recognise the verified hypothesis as false and we reject it, although the hypothesis is a true one. Similarly, we may make a mistake by accepting a hypothesis which an untrue one. This means that two possible errors can be made during statistical inference. The relationship between the property of the hypothesis—true or false—and the decisions made during statistical inference is presented in the table below.

	Hypothesis H_0	
Decision	True	False
Reject	I type error	√
Accept	√	II type error

The **probability** that an **error of the first type** will be made is given by the pattern:

$$P(W_n \in \mathfrak{L} \,|\, H_0) = \alpha(\mathfrak{L}) \tag{1.98}$$

whereas the **probability** that an **error of the second type** will be made is given by the pattern:

$$P(W_n \in (\mathfrak{Z} - \mathfrak{L}) \,|\, H_1) = \beta(\mathfrak{L}) \tag{1.99}$$

The best test would be one which ensures a minimum of both errors. Unfortunately, there is no way to attain the simultaneous minimisation of both probabilities. If the probability of making an error of the first type is zero, then the rejection region is an empty set. Thus, the acceptance region overlaps with set $\mathfrak{Z}$, and for this reason, the probability is that relation $W_n \in \mathfrak{L}$ will be 1 for all of the hypotheses. This also means that $\beta(\mathfrak{L}) = 1$ for hypothesis H_1.

In the theory of the verification of statistical hypotheses, tests are constructed in such a way as to minimise the probability of making a type II error presuming that the probability of making a type I error is constant and appropriately low. Such **tests** are called the **most powerful** ones. A certain probabilistic measure is associated with these tests—the probability that a false hypothesis will be rejected and the alternative hypothesis which states the truth will be accepted. This measure is called the **power of a statistical test**.

Therefore it can be written as:

$$P(W_n \in \mathfrak{L} \,|\, H_1) = M(\mathfrak{L}) \tag{1.100}$$

where $M(\mathfrak{L})$ is the **power of the test**.

The relationship between the power of the test and the probability that a type II error will be made is given by the relationship:

$$\beta(\mathfrak{L}) = 1 - M(\mathfrak{L}) \tag{1.101}$$

The task of the general theory of testing statistical hypotheses is to formulate methods for the construction of the best tests, i.e. the most powerful tests. However, in some cases such tests do not exist. Thus, the further task of the theory is to indicate what to do when there is no most powerful test.

Look more carefully at formula (1.98), which provides information about the probability of making a type I error. If the verified hypothesis is rejected due to the information contained in the sample, then it can be assessed that a rare event occurred because the probability α is small. Moreover, if the event was $W_n \in \mathfrak{L}$, then the assumption on the truthfulness of the null hypothesis was wrong. In a case when the event will not happen, we can say that we have no ground to discard the hypothesis. Notice, that we have no basis for evaluating the hypothesis as a true one because true hypothesis can be different. Such a property of statistical tests characterises **tests of significance**. These tests allow the verified hypothesis to be rejected with a high probability when it is false. However, they do not allow the problem of whether the null hypothesis is a true one to be resolved. The probability α is called the **level of significance**. Thus, if this level is assumed to be 0.05 (this is the most frequently presumed level of significance in engineering investigations), it means that taking 5 out of 100 cases on average, the verified hypothesis will be rejected—based on the sample taken—despite the fact that the hypothesis is a true one. The reason for the rejection is connected with the information contained in samples and is not associated with the statistical procedure conducted.

Tests of significance are most frequently used in practice not only in the engineering field and they are very simple in application.

Nonetheless, it should be noted that there is a certain freedom in the construction of tests of significance and this freedom is not only connected with the arbitrary presumed level of significance. Basically, there is independence in the selection of the statistic (estimator) which will be used to estimate the unknown parameter of the general population in parametric tests or in a different type of inference (in nonparametric tests of significance). Practically, the method of the selection of the level of significance and the selection of the statistic alone is usually imposed by what would be found in the literature on the subject. As was stated, the level of significance is usually assumed to be $\alpha = 0.05$ and it is a rare event when this level is higher or lower than that.

There is also a rule that is only one hypothesis is articulated presuming silently that the alternative hypothesis is its negation in tests of significance. In addition, nothing is stated about the level of the probability of a type II error.

When a parametric test of significance is applied, its procedure is as follows.

1. Formulate the basic hypothesis (null one) H_0 stating that parameter $q = q_0$, which means it is suspected that the population of interest has a parameter q of q_0 value
2. Take a sample of size n from the general population
3. Construct the statistic Q, which is a gauge (tester) of the verified hypothesis
4. Determine the probability distribution of the tester and find the critical region $\mathfrak{L}$ for the presumed level α of significance; a condition to fulfil is:

$$P\{Q \in \mathfrak{L} \mid H_0\} = P\{|Q - q_0| > k\} = \alpha \quad k > 0, \quad \text{(Figure 1.23)} \quad (1.102)$$

5. Make a decision based on the result of the investigation: reject the verified hypothesis or there is no basis to discard it.

A different case can also be considered. If an estimation of the unknown parameter is obtained using the information contained in the sample taken and its value is $\hat{q}$, then the corresponding probability can be calculated from the formula:

$$P\{|Q - q_0| > \hat{q}\} = p \quad (1.103)$$

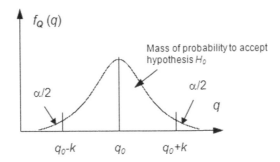

Figure 1.23. Graphical illustration of the critical region determined by formula (1.102).

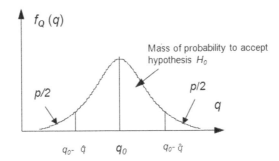

Figure 1.24. Graphical illustration of the critical region determined by the pattern (1.103).

If the following inequality holds:

$$p > \alpha \qquad (1.104)$$

then there is no basis to reject the verified hypothesis H_0 (Figure 1.24).

1.3.5 *Probability distributions for the verification of statistical tests*

Coming to the end of the consideration of the basic terms used in mathematical statistics, let us briefly review the most important probability distributions that are applied in the verification of the statistical hypotheses that are frequently used in engineering practice, not only in mining.

Many tests are based on the normal distribution when the hypotheses concern the parameters of random variables, especially in cases when some parameters are known or when the sample taken is large. However, when conditions of application of the normal distribution are not fulfilled then several different probability distributions can be applied. These are:

- the Student's *t* probability distribution
- the χ^2 (chi-squared) probability distribution
- the *F* Snedecor's (Fisher-Snedecor) probability distribution.

In some cases (see for instance in the chapter concerning prediction), the Fisher's *z* distribution is useful.

We say that the random variable t_r has a **Student's probability distribution** with r degrees of freedom (this is a parameter of the random variable) when its probability distribution is given by the formula:

$$S_r(t) = P\{t_r \le t\} = \frac{2}{\sqrt{r}\,B(1/2, r/2)} \int_{-\infty}^{t} \frac{dt}{(1+t^2/r)^{(r+1)/2}}, \quad t \in (-\infty, \infty) \qquad (1.105)$$

Notice that there is a function beta defined by the pattern (1.62) in the denominator here.

Example plots of the probability density function of the Student's distribution are shown in Figure 1.25.

The mode and median are zero and the expected value provided that $r > 1$, otherwise is undefined.

The density of this random variable is symmetric with regard to point $t = 0$ and for this reason:

$$S_r(-t) = 1 - S_r(t)$$

Let there be n independent measurements of a random variable X. Let $\bar{x}$ denote the mean from the sample and $\hat{s}(X)$ is the unbiased estimator of the standard deviation (1.95) in the sample. A random variable defined by the formula:

$$t_r = (\hat{s}(X))^{-1}(\bar{x} - m)\sqrt{n} \qquad (1.106)$$

where m is the expected value of the random variable of interest is the Student's random variable with $r = n - 1$ degrees of freedom[34]. The random variable which probability distribution function is given by formula (1.105) is the same as the random variable determined by the formula:

$$t_r = (s(X))^{-1}(\bar{x} - m)\sqrt{n-1} \qquad (1.106a)$$

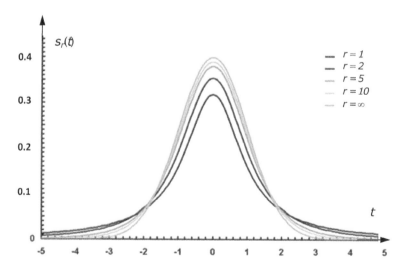

Figure 1.25.   Student's probability density functions.

---

[34]The Student's *t*-distribution is a special case of the generalised hyperbolic distribution.

where $s(X)$ is the biased estimator of the standard deviation.

A number $t(\alpha, r)$ which fulfils the equation:

$$P\{|t_r| > t(\alpha, r)\} = \alpha \qquad (1.107)$$

is called a critical value of the distribution t_r.

For a large r the Student's random variable has approximately the normal standardised probability distribution $N(0, 1)$[35].

Tables of the critical values $t(\alpha, r)$ are given at the end of this book—Table 9.3.

Random variables defined by patterns (1.105) and (1.105a) are applied in the interval estimation of the mean and in the verification of some statititical hypotheses. This will be a point of interest in a further part of the book.

A random variable defined as the sum of squares of n independent random variables X_i, $i = 1, 2, ..., r$ having the identical normal distribution $N(0, 1)$ is the random variable of the **chi-squared distribution** with r degrees of freedom. Thus, it can be written as:

$$\chi_r^2 = X_1^2 + X_2^2 + \cdots + X_r^2 \qquad (1.108)$$

The random variable χ_r^2 has the chi-squared probability distribution with r degrees of freedom if its cumulative function is given by:

$$F_r(\chi^2) = P\{\chi_r^2 \leq \chi^2\} = \frac{1}{2^{r/2}\Gamma(r/2)} \int_0^{\chi^2} u^{(r/2)-1} e^{-u/2} du, \quad \chi^2 > 0 \qquad (1.109)$$

Typical charts of the probability density function of the chi-squared random variable are shown in Figure 1.26.

The expected value and the variance are determined by the patterns:

$$\begin{cases} E(\chi_r^2) = r \\ \sigma^2(\chi_r^2) = 2r \end{cases} \qquad (1.110)$$

For large values of r, the random variable $(2\chi_r^2)^{1/2} - \sqrt{2r-1}$ has approximately the standardised normal distribution $N(0, 1)$.

A number $\chi_\alpha^2(r)$ that satisfies the equation:

$$P\{\chi_r^2 \geq \chi_\alpha^2(r)\} = \alpha \qquad (1.111)$$

is called a critical value of the chi-squared distribution; $\chi_\alpha^2(r)$ is the quantile of the order $(1 - \alpha)$ of the chi-square distribution with r degrees of freedom.

If r is large, the critical values can be calculated using the following approximation:

$$\chi_\alpha^2(r) \cong r\left(1 - \frac{2}{9r} + u_{1-\alpha}\sqrt{\frac{2}{9r}}\right)^3 \qquad (1.112)$$

---

[35] The Student's t-distribution, especially in its three-parameter (with a location-scale) version, arises frequently in Bayesian statistics as a result of its connection with the normal distribution. Whenever the variance of the normally distributed random variable is unknown and a conjugate prior placed over it that follows the inverse gamma distribution, the resulting marginal distribution of the variable will follow the Student's t-distribution.

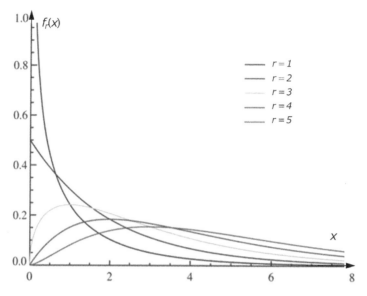

Figure 1.26.   The Chi-squared probability density functions.

or

$$\chi_\alpha^2(r) \cong \frac{1}{2}\left(\sqrt{2r-1} + u_{1-\alpha}\right)^2 \qquad (1.112a)$$

where u_α is the quantile of the order α of the standardised normal distribution (see Table 9.2).
Notice that the random variable:

$$t_r = U\sqrt{r}(\chi_r^2)^{-1/2} \qquad (1.113)$$

where U is the standardised normal random variable is the Student's random variable with r degrees of freedom.

Tables of the critical values $\chi_\alpha^2(r)$ are given in Table 9.4.

The chi-squared distribution is one of the most widely applied probability distributions in inferential statistics, e.g., in hypothesis testing or in the construction of confidence intervals. The chi-squared distribution is used in the common chi-squared tests for the goodness-of-fit of an observed distribution to a theoretical one (Chapter 4.2), the independence of two criteria of the classification of qualitative data (Chapter 5.1) and in the estimation of the confidence interval for a population standard deviation of a normal distribution from a sample standard deviation. Many other statistical tests also use this distribution, like the Friedman's analysis of variance by ranks.

We say that a random variable F_{r_1,r_2} has **the Snedecor's F** distribution (or F-distribution or the Fisher-Snedecor distribution, after R.A. Fisher and G.W. Snedecor) with (r_1, r_2) degrees of freedom if its distribution function is determined by the pattern:

$$P\{F_{r_1,r_2} \le x\} = \frac{\Gamma\left(\dfrac{r_1+r_2}{2}\right)}{\Gamma\left(\dfrac{r_1}{2}\right)\Gamma\left(\dfrac{r_2}{2}\right)} r_1^{r_1/2} r_2^{r_2/2} \int_0^x y^{(r_1/2)-1}(r_2 + r_1 y)^{-(r_1+r_2)/2} dy, \quad x > 0 \qquad (1.114)$$

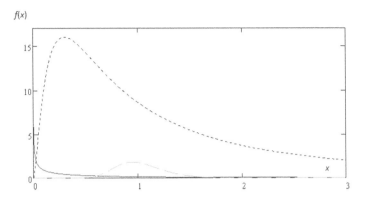

Figure 1.27. The Snedecor's **F** probability density functions.

A few probability density functions of the Snedecor's distribution are shown in Figure 1.27.

The number $F_\alpha(r_1, r_2)$ that satisfies the equation

$$P\{F_{r_1, r_2} \geq F_\alpha(r_1, r_2)\} = \alpha \tag{1.115}$$

is called a critical value of the **F** distribution; this is the quantile of the order $(1 - \alpha)$ of this distribution.

The critical values satisfy the equation:

$$F_\alpha(r_1, r_2) F_{1-\alpha}(r_1, r_2) = 1 \tag{1.116}$$

and also

$$F_{r_1, r_2} F_{r_2, r_1} = 1 \tag{1.117}$$

If the random variables χ_r^2 and χ_s^2 are of the chi-sqaured distribuiton with r and s degrees of freedom, respectively, then the random variable:

$$F_{r,s} = (\chi_r^2/r) : (\chi_s^2/s) \tag{1.118}$$

has the Snedecor's **F** distribution of (r, s) degrees of freedom[36].

Let $S_r(t)$ be the cumulative function of the Student's distribution with r degrees of freedom and let $t(\alpha, r)$ be the critical value in this distribution. Then

$$S_r(t) = \frac{1}{2}[1 + (\text{sign } t)P\{F_{1,n} < t^2\}] \tag{1.119}$$

$$F_\alpha(1, r) = t^2(\%, r) \tag{1.120}$$

$$F_\alpha(r. r) = 1 + \frac{2t^2(\alpha, r)}{r}\left[1 + \text{sign}(0.5 - \alpha)\sqrt{1 + \frac{r}{t^2(\alpha, r)}}\right] \tag{1.121}$$

where $\text{sign } x = -1$ for $x < 0$, 0 for $x = 0$ and $+ 1$ for $x > 0$.

---

[36] Because the chi-squared distribution is a particular case of the gamma distribution then the random variable of the **F** Snedecor distribution can be treated as the quotient of two independent random variables of gamma distribution.

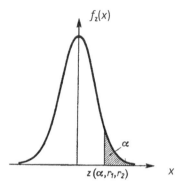

Figure 1.28.   The Fisher's z probability density function.

Tables 9.5–9.7 contain the critical values $F_\alpha(r_1, r_2)$.

We say that a random variable has a z **Fisher's distribution** with (r_1, r_1) degrees of freedom if its distribution function is given by the formula:

$$P\{\boldsymbol{Z}_{r_1, r_1} < x\} = \frac{2r_1^{r_1/2}r_2^{r_2/2}}{B(r_1/2, r_2/2)} \int_{-\infty}^{x} \frac{e^{zr_1}dz}{(r_1 e^{2z} + r_2)^{(r_1 + r_2)/2}} \quad (1.122)$$

An example of the probability density function of the Fisher's z distribution is sown in Figure 1.28.

The number $z(\alpha, r_1, r_2)$ that satisfies the equation

$$P\{\boldsymbol{Z}_{r_1, r_1} \geq z(\alpha, r_1, r_2)\} = \alpha \quad (1.123)$$

is called a critical value of the z distribution.

If the random variable $\boldsymbol{F}_{r,s}$ has the distribution F with (r, s) degrees of freedom then the random variable:

$$\boldsymbol{Z}_{r,s} = \tfrac{1}{2} \ln \boldsymbol{F}_{r,s} \quad (1.124)$$

has the z Fisher's distribution with (r, s) degrees of freedom.

When the degrees of freedom become large $(r, s \to \infty)$, the Fisher's distribution approaches normality with the mean and variance determined by the formulas:

$$E(\boldsymbol{Z}_{r,s}) = \frac{s^{-1} - r^{-1}}{2} \quad \sigma^2(\boldsymbol{Z}_{r,s}) = \frac{s^{-1} + r^{-1}}{2} \quad (1.125)$$

This book is primarily intended to present the relationships between mathematical statistics and mining engineering practice and for this reason, the next chapter presents a description of an exploitation (operation) process of a technical object and the problems of a statistical nature that are associated with it.

CHAPTER 2

Some areas of the application of mathematical statistics in mining

In English the term 'mining' usually means[1]:

a. The act
b. A kind of process
c. A branch of industry

all of which are connected with the extraction of useful mineral[2]. Here, we neglect its association with the military.

Mining engineering is an engineering discipline that involves the practice, the theory, the science, the technology and the application of extracting, hauling (sometimes dumping) and processing minerals from a naturally occurring environment. Mining engineering also comprises the processing of minerals for additional value. Thus, in mining engineering the points of interest are the identified processes and their properties and also the properties of objects, understood here in a broad sense.

These processes are of different natures. They are connected with methods of winning rocks and the identification of mineral deposits, their extraction, haulage and ore dressing, dumping of overburden etc. Different processes are connected with the operation of the equipment involved in mining development and the point of interest here is the exploitation process of pieces of equipment, their parts and assemblies and also entire machinery systems. In mining, the interesting processes are those that accompany mining development, i.e. the displacement of rocks due to rock extraction and all of the repercussions connected with this process.

The objects of interest are mainly of two kinds:

- Pieces of equipment that are a part of mine development and
- Surrounding rocks near a mine.

Changes in the properties of these two kinds of objects during a mining operation require the greatest attention of mining engineers.

Many problems considered in this book concern technical objects, and whether this object means a single item or a system does not matter. For this reason, it seems worth considering some aspects of the properties and life course.

Each technical object basically has three characteristic phases of its life.

It is presumed that the source of the birth of any technical object is 'a need'[3]. There is a certain technological process and there is a need to design an object that will be able to realize this process or that will be of service in order to realize some phases of the process. In some cases 'a need' may be different. There is a piece of equipment which properly fulfils its duties but its parameters are not so advantageous compared to modern standards. This may concern the output that is usually attained by this item, its low effectiveness, the average level of its reliability and so on. All of these indicate that there is 'a need' to create a new item

---

[1] In English books, one may find slightly different definitions, e.g. SME Mining Engineering Handbook (1973, 2011).
[2] In some other languages the term 'mining' only means (b) and (c) (e.g. in Polish).
[3] A *need* here is a concept of primary (primitive notion) as in economics. One can find a definition of a need in psychology, e.g. it is such a state of an individual that is a deviation from an optimal state.

with better parameters. Sometimes, a competitive company has just presented a new piece of equipment with better characteristics than our product. Finally, pressure sometimes appears to get better achievements. Our factory has modern machines, well-trained personnel with the knowledge and experience to produce a given piece of equipment cheaper. There is 'a need' to create a new item.

In the first stage the object does not exist physically. It comes into being. It commences its existence in concepts, models, drafts, notes and in virtual notations. Later on it becomes more concretized; it appears as a list of parts and assemblies, a description of their mechanical and electrical connections and comes into being in a calculation procedure. This stage is ended when the design and construction documentation is completed. Actually, this is a virtual version. The non-existent object has got its properties—forecasted properties.

The second stage of an object's life is its production. The item is formed physically. Its final properties are created during the production process, i.e. the features of the object which will characterize it in the third phase of its life. Properties—these real ones—are usually rather different than those given in the first stage of its life by designers and constructors; the production process is not an ideal realisation of their intentions. This stage is finished when the object exists physically and is ready to be transferred to the user.

When the object is purchased by its user, the third phase of its life usually begins[4] and its usage commences. This is a process that continues over time and for the majority of objects is accompanied by the process of its maintenance[5]. These two processes interlace each other. The object realizes the purpose of its existence. Also, the process of changes in its properties commences. Elements of the object begin to show wear and tear. Some of these occur in a significant way and failures occur. The object becomes more and more degraded. However, for objects that can be renewed periods of maintenance occur (repairs, prophylactic actions, adjustments) and the process of degradation is reduced. Periods of maintenance happen in either a random way or in a deterministic way if planned. The process of changes in an object's properties used to be termed the **exploitation process**[6].

There are two essential terms of exploitation theory associated with the term 'exploitation process'. These are: the *state of the object* and the *exploitation events*.

During the object's exploitation, i.e. during the process of the object's utilisation and maintenance, the properties of the object change. For some features these changes will be of a continuous type, sometimes slow, sometimes transitional and sometimes drastic. Therefore, an object at a given moment in time is not identical to the object at a different moment in terms of its properties. In order to describe the process of these changes the term *state* is applied.

When defining a set of an object's essential properties $\mathfrak{C}$, $\mathfrak{C} = \{c_1, c_2, ..., c_m\}$, we can say that the state of the object at time t is determined by a certain function:

$$\mathfrak{G}(t) = f[\mathfrak{C}(t)] = f[c_1(t), c_2(t), ..., c_m(t)].$$

Kaźmierczak (2000, p. 119) gave a similar assessment of the term 'state': *under the term state of object we are going to understand here a 'photograph' of values of object properties in a given moment of time.*

In practical applications this function is not considered to be a continuous one. Discretisation occurs regularly and states are named. These names are usually associated with the

---

[4] In some considerations more than three phases of an object's life are taken into account, e.g. storage or montage in the final operation place.
[5] There is a class of technical objects that cannot be renewed, e.g. hoist head ropes or balance ropes.
[6] Some researchers are of the opinion that 'the exploitation process of an object' is everything that happens with the object from the moment of the end of its production until the moment of its final withdrawal from utilisation (Kaźmierczak 2000, p. 156).

physical nature of the state, e.g. repair state, work state, stand-still state and so on. Notice that a simple conclusion can be made here: The exploitation process of an object can be understood as the sequence of the states of the object or—the usual formulation—as the process of the changes of states.

As a result of this discretisation, at each moment when a change of state occurs, an *exploitation event* takes place. Sometimes, such events are *visible* and to some extent *perceptible*, e.g. a certain element of the object fails and the machine ceases its operation. Sometimes events are conventional ones—nothing physically happens apart from the fact that a certain object parameter exceeded its assumed limited value, e.g. a brake lining worn excessively. At this moment, it is assumed that the object is in a different state.

Based on the consideration above, different theoretical models of the exploitation processes of technical objects can be constructed.

The simplest model is the one that describes the process of impulses (Figure 2.1). An object operates and as a parameter of the process a time is taken into account. At moments t_1, t_2, t_3, ... interesting exploitation events occur, e.g. failures. In the process considered, only **one state** is distinguished and the process has **one type of exploitation events**. At first glance, this model looks very simple. However, when considering it more carefully, many significant problems arise.

Many essential questions connected with this process can be formulated; essential for the object's user and for the object's constructor and producer. Some of these questions are as follow:

– What kind of statistical properties does the observed sequence of times $\{t_{i+1} - t_i\}$, $i = 1, 2,$ 3, ... have?
– If this sequence is a stationary one and it has no peculiar properties, then what kind of probability distribution can be used to associate it with the random variable: 'time between neighbouring impulses'?
– If there is a possibility of two or more failures occurring at the same moment of time, then is this possibility stable or not?
– If the probability of the occurrence of failures is independent of time, then what is the probability distribution that describes the number of failures that can occur at a given moment in time?
– Until what moment the analysis of the course of this process makes sense?

To obtain answers to the above questions it is necessary to have a knowledge of the physical nature of the object as well as to have the knowledge and skills to conduct a proper mathematical analysis. It is necessary to use the appropriate mathematical tools from the probabilistic area as well as from the field of mathematical statistics. All of these should be undertaken in order to identify the properties of the object exploitation process that is described by the theoretical model just presented.

Secondly, the most frequently used theoretical model is the model of the process of changes of states that is illustrated in Figure 2.2.

The object is used and in moments t_1, t_2, t_3, ... interesting exploitation events occur— changes of states. This model is more complicated than the previous one. The **exploitation**

Figure 2.1.   Exploitation process as a process of impulses.

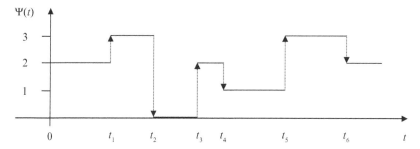

Figure 2.2.   Model of an exploitation process as the process of changes of states.

repertoire 𝕽𝕮 is associated with this model. It is determined by the set of states of the object that can happen during the operation of the object:

$$\mathbf{\mathfrak{RC}} = <\mathbf{s}_1, \mathbf{s}_2, ..., \mathbf{s}_m> = <\mathbf{s}_i; i = 1, 2, ..., m>$$

The set **𝕾𝕻** of possible transitions between states is associated with this set:

$$\mathbf{\mathfrak{SP}} = <\lambda_{ij}; i, j = 1, 2, ..., m>$$

where λ_{ij} are the intensities of the transitions between states. For some stochastic processes the probabilities are considered instead of the intensities.

The determination of both sets can be achieved by logical analysis bearing in mind the operational reality. Notice that a special case of the above process is the process of changes in states: work-repair type, which is very well known in reliability theory (e.g. Gnyedenko et al. 1969, Kopociński 1973). The estimation of measures associated with the individual nonzero elements of the sets requires an extensive analytical procedure and here again one may formulate a list of questions that are worth answering.

– What kind of statistical properties does each sequence of times of states have?
– If a given sequence is stationary and has no peculiar properties, then what kind of probability distribution may be used to describe *well*, in a statistical sense, the random variable 'time of given state'?
– Are times of states independent of each other?
– If some states are stochastically dependent, then what kind of random dependence is it?
– Until what moment the analysis of the course of this process makes sense?

The above list is similar to the previous one. Nevertheless, the necessary analysis connected with a trial to get answers to these questions is more complicated and more comprehensive.

In both lists of questions is the presumption stating 'if this sequence is a stationary one and it has no peculiar properties then ...'. Let us explain what this means.

When a statistical observation has been made and a sample has been taken, one obtains a sequence of the times of a given state $\{t_1, t_2, ..., t_n\}$. And here a fundamental question should be formulated: What kind of stochastic properties characterize the sequence? These properties contain rich information about the running physical processes in the object being investigated and about the repercussions of these processes. An exploitation reality is determined by the particular realisation of a given sequence. It is worth remembering that three elements determine the course of the exploitation process of a technical object, namely:

• Properties of the object given by the designer, the constructor and set up by the producer
• Properties of the surroundings of the object (recall: surroundings of an object includes everything that is around it and that remains in a certain interaction with it)
• Executed policy of use and maintenance of the object.

Each sequence observed contains encoded information. By performing an adequate analysis and using suitable statistical tools in an appropriate sequence, we are able to decode it and translate it into engineering language. Usually, we are looking for answers, among other things, to the following questions:

– Is the observed sequence homogeneous or does it perhaps have untypical elements that distinctly differ from the others?
– Is the observed sequence stationary or is there a trend in it?
– Is there a stable dispersion of values in the sequence or does it depend on time?

Obviously, the above list can be extended according to our needs.

A trial to answer the above listed questions is the subject of consideration in the next chapter, which is the first part of an analysis conducted in the book. The next step will be the statistical synthesis: The estimation of the parameters of random variables that are being investigated as well as finding the theoretical probability distributions that describe the empirical distributions *well*[7].

A further part of the considerations will concern a case in which two or more random variables are observed. Two problems are important here—the investigation of whether random variables are independent of each other and—if not—an examination of the interdependence (in the shape of a correlation) between the variables. The second problem comprises a much broader scope of analysis than the first one.

The next part, in turn, comprises the second stage of the statistical analysis. If there is information that random variables are stochastically dependent, the problem is: What kind of relationship exists between them? This part will present a description in the form of a function illustrating this stochastic interdependence. Consideration commences from a simple linear regression analysis and linear transformations up to multidimensional models. Next, more advanced models will be presented starting from autocorrelation and autoregression models, through classical linear regression for many variables and regressions when errors in the values of random variables are traced. This part of the consideration concludes by taking into account that in some cases there is additional information on the random variables examined and this information should be included in the study in order to improve statistical inference conducted.

Chapter 7 contains a special topic—statistical prediction. There are many problems connected with any inference about the future. They concern terminology, definitions in use, the areas of study and so on. In this chapter some order is presented in this regard and a few examples are presented based on data taken from practice.

The penultimate chapter is a supplement where basic statistical terms are defined in order to better understand the considerations presented in the book.

The book concludes with Chapter 9 in which a set of tables to carry out statistical inference is included.

---

[7] In this book the phrase '...describe *well* ...' will be found in many places. This term 'well' does not have the commonly understood meaning in this context but it is the use of 'well' in a statistical sense. This means further that the statistical investigation was conducted and a positive result was obtained. Thus, we are authorized to state that, for example: 'This model describes *well* the empirical data.'

CHAPTER 3

Analysis of data

The subject of consideration in this chapter is the study of the properties of the observation outcomes noted during the realisation of a statistical investigation. It is presumed that the recorded data has the form of a sequence of numbers. In order to translate this into statistical language—a sample is taken and it is necessary to make a primary analysis; this is the primary because a further part of the analysis depends on the information obtained from it. This information can be of different natures and this will determine the direction of the later investigations into different areas of the statistical consideration.

3.1 TESTING OF SAMPLE RANDOMNESS

When the decision is made to observe only a certain part of the population that was the point of interest, we must be sure that this separated representation will have all significant properties of the population. Practically, a point of interest is one feature or a few features, and it is expected that this separated part in the form of a sample will characterise the whole population *well*, i.e. the **sample** will be **representative**. A long time ago, it was stated that a sample is representative when it is taken in a random way[1]. And that this is a necessary condition. Therefore, the subject under consideration here will be a certain property of a sample—its randomness.

Neglect here the problem of the definition of randomness. By studying some papers concerning randomness (starting from Kendall and Smith 1938 up to Wolfram 2002 p. 1067), some subtle differences can be noticed. It is also dependent on whether the consideration is in the area of mathematical statistics or whether attention is being paid to engineering problems. Our approach to randomness is a typical engineering one—a sample was taken and we want to know whether it is random. In order to resolve this problem it is necessary to choose the appropriate statistical tool—a test. This problem can be solved in the area of the theory of statistical hypotheses.

It is said that there are not many tests that can be used in this case[2]. Here our analysis will be based on a number of series (runs) in a sample taken using the median of the variable that is being investigated[3].

The procedure of the test is as follows.

3.1.1 *The test procedure*

The sample is in the form of a sequence of numbers that are successively noted according to their occurrence. One should order the sequence monotonically in order to estimate the median. The number in the middle is the estimation of the unknown median from the

---

[1] If all of the samples of the same size have an equal chance of being selected from the general population, we say that the samples have a random character.

[2] By the way, some tests for the randomness of a sample—for example 'the rank correlation test for the randomness of a sample' presented by Gopal (2006, test 71)—do not test the randomness of a sample. Gopal's test is, in fact, a test for the stationarity of a sequence. Similar examples of improper statements can be found in the literature related to the subject.

[3] Recall, the median was defined by formula (1.34).

population. When the sample has an even number of items, then the arithmetic mean of the two middle numbers is an estimate of the median. Now, the original sequence of numbers translates into a sequence of plus and minus signs; each number will have a sign. A plus sign will be given to all numbers greater than the median and a minus sign will be associated with all numbers lower than the median. Any numbers equal to the median should be rejected. The next step in the test is the calculation of the number of series, i.e. the number of monomial signs. Denote the number of + signs by n_+ and the number of − signs by n_-.

Now, the statistical hypothesis H_0 is formulated, which proclaims that the elements of the sample were selected in a random manner, whereas the alternative hypothesis H_1 rejects H_0. In order to verify the premise H_0, one compares the number of series in the sample with the critical value that is taken from the statistical table for the given number of signs and a presumed level of significance α. The critical region consists of two sub-regions: the left side and the right side, which means that there are two limited values: the minimum number of series (the critical region for α/2) and the maximum number of series (the critical region for $1 - \frac{\alpha}{2}$). If the number of series falls between these two critical values, we have no ground to reject the null hypothesis. Otherwise, the alternative hypothesis H_1 is the true one. This means that the rejection of the verified supposition is a consequence of the fact that there are either too many series in the sample or too few series.

Consider an example.

■ Example 3.1

A reliability investigation of selected machines was carried out in an underground copper mine. The sequence of the repair times of one LHD machine was noted:

2.5; 1.4; 4.3; 0.8; 3.2; 0.4; 2.2; 3.4; 5.4; 7.2; 0.9; 2.8; 2.9; 1.8 h.

Verify the hypothesis that the observed sequence is random.

By arranging the sequence monotonically we have:

0.4; 0.8; 0.9; 1.4; 1.8; 2.2; 2.5; 2.8; 2.9; 3.2; 3.4; 4.3; 5.4; 7.2

The sample contains 14 elements. Calculate the median of the sample:

$$Me = \frac{2.5 + 2.8}{2} = 2.65$$

Convert the original sample into a sequence of signs. We have:

$$- - + - + - - + + + - + + -$$

The number of the series is 9, the number of signs $n_+ = n_+ = 7$. Presuming a level of significance α = 0.05 and using Table 9.8, we have:

$$K_{\alpha/2}(7, 7) = 3 \quad \text{and} \quad K_{1-\alpha/2}(7, 7) = 12$$

The empirical number of the series fulfils the inequality $3 < 9 < 12$, thus we have no ground to reject the verified hypothesis H_0. We can now agree with the statement that the sample has a random property.
◄

The statistic that is the number of series also has an application in the verification of the hypothesis that proclaims that the two samples are from the same population.

As a rule, tables of the critical values for a series comprise up to 20 signs so the problem arises of what to do when a sample size is greater than 20. For large n_+ and n_- the series number distribution can be satisfactorily described by the normal distribution $N(m, \sigma)$ of the parameters determined by the formulas:

$$m = \frac{2n_+n_-}{n_+ + n_-} + 1 \tag{3.1}$$

and

$$\sigma = \sqrt{\frac{2n_+n_-\left(2n_+n_- - n_+ - n_-\right)}{\left(n_+ + n_-\right)^2\left(n_+ + n_- - 1\right)}}. \tag{3.2}$$

The above relationships can be used for approximate calculations.

3.1.2   *Results of a randomness investigation*

The finding that a given sample is non-random in mining practice does not occur frequently. This regularity is undoubtedly associated with the fact that studies are usually prepared with certain insight and diligence bearing in mind observations of the conditions for proper investigations. Nevertheless, there are some realisations of random variables in mining engineering that are non-stationary ones, e.g. the total number of wire breaks versus the time in the hoist head ropes (or better—versus the number of hoist cycles executed). In this case, one observes the realisation of a non-stationary random process and randomness testing makes no sense.

If—as the result of statistical testing, the non-randomness of the sample was stated—we cannot make any further statistical inference concerning the random variable except to trace why this regularity has been noticed.

There are many reasons for such a set of circumstances. One possibility is the existence of a cyclic component in the realisation of the observed random variable. The operation of many pieces of equipment in mining has a cyclic character and this periodicity can generate a cyclic component in the process of their exploitation. A stream of rock that is being excavated—it does not matter whether it is continuous or discrete—has a periodic character because of the cyclic character of a mining operation. And again, this can have an influence on processes that are running in mining. Another possibility is that during the repair of a technical object, a certain assembly has been replaced by an assembly from a different machine. As a rule, this new item is much more sensitive to failures than the original one. It can be much more susceptible to the periodic character of the operation. Generally, all these 'abnormal' events can generate non-randomness in the data observed.

For a researcher carrying out an investigation, information about the non-randomness of the sample should be a clear signal that something untypical was noticed. Finding the reasons for this untypical regularity is by all means recommended. It may be the source of significant information on the object being investigated and it does not matter whether it is a technical item, a process or a property of the surrounding rocks. Sometimes, the reason can be prosaic—an informatics error in the system that is collecting the data.

3.2   AN OUTLIER IN A SAMPLE

This problem can be included in the category of homogeneity investigations. Homogeneity alone can be defined in different ways depending on the subject of the research and in what sense it is understood. An example in this regard are studies on the homogeneity of a shovel-truck system that were presented in Czaplicki's book (2010, Chapter 10). The basic statistical measure of the homogeneity of the machinery system was the standard deviation and also the probability distribution that was the result of the calculation scheme. Homogeneity, which is the point of interest here, is understood as a property of a random variable being investigated. This property relies on the possibility of describing data by using a model that is the probability distribution function.

When a sample is taken, there is sometimes one outcome, very rarely two, that does not fit the sample at first glance due to the fact that this observed value clearly differs from the others. The outcome either has a very high or a very low value compared to the other data. In mathematical statistics such an outcome is termed an *outlier*[4]. Immediately the question arises as to whether this number belongs to the sample or whether it has been put into the data due to a wrong decision, a mis-recording or an error in the measurement. Outliers are often easy to spot in histograms.

A **histogram** is an important tool in statistics and is a graphical representation that shows a visual impression of the distribution of data. It is an estimate of the probability distribution of a continuous variable and was first introduced by Karl Pearson (1895). A histogram consists of tabular frequencies, which are shown as adjacent rectangles that are erected over discrete intervals (called bins) with an area equal to the frequency of the observations in the interval. The height of a rectangle is also equal to the frequency density of the interval, i.e. the frequency divided by the width of the interval. The total area of the histogram is equal to the number of the items of data. A histogram may also be **normalised** in order to display relative frequencies. It then shows the proportion of cases that fall into each bin/category with the total area equalling 1. The categories are usually specified as successive, non-overlapping intervals of a variable. The bins must be adjacent, and often are chosen to be of the same size. The rectangles of a histogram are drawn so that they touch each other to indicate that the original variable is continuous. An example of a histogram with an outlier being far to the right is shown in Figure 3.1.

An outlier is also easy to trace in a conventional x–y diagram when the data gathered concern the relationship between these two variables. A set of collected data shown in the coordinate system and an outlier visible far beyond main course of the rest of data are presented in Figure 3.2.

The main problem that arises when an outlier is noticed in a sample is whether this extraordinary number belongs to the probability distribution that describes data observed *well* or whether it does not.

Ignoring the problem that authors have given different definitions of an outlier (see, for example, Grubbs 1969, Zeliaś 1996, Moore and McCabe 1999, Czekała 2004, www.math-words.com/o/outlier.htm), it is worth noting that mathematical statistics has a variety of tests that allow the hypothesis that states that such an outcome belongs to a given sample to be verified. Literature from this field is rich (cf. Barret and Lewis 1994) and particular problems of untypical observations have been considered by many authors for years (see, for example, Fisher 1929, Gnyedenko et al. 1965, Czaplicki 2006).

Let us look at the problem of outliers from a practical point of view. Sometimes in the reliability investigations of pieces of equipment, a repair occurs which has evidently taken a

---

[4] One can find some further definitions that are not entirely statistical ones: a person whose residence and place of business are at a distance or something (such as a geological feature) that is situated away from or classed differently from a main or related body.

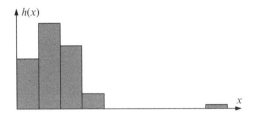

Figure 3.1.   Histogram with an outlier.

Figure 3.2.   Empirical relationship between two variables and an outlier.

longer time to clear than the other data. Ignoring the problem of what are the causes of such observation, we can quickly come to the conclusion that the mean time of repair becomes significantly longer than before the outlier appeared and especially the corresponding standard deviation.

In such a case, the first reaction should be to perform a penetrating analysis of the reasons for the occurrence of such an event—**an out-of-statistics analysis**.

Before one commences this type of analysis, it is worth being aware of a certain regularity. There is a large class of technical objects that can be seriously damaged during their operation. The environment of the objects can also be devastated and the threat to the lives and health of personnel can be observed as well. For example, when a hoist conveyance over-winds beyond its capacity (a catastrophic type event), it can strike the crush beams and a rupture can occur in the hoist head rope. The intensity of failures of this type is completely different than other regular failures of the object. In addition, the time required to clear the failure is significantly longer. This means that these two types of exploitation events cannot be analysed together. The intensity of failures and the times of repair (if repair is possible) are completely different, but repair time, for instance, is not an outlier in a statistical sense. The exploitation process of an object of this type is a superposition of two processes; regular failures occur in one process whereas in the second one, rare events occur.

A similar approach to events that occur can be repeated in different areas. When registering the shocks that appear in the rocks around underground excavations a sudden great earth tremor can sometimes be observed. In reality, it is obvious that these unusual events should be analysed separately and they are not outliers compared to the tremors that are registered day-to-day in a mine[5].

---

[5]Often, it is additionally presumed that extraordinary shocks and regular ones are independent of each other. Fortunately, in some cases this assumption does not hold; closely followed events in this regard allowed such an event that is a precursor of a coming great shock to be registered (Sokoła-Szewioła 2011). This information is important for the safety of mining operations.

Thus, let us presume at the very beginning of our consideration that we are able to recognise and exclude events which are untypical and are caused by the nature of a complex phenomenon from further study.

However, if there is no indication as to why an outlier has occurred in the sample, we have to make use of statistical tools.

Consider the following statistical scheme.

Presume there is a sample that was gathered from the elements of two different populations. The variables connected with these populations differ from each other considerably in regard to the mean value. Moreover, one population is very large in number in relation to the second one. In such a situation, we can expect that a number may occur in the sample taken that is significantly different from the other elements of the sample. The occurrence of such an item allows a hypothesis to be formulated stating inhomogeneity with regard to the feature that is the objective of the investigation. In order to formalise it, one can write it down.

There is a given sample:

$$\{X_1, X_2, ..., X_k, ..., X_n\}$$

There is a supposition that the k-th element of the sample is untypical—an outlier.

Assume the mean value of the rest of the elements is:

$$E(X_{i, i \neq k}) = \mu_i = \mu.$$

Formulate a null hypothesis $H_0 : E(X_i) = \mu$. This supposition says that all of the elements are taken from one population. An alternative hypothesis is:

$$H_1 : \mu_k \neq \mu \wedge \mu_{i, i \neq k} = \mu$$

which means that the k-th element has a significantly different mean value.

A further part of the analysis depends on the kind of information that is available. In mining practice two probability distributions are applied most frequently, namely, exponential distribution and Erlang distribution. (Gaussian distribution is also applied often, especially when the times of the work cycle of machines operating periodically is concerned.) Let us also discuss a case when there is no information on the distribution of the random variable being investigated.

Commence consideration when the random variable being investigated can be described by an exponential distribution.

3.2.1   *Exponential distribution*

Note at the very beginning that when an exponential distribution is concerned, we are in a peculiar situation. The intensive development of reliability theory in the sixties and seventies of the previous century as well as exploitation theory later on initiated the development of mathematical tools from the field of statistics in order to investigate the properties of technical objects that are manifested during the operation of an object. For this reason, there are several tests on the homogeneity of a sample and the tracing of outliers.

Presume that the exponential distribution is of the parameter λ. It is a well-known fact that the product $2\lambda \Sigma_{i=1, i \neq k}^{n} X_i$ has the probability distribution χ^2 with $(2n - 1)$ degrees of freedom because the sum of the random variables that are exponentially distributed has the Erlang distribution of $(n - 1)$ degrees of freedom and multiplying it by the product 2λ, one obtains the Chi-square random variable with $(2n - 1)$ degrees of freedom. Denote it conventionally:

$$2\lambda \sum_{i=1, i \neq k}^{n} X_i \quad \text{and} \quad 2\lambda \sum_{i=1}^{n} X_i : \chi^2(2(n-1)) \tag{3.3}$$

By introducing the estimator of the expected value:

$$m_{n-1} = \frac{1}{n-1} \sum_{i=1, i \neq k}^{n} X_i \qquad (3.4)$$

one can rearrange pattern (3.3) into

$$2\lambda(n-1) m_{n-1} : \chi^2(2(n-1))$$

Presuming that the null hypothesis is a true one and considering the ratio of quotients, it is easy to perceive that

$$m_n \frac{2\lambda(n-1)}{2n} : \frac{2\lambda n}{2(n-1)} m_{n-1} : F(2n, 2(n-1))$$

where m_n is the estimator of the mean taking into account the outlier and the symbol $F(2n, 2(n-1))$ denotes the F-Snedecor's random variable with $2n$, $2(n-1)$ degrees of freedom. By simplifying one can get

$$(m_n : m_{n-1}) : F(2n, 2(n-1)) \qquad (3.5)$$

Now we are able to verify the null hypothesis. It should be rejected if—for the presumed level of significance α—the quotient of the left-hand side of this relationship is greater than the quantile of the order α of the distribution F, that is

$$(m_n : m_{n-1}) > F_\alpha(2n, 2(n-1)) \qquad (3.6)$$

A slightly different approach for the verification of an outlier was presented by Fisher in 1929 (the problem of outliers in those days was almost non-existent in the formal sense). He presented a test for the simultaneous verification of the hypothesis of the exponential distribution of the random variable being tested together with the existence of an outlier that was significantly greater than the rest of the elements in the sample in his paper. Fisher proposed that the following formula be considered:

$$\eta = \frac{X_k}{\sum_{i=1, i \neq k}^{n} X_i} \qquad (3.7)$$

If the random variables X_i; $i = 1, 2, …, n$ are stochastically independent and have the same exponential distribution, then the probability distribution of statistic η does not depend on λ.

The critical values for this function were given in the paper cited, thus allowing for the verification of the formulated hypothesis (see also Gnyedenko et al. 1965 or Gnyedenko 1969).

■ **Example 3.2**

During a reliability investigation carried out in an underground coal mine, the following sequence of work times of a belt conveyor was observed:

290, 880, 670, 1420, 590, 380, 6220, 110, 280, 410, 770, 480, 60, 1330, 2070, 860, 1190, 1610, 820, 180 h

The seventh element of the sample looked strange at first glance. A suspicion was formulated that possibly a gap in recording took place.

Perhaps, a failure occurred at that time and was not recorded[6].

Begin from a randomness testing of the sample. Presume that the outlier belongs to the sample. If during a further part of the research it will be proved that this does not hold, our analysis will be repeated. The sample consists of $n = 20$ elements. If so, the median of sample is:

$$Me = \frac{670 + 770}{2} = 720 \text{ h}$$

The sequence of signs that depend on the fact that the given value is greater or lower than the median is as follows:

$$- + - + - - + - - - + - - + + + + + + -$$

The number of series in the sequence is 11. The number of plus and minus signs is identical and equals 10. Presume a basic hypothesis H_0 that proclaims the randomness of the sample versus the alternative supposition that it does not hold. One can read the critical values 6 and 15 from Table 9.8. Because the number of series in the sample is 11, there is no basis to reject the basic hypothesis. We can assume that the sample has a randomness character.

Let us calculate two mean values: one for all of the elements but without the outlier and the second one including it. We have:

$$\overline{X}_{n-1} = 1031 \text{ h}, \quad \overline{X}_n = 1335 \text{ h}$$

The difference between them is significant. (A greater difference is visible in the corresponding standard deviations: 552 h and 758 h for this obvious reason).

As many empirical investigations have proven, in the majority of cases the probability distributions of the work time for belt conveyors can be satisfactorily described by an exponential distribution.

Calculate the ratio of the mean values:

$$\frac{\overline{X}_n}{\overline{X}_{n-1}} \cong 1.3$$

Formulate a null hypothesis stating that the data are homogeneous and the quotient of means is not significantly different from the alternative hypothesis that proclaims that the ratio of the average values is significantly different from unity.

---

[6] There are many such cases in mine practice and it is not a mining 'specialisation'. In their book Gnyedenko et al. (1969) described a case where a member of staff who was responsible for registering failures took a break and nothing was recorded during this period.

The critical value is taken from the *F*-Snedecor's tables for degrees of freedom for this example (Table 9.6) and it is presumed that $\alpha = 0.05$ is:

$$F_{\alpha=0.05}(40, 38) = 2.12$$

Because the empirical value does not exceed the critical one, there is no basis to reject the hypothesis, i.e. there is no ground to remove the outlier from the sample. ◄

3.2.2 *Erlang distribution*

The above consideration can be expanded for a case where the values of the sample can be satisfactorily described by the Erlang probability distribution. It is enough to notice that the product $2\lambda\Sigma_{i=1,i\neq k}^{n}X_i$ has a Chi-squared distribution with $2k(n-1)$ degrees of freedom. It can be noted conventionally as

$$2\lambda \sum_{i=1,i\neq k}^{n} X_i : Ga\left(k(n-1), \frac{1}{2} \right) \tag{3.8}$$

Applying analogical reasoning one obtains:

$$(m_n : m_{n-1}) > F_{\alpha}(2kn, 2k(n-1)) \tag{3.9}$$

which means that the null hypothesis should be rejected when for the presumed level of significance α the quotient visible on the left side of the inequality (3.9) is greater than the quantile of the order α of the *F* probability distribution.

- **Example 3.3**

The operation of a suspended loco used in an underground coal mine to refurnish long-wall faces was investigated. A sequence of repair times was recorded giving the following data:

160, 60, 175, 320, 100, 360, 120, 70, 210, 45, 590, 250 120, 230, 30, 140, 55, 510, 60, 95, 180, 310, 190, 150, 2220, 200 min.

Attention was paid to element number 25, which was significantly greater than the other times. There was no description of what had happened and why this time was so long. There was also a suspicion that something had broken and there was no spare part in the mine's warehouse. But in such a case there should be proof that a special order was submitted to the producer of the loco on that date. There was no such proof.

Investigate whether the sample has randomness character.

The median of the sample is:

$$Me = \frac{150 + 160}{2} = 155 \text{ min}$$

and the sequence of signs is as follows:

$$+ - + + - + - - + - + + - + - - - + - - + + + - + +$$

The series number is 17. The number of plus and minus signs is the same and equals 13. The critical numbers for the presumed level of significance $\alpha = 0.05$ are: 8 and 19 (see Table 9.8). Conclusion: there is no basis to reject the null hypothesis that proclaims the randomness of the sample that has the outlier in it.

Discuss now the existence of the outlier in the sample from a statistical point of view.

An earlier reliability investigation gave grounds to conclude that the probability distribution of repair times could be satisfactorily described by the gamma distribution with the shape parameter just below two, and it was assumed that $k \cong 2$. Thus, it was assumed that the Erlang distribution described the statistical data *well*.

Two average times were calculated for the loco based on the sample; one presuming that all of the elements come from one homogeneous population and that the second one excludes the outlier. Results are

$$m_n = 267.3 \text{ min} \quad \text{and} \quad m_{n-1} = 189.2 \text{ min};$$

(the standard deviations are 428.8 and 140.5 min, respectively).

The ratio is:

$$\frac{m_n}{m_{n-1}} = 1.41$$

Formulate a hypothesis H_0: the data are homogeneous and the quotient of the average values is non-significantly different than the one against the hypothesis H_1, which states that the ratio of the average values is considerably greater than one, which means that the outlier does not belong to the sample.

The quantile of the order $(2kn, 2k(n - 1))$ taken from the appropriate table (*F* Snedecor, Table 9.6) for the presumed level of significance $\alpha = 0.05$ is:

$$F_{0.05}(104, 100) \approx 1.39$$

The empirical value is greater than the critical one and for this reason we have ground to reject hypothesis H_0. The sample is not homogeneous and the outlier should be excluded from further analysis.

The above result indicates that our previous inference on the randomness of the sample has to be repeated. If so, calculate the median of the sample which is now 160 min. The sequence of signs is as follows:

$$- + + - + - - + - + + - + - - - + - - + + - +$$

(The number of signs was reduced by 2 because there is no outlier and the element that equals the median was also rejected).

The number of series is now 16. The number of – signs is 12. The critical values for the same level of significance are: 7 and 18 (Table 9.8). Corollary: there is no basis to reject the hypothesis that proclaims the randomness of the sample.

Note. Further inquiry in this regard permits information about what happened to be obtained. Two bolts fixing the rail to the roof had been slackening due to the displacement of the surrounding rocks. The displacement was small but was strong enough to weaken the bolts. When the heavy unit arrived at the loosened rail part, the bolts slipped out and the loco went off the rail. Some parts of the loco were broken. We can assess this event as a rare one, and it can be counted among catastrophic ones. Observations of the operation of several similar locos at that mine indicated that there was no evidence of any comparable event.   ◄

3.2.3   *Other distributions*

Consider now when the data contained in a sample can be satisfactorily described by a Gaussian distribution.

As we know, the domain of determinacy for a random variable described by this distribution is the whole axis of real numbers. This is not accepted in engineering applications. All physical magnitudes used in the field of engineering have their own limits.

However, such an approach in which the values of a random variable are supported over the whole real axis should be treated as a margin model. Very often the left-side natural limit for physical magnitude is zero. Not many physical magnitudes are supported over a certain part of a negative value axis. These are, first of all, load, stress, braking force, deceleration or the dynamic moment.

Magnitudes that are considered in reliability analyses are determined over a nonnegative set of values as a rule. This set contains real number values when times of states are concerned. The effect of the existence of a machine work state is the specific job executed by this machine and in mining engineering may have the form of an excavated, hauled, dressed, dumped or stored mass of rock that has been won. This set can also contain natural numbers only if the point of interest is the number of pieces of equipment that are in a particular state at a given moment of time, the number of repair stands, spare parts etc. Where transport means are concerned, the proper determination of the right-side limit for random variable values determines whether the results of the calculation of their basic parameters will be adequate to the reality. Many practical examples testify that in the world of engineering and especially in mining the Gaussian distribution should have both side limits, which means that the probability distribution is truncated. In a case where the distribution has only one boundary or it has no limits, it is necessary to evaluate any error generated by such an assumption. If its value is small, it may be approved.

Consider a homogeneity analysis. A hypothesis H_0 that is verified is a statement that all of the elements of the sample are taken from one population in spite of the fact that one element differs considerably from the others. This formula can be used to check whether the supposition H_0 is true:

$$T_k = \sqrt{\frac{n}{n-1}} \frac{X_k - m_{n-1}}{S_{n-2}} \tag{3.10}$$

where:

$$S_{n-2} = \sqrt{\frac{1}{n-2} \sum_{i=1, i \neq k}^{n} (X_i - \bar{X}_{n-1})^2} \tag{3.11}$$

is the estimator of the standard deviation of the variable being observed without the outlier. If the verified hypothesis is true, the random variable T_k has a Student's *t*-distribution with $(n-2)$ degrees of freedom. If the alternative hypothesis H_1 is true, the random variable T_k has a non-central Student's *t*-distribution[7]. The hypothesis that is verified should be rejected for the presumed level of significance α if the following inequality holds:

$$T_k \geq t_\alpha(n-2) \tag{3.12}$$

where $t_\alpha(n-2)$ is the quintile of the order α of the Student's *t*-distribution with $(n-2)$ degrees of freedom.

■ **Example 3.4**

The unloading times of hauling trucks were recorded in a certain open pit mine. The following sequence was noted:

  1.02; 1.10; 1.12; 0.98; 1.21; 1.09; 0.94; 1.05; 0.93; 1.66; 1.22; 1.08; 1.02; 0.88; 1.03; 1.09; 1.00; 0.86 min.

  The 10-th time looks strange (1.66 min) so it is excluded from further analysis. Verify whether this decision was a proper one.

  Commence from testing whether the sample has a randomness property.
  The sample consists of 18 elements. If the outlier is rejected, the sample has an odd number of items and for this reason—after arranging the sequence monotonically—the element that is in the middle is the estimate of the unknown median. This means that the time of dumping, which equals 1.03 min, is the estimate of the median. There is such an element in the sample,

---

[7] In probability and statistics, the **non-central *t*-distribution** generalises the Student's *t*-distribution using a non-centrality parameter. Like the central *t*-distribution, the non-central *t*-distribution is primarily used in statistical inference, although it may also be used in robust modelling for data. In particular, the non-central *t*-distribution arises in power analysis. The cumulative distribution function with v degrees of freedom and non-centrality parameter μ can be expressed as:

$$F_{v,\mu}(x) = \begin{cases} \dfrac{1}{2}\displaystyle\sum_{j=0}^{\infty}\dfrac{1}{j!}\left(-\mu\sqrt{2}\right)^j e^{-\frac{\mu^2}{2}}\dfrac{\Gamma\left(\dfrac{j-1}{2}\right)}{\Gamma(1/2)}B\left(\dfrac{v}{v+x^2};\dfrac{v}{2},\dfrac{j+1}{2}\right) & \text{for } x \geq 0 \\[3em] 1-\dfrac{1}{2}\displaystyle\sum_{j=0}^{\infty}\dfrac{1}{j!}\left(-\mu\sqrt{2}\right)^j e^{-\frac{\mu^2}{2}}\dfrac{\Gamma\left(\dfrac{j-1}{2}\right)}{\Gamma(1/2)}B\left(\dfrac{v}{v+x^2};\dfrac{v}{2},\dfrac{j+1}{2}\right) & \text{for } x < 0 \end{cases}$$

where $B(x;a,b)$ is the regularised incomplete beta function, i.e. $B(x;a,b) = \int_0^x u^{a-1}(1-u)^{b-1}du$

so if one rejects it, the sequence of plus and minus signs (elements greater than the median and elements lower than the median) is as follows:

$$- + + - + + - + - + + - - + - -$$

The number of series is 11 and the number of plus signs is equal to the number of minus signs, which is 8. The critical values for the level of significance $\alpha = 0.05$ are: 4 and 13 (Table 9.8). Corollary: because the empirical value falls between the critical boundaries there is no basis to reject the hypothesis that states the randomness of the sample.

Now evaluate whether the decision to reject the outlier was a correct one.

Calculate the average values for the sample with the outlier and without it. Here we have:

$$m_{n-1} = 1.04 \text{ min} \quad \text{and} \quad m_n = 1.07 \text{ min}$$

The corresponding standard deviations are:

$$S_{n-2} = 0.10 \text{ min} \quad \text{and} \quad S_{n-1} = 0.18 \text{ min}.$$

Now calculate the value of the statistic T_k—patterns (3.10) and (3.11): $T_k = 6.36$.

Presume the level of significance $\alpha = 0.05$. The critical value for this level is $t_\alpha(n - 2) = 2.12$ (Table 9.3). The empirical value significantly exceeds the critical value. The outlier has to be rejected from further considerations.

Note. This extraordinary dumping time was caused by a malfunction of the hydraulic system that lifted and lowered the loading box. □

It is time to consider a case in which there is no information on the random variable being investigated, and thus we are not able to formulate a statistical hypothesis in this regard in addition to the fact that we know the sample that has just been taken. This may be the case when the sample is small or when the test that permits the probability distribution to be identified gives ambiguous information.

In such a case, the oldest principle that is used is the rule of 3σ. According to this principle, the rejection region is determined by the inequality given by formula:

$$X > \bar{X}_{n-1} + 3S_{n-2} \tag{3.13}$$

where the first component of the right side of the equation is the mean based on the $(n - 1)$ sample whereas the second is the product of three times the standard deviation of the sample.

By the way, let us evaluate how this principle holds for the case just discussed—example 3.4. Applying the estimates that were found during the analysis, one obtains

$$X > 1.04 + 3 \times 0.10 = 1.34 \text{ min}$$

The outlier (1.66 min) is above this value. Notice that the dispersion of the random variable being investigated is comparatively low[8]. ◄

---

[8] A basic measure of the relative dispersion of a random variable is the **Coefficient of Variation** (CV). It is a measure of the differentiation of the distribution of a random variable. It is the ratio of the standard deviation to the expected value of the random variable provided that the mean is different than zero. Its estimator—both parameters assessed based on the sample—is consistent but biased in a general case. CV is usually given in percentages.

Although it is difficult to determine unequivocally what is the probability that the above inequality holds (it depends on the type of probability distribution that we are dealing with), we can be sure that this probability is very small, assessed as a few per cent (in the case of an exponential distribution for instance, it does not exceed 2%). A rather stronger statement in this regard is the limited estimation that is determined by Cheybyshev's inequality theorem, which is as follows.

Let X be a random variable with a finite expected value and a finite non-zero variance. Then, for any real number $k > 0$, the following inequality holds:

$$P\{|X - E(X)| \le k\sigma_x\} \ge 1 - \frac{1}{k^2} \tag{3.14}$$

However, only the case $k > 1$ provides useful information. When $k < 1$, the right side of the inequality is greater than one, so it becomes vacuous because the probability of any event cannot be greater than one. When $k = 1$, it simply means that the probability is less than or equal to one, which is always true. If for instance $k = 3$ then, keeping in mind pattern (3.14), the lower estimation is approximately 0.9.

Several further tests concerning outliers are applied in statistics. One of them is the so-called Q test.

For verification of the statistical hypothesis H_0 that proclaims that an outlier is necessary in a sample to order the sample monotonically, presume that the elements of the sample were presented in a non-decreasing way, i.e.

$$\{y_1 \le y_2 \le \cdots \le y_n < X_k\} \tag{3.15}$$

Define the statistic:

$$Q = \frac{X_k - y_n}{X_k - y_1} \tag{3.16}$$

The formula (3.16) determines the empirical value of the statistic Q. It should be compared with its counterpart $Q_\alpha(n)$, which is a critical value. The critical values are given in Szepke's paper (1967). If the following inequality holds:

$$Q \ge Q_\alpha(n)$$

then the verified hypothesis H_0 should be rejected at the level of significance α. It is obviously understood that the alternative supposition is accepted; the hypothesis that proclaims that the outlier belongs to the sample.

At the conclusion of the considerations on outliers, note the following regularity. There are many probability distributions that are used in mathematical statistics and they have various properties. One such property is **robustness on the outliers' appearance**. Although the problem of robustness in statistics is much wider[9], we are discussing the possibility of the occurrence of untypical outcomes in a sample here. Literature on the subject of robustness appeared mainly in the late seventies and eighties of the previous century (e.g. Huber 1981, Rousseeuw and Leroy 1987), and many probability distributions were tested in this regard. The Gaussian distribution

---

[9] There is a term 'robust statistics' which provides an alternative approach to standard statistical methods, such as those for estimating some parameters of a random variable. The main point is to make such estimators which are not unduly affected by outliers or small departures from model assumptions.

proved to be robust on the appearance of outliers while gamma and Weibull did not. This means that an extraordinary outcome may appear and the sample will be homogeneous.

Generally, it is worth remembering that if an untypical outcome is in a sample, it is necessary to verify the source of information in order to check the possible physical reasons for its appearance etc. And, if all of these non-statistical potential reasons are attested and no motive is found then it is time to apply statistical tools.

As has been shown, the presence of an outlier in the sample generates a problem. If further analysis is carried out and some statistical measures are estimated—or, when for example, performing least squares fitting to the data—it is often the best solution to discard the outlier before computing. This should even be done when the appropriate statistical test gives no grounds to reject the hypothesis that the outlier belongs to the given sample characterised by a certain probability distribution. It is worth noting that if a sample is large, the rejection of one element does not cause a great loss. If, however, a sample is small, the value of the outlier has an enormous influence on the values of the calculated parameters. Estimates will be distorted and therefore it is better to reject it.

3.2.4 *Result of an outlier analysis*

This provides unequivocal information about which part of the gathered data should be taken into further investigations.

In a case where an untypical element belongs to the sample, it is noteworthy information that during the operation of the object such event can sometimes occur.

3.3 STATIONARITY TESTING OF SEQUENCES

In our analysis, two steps were done first—we checked the randomness of the sample taken and rejected the outlier (if there was one). The next step should be an examination of whether the observed sequence of outcomes is stationary or not. The result of this examination has a significant influence on further parts of the statistical analysis. This is because if a sequence is a non-stationary one (which means that the values of the sample are increasing or decreasing, on average), it indicates that we have observed the realisation of a certain stochastic process of the monotonically expected value of the variable. If so, our further analysis should be done in the area of the theory of stochastic processes. However, if the sequence is stationary, we can presume that our sample was the realisation of a certain random variable and the scope of further consideration should be on the analysis of the field of the random variables. Such analysis is much simpler than an analysis of stochastic processes.

Generally, **stationarity** means that something is fixed in a position or mode that is immobile or unchanging in condition or character. Stationarity in connection with the exploitation process of a technical object is associated with the way in which the process is realised. Therefore, it is a kind of **property of a random process**.

Let us ignore the subtleties connected with the different kinds of stationarity that are considered in the theory of stochastic processes (see, for example, Feller 1957, Kovalenko et al. 1983, Ross 1995). Instead, we will approach the problem of stationarity from a practical engineering point of view.

Each sequence of the times of a given state should be tested to see whether the observations increase or decrease on average over time. Because the parameter of the process is a time, this verification concerns stationarity. Nevertheless, in engineering practice there are some operations for which the most important process parameter is not time. For a mine hoist installation, the number of winds to be executed is much more important. Similarly, for a transportation system, the number of tons of mass to be displaced is critical. For many machines the number of work cycles to be performed is vital. For these reasons, the point of interest is a defined stochastic property of the sequence of values being observed. However, in order to make further considerations more communicative, it will be presumed that the process parameter is time.

There are several statistical tests that allow this property to be checked for a given sequence, but the one that is most frequently used in econometrics and technometrics is the test that applies the Spearman's rank correlation coefficient (see, for example, Hollander and Wolfe 1973). **Correlation** is a kind of stochastic relationship between random variables. It is the statistical proportionality of the results of the measurements of different phenomena. It is sometimes stated that it refers to the departure of variables from independence[10]. A basic measure of correlation is a correlation coefficient, however several other correlation coefficients are used in different investigative situations.

The procedure for the Spearman rank correlation test is as follows.

3.3.1   *The test procedure*

A natural number that follows the sequence of the occurrence of the elements in time is assigned to each element of the sample. These natural numbers are called ranks[11]. Next, the sample is ordered monotonically. Natural numbers going up or down are assigned to all of the elements of the new sequence. (For general purposes, it does not matter if these numbers are assigned going up or down, as this only generates a change in the sign of the coefficient.) If a situation occurs in which a few values are identical, a rank that is the arithmetic mean of their ranks should be assigned. In such a way, a second set of ranks, v_i, $i = 1, 2, ..., n$ is obtained. These two sets of ranks create a matrix:

$$\begin{bmatrix} v_1 & v_2 & ... & v_n \\ 1 & 2 & & n \end{bmatrix}$$

To assess the Spearman's rank correlation coefficient, the following statistic is applied:

$$r_S = 1 - \frac{1}{n(n^2 - 1)} 6 R_n \tag{3.18}$$

where:

$$R_n = \sum_{i=1}^{n} (v_i - i)^2 \tag{3.19}$$

This correlation coefficient is a normalised measure that is determined over $[-1, 1]$ interval and it is created based on the Pearson's linear correlation coefficient. If the investigated sequences are independent of each other, the rank Spearman's correlation coefficient equals zero. However, if these sequences are functionally dependent, then this coefficient equals 1 in the modulus. A value near zero indicates that these sequences are not correlated; there is no interrelationship between the value that appears in the sequence and its order in the sequence, i.e. it testifies to the stationarity of the sequence being investigated.

Formally, the procedure of statistical investigation is as follows. A statistical hypothesis is formulated H_0: $\rho = 0$ (where ρ is the correlation coefficient in the entire population), which states that there is no dependence between the values with respect to time. This hypothesis is

---

[10] For more on correlation see Chapter 6.
[11] **Rank** refers to the relative position, value, worth, complexity, power, importance, authority, level etc. of a person or object.

set against the hypothesis $H_1 : \rho \neq 0$, which states that the values of a variable depend on time. If the following inequality holds:

$$|r_S| \geq r_S(\alpha, n) \tag{3.20}$$

where $r_S(\alpha, n)$ is the critical value (taken from Table 9.14) for the given level of significance α and the sample size n, the hypothesis H_0 must be rejected. Paying attention to the sign of the coefficient and confronting it with the way of assigning ranks, we can conclude what kind of tendency is in the sequence: increasing when the sign is plus or decreasing when it is minus.

The estimator (3.18) had been applied in statistical investigations for a long period of time up to a moment when the problem of the influence of the ranks associated with the same values (called **tied ranks**) was taken into consideration. Due to the fact that in mathematical statistics there are many tests that apply ranks (e.g. Mann-Whitney test or Kruskal-Wallis test), the problem was crucial. Researchers easily came to the conclusion that such ranks have an effect on the statistical inference being conducted and that this effect is negative. Generally, the greater the number of tied ranks, the greater their effect is. In the test just presented, they decrease the value of the sum of squares of the deviations of natural numbers from their mean. Thus, it is necessary to introduce a correcting sum:

$$T_\Sigma = \sum_{j=1}^{g} (w_j^3 - w_j) \tag{3.21}$$

where g is the number of groups of tied ranks, while w_j is the number of tied ranks in the j-th group. The final pattern of the estimator for the rank correlation coefficient is:

$$r_S' = \frac{(n^3 - n) - 6R_n - \frac{1}{2}T_\Sigma}{\sqrt{(n^3 - n)^2 - T_\Sigma(n^3 - n)}} \tag{3.22a}$$

or

$$r_S' = \frac{12v_i i - 3n(n+1)^2}{\sqrt{(n^3 - n - T_\Sigma)(n^3 - n)}}. \tag{3.22b}$$

Generally, the application of the test for an investigation of the rank correlation coefficient is possible when $n \geq 4$ for the presumed level of significance $\alpha = 0.05$. If $\alpha = 0.01$, the sample size $n \geq 5$. For a larger n, in practice for $n > 10$, this correlation coefficient has approximately the normal distribution $N(0, \sqrt{(n-1)^{-1}})$, while statistic $\boldsymbol{R}_n$ has approximately the normal distribution:

$$N\left(\frac{n(n^2 - 1)}{6}, \frac{n(n+1)}{6}\sqrt{n-1} \right)$$

For this reason if n is large (practically, already if $n > 10$), the critical value can be calculated from the pattern:

$$r_S(\alpha, n) \approx \frac{u_{1-\alpha}}{\sqrt{n-1}}, \tag{3.23}$$

where $u_{1-\alpha}$ is the quantile of the order $(1 - \alpha)$ of the standardised normal distribution $N(0, 1)$.

■ **Example 3.5**

Reliability investigations concerned armoured flight conveyors operating in underground coal mines where the longwall method was being used. For one conveyor the sequence of repair times was as follows:

110, 40, 60, 10, 100, 70, 30, 50, 155, 65, 170, 35, 90, 85, 230, 40, 35, 95, 65, 100, 45, 180, 80, 60, 30, 140, 170, 120, 25, 80, 130, 105, 70, 20, 15 min.

A question to answer is whether this sequence is stationary.

Firstly, investigate the randomness of the sample. The size of the sample is 35. Arranging the sequence monotonically, one can find that 18-th element is 70 min. This is the median of the sample. By transforming the sequence into the sequence of signs, one has:

$$+ - - - + - - + - + - + + + - - + - + - + + - - + + + - + + + - -$$

The number of series of signs is 20. The sequence now has 33 elements because two elements were rejected (they were identical to the median). The number of plus signs is 17 whereas the number of minuses is 16. We next verify the basic hypothesis that states the randomness of the sample versus the alternative supposition that rejects it. Presuming the level of significance $\alpha = 0.05$ and bearing in mind parameters of the sample, the critical values are: 11 and 23 (see Table 9.8). Therefore, we have no ground to reject the null hypothesis—we assume that the sample has a randomness character.

Because the sample has no outlier, let us check whether the sequence observed is stationary. Making all of the necessary calculations (Table 3.1), one gets:

$$r_S = 1 - \frac{6 \times 6957.5}{n(n^2 - 1)} = 1 - \frac{6 \times 6957.5}{35(35^2 - 1)} = 0.026$$

Immediately a conclusion can be formulated that such a small value should not be significant. Formally, a null hypothesis is stated that proclaims the stationarity of the sequence given by the sample versus an alternative supposition rejecting it. The critical value can be calculated by applying the formula (3.23). Here we have:

$$r_S(\alpha, n) = \frac{1,645}{\sqrt{n-1}} = 0.282$$

Looking at this result one can state that there is no basis to reject the null hypothesis that proclaims the stationarity of the sample.

Some readers have probably noticed that there are several values in the table that are repeated and obviously their ranks also. Let us investigate the influence of these ranks on our statistical inference—on the value of the correlation coefficient. Calculate the correcting sum as determined by formula (3.21). In the case being considered, $g = 9$ (the number of groups) and the sum is:

$$T_\Sigma = 132$$

By applying pattern (3.22a), one calculates the corrected value of the rank correlation coefficient, which is now:

$$r'_S = -0.037.$$

As you can see the difference is small, negligible.

Table 3.1.   Auxiliary calculations.

Value	Rank v_i	# group	Number i	$(v_i - i)^2$
110	27		1	676
40	9.5	III	2	56.25
60	13.5	IV	3	110.25
10	1		4	9
100	24.5	VIII	5	380.25
70	17.5	VI	6	132.25
30	5.5	I	7	2.25
50	12		8	16
155	31		9	484
65	15.5	V	10	30.25
170	32.5	IX	11	462.25
35	7.5	II	12	20.25
90	22		13	81
85	21		14	49
230	35		15	400
40	9.5	III	16	42.25
35	7.5	II	17	90.25
95	23		18	25
65	15.5	V	19	12.25
100	24.5	VIII	20	20.25
45	11		21	100
180	34		22	144
80	19.5	VII	23	12.25
60	13.5	IV	24	110.25
30	5.5	I	25	380.25
140	30		26	16
170	32.5	IX	27	30.25
120	28		28	0
25	4		29	625
80	19.5	VII	30	110.25
130	29		31	4
105	26		32	36
70	17.5	VI	33	240.25
20	3		34	961
15	2		35	1089
			Σ	6957.5

□

It is worth noting that the non-stationarity of the sequence examined here is not frequently observed. However, there are some processes in mining engineering that have a dependence on time or another process parameter. There are three basic areas in which the non-stationarity of data can be perceived in mine mechanisation.

When a piece of equipment commences to fulfil its duties, a period of running-in (grinding-in) often begins. Some structural elements that work together begin to fit better. The cooperation becomes smoother over time. If this is so, any potential failures should occur less frequently. For this reason, the mean work time between failures increases. This process has a fading character. At the very beginning, the intensity of failures, which is relatively high, decreases toward its characteristic appropriate level that is associated with the properties of the object. When this level is reached, the regular exploitation process of the object begins. Producers of high-quality machines eliminate this period by performing the running-in period when the production of the machine is finished in their factory. The purchaser of the machine obtains an object in which the elements already work together smoothly. Some other machine manufacturers whose products have a running-in period during regular operations guarantee the replacement of failed elements through this special period free of charge or guarantee a low price on the items necessary to replace those that have failed. This period is warranted by the producer.

A second special period in which the intensity of failures increases is the period during which a piece of equipment becomes worn—aging. Some signs of wear and tear are clearly visible in some machine elements. Elements begin to work together less efficiently over time. The number and intensity of failures increases. The exploitation process of the machine shifts from the regular period (II) to a third period—a period of significantly increasing degradation. The end of the working life of this object is approaching. Users of the machine are interested in getting rid of it in order to avoid operational problems and economic losses. However, nowadays, the economic end of a machine occurs rather than its mechanical death. A problem that sometimes faces users of machines is the appearance on the market of a new machine that has better parameters and better properties. This problem raises the question of whether it may be more profitable to withdraw the machine from use, sell it and purchase a new machine.

There is also a different source generating non-stationarity in data.

Mining practice has shown that sometimes unusual situations arise. The process of the generation of a stream of rock that is being excavated is running with a high degree of intensity when suddenly a failure occurs in a machine that is important for production. There is *time pressure*—this machine should be repaired as quickly as possible. And, unfortunately, there is no spare element to replace the one that has failed. Miners sometimes try to solve this problem by making use of a *similar* element (as a direct replacement or after some modification) during the repair. The machine operates again but the intensity of failures changes. The modified element (sometimes it is an entire assembly) fulfils its duties but not entirely in the way that is expected. This period when the machine begins to perform its duties and the intensity of failure increases is a non-stationary one, and this means that the work time between two neighbouring failures is shortened and the intensity of the failures increases. This type of repair practice is obviously not recommended but unfortunately it can sometimes be observed in mining reality.

Be aware that phases of operation of machines described above concern objects that can be repaired. However, there are some other technical items that are non-repairable and that fail during their usage and that some periods of non-stationarity occur when this happens. A hoist head rope in a hoist can be given as an example. Its process of wearing, which is usually understood as the process of breaks in the wires, results in breaks after a period without showing any signs of fatigue (stationary period) thus entering into a non-stationary phase with an accelerating accumulation of breaks.

■ **Example 3.6**

The use of an underground suspended loco was investigated. The times of its work were noted as well as any failures that occurred. Next, the total work time between two neighbouring failures was calculated. In such way a sequence of 13 times of the use of the loco was obtained:

241.1; 205.4; 26.1; 25.0; 68.1; 41.0; 9.2; 17.1; 5.2; 69.1; 1.0; 4.9; 0.5 h

Looking at this sequence it is easy to see that the total work time from failure to failure decreases stochastically. Let us verify this supposition. Apply a test based on the Spearman's rank correlation coefficient.

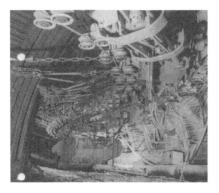

By assigning ranks and making all of the necessary calculations, one gets the data shown in Table 3.2.

Calculate the coefficient. Here we have:

$$r_S = 1 - \frac{6 \times 662}{n(n^2 - 1)} = -0.82$$

Formulate a null hypothesis stating that the sequence is a stationary one. Compare this value with the corresponding critical one taken from Table 9.14:

Table 3.2. Auxiliary calculations.

Value	Rank v_i	Number i	$(v_i - i)^2$
241.1	13	1	144
205.4	12	2	100
26.1	9	3	36
25	8	4	16
68.1	10	5	25
41	7	6	1
9.2	5	7	4
17.1	6	8	4
5.2	4	9	25
69.1	11	10	1
1	2	11	81
4.9	3	12	81
0.5	1	13	144
		Σ	662

$$r_S(\alpha = 0.05;\ n = 13) = 0.53.$$

The empirical value is significantly greater in modulo than the critical one. This gives the ground to reject the null hypothesis. Taking into consideration how the ranks were ascribed and the sign of r_S, we can conclude that the total work times from failure to failure decreases and not incidentally, significantly in a statistical sense.

Remark. There was no way to obtain information about what the physical reason for such a phenomenon was but when the last failure occurred the loco went under a general repair that lasted for quite some time.   ◄

There are some further different areas of mining engineering where the data indicate non-stationarity. Figures 3.3–3.6 illustrate the statistical relationships between:

- The number of tonnes of hard coal sold in Poland on a monthly basis in 1995–2007
- The lost time incident rate for the mining industry in the USA
- Vertical displacements of rocks over the time interval connected with tremor recorded in the 'Rydułtowy-Anna' coal mine, Poland
- The theoretically determined subsidence $w(t)$ of point S and the observed vertical displacement $w_p(t)$ in an area of the 'Halemba' underground coal mine, Poland.

All of these plots contain important and useful information although in some cases further statistical analysis is difficult and requires the application of sophisticated statistical tools.

Make some comments on the problem of stationarity.

Let us repeat at the very beginning, due to importance of this material, that information on non-stationarity has great repercussions for any further statistical analysis that is conducted. This information is also vital if the analysis concerns the problems of mine mechanisation for instance. If the data collected refers to the realisation of the exploitation process of a technical object (and it does not matter whether it is a single item or the entire system), the information on non-stationarity should be a clear signal to observe the object carefully. The main point of interest should be the answer to two questions:

- What is the physical source of such regularity?
- What will probably happen to the object in the near future?

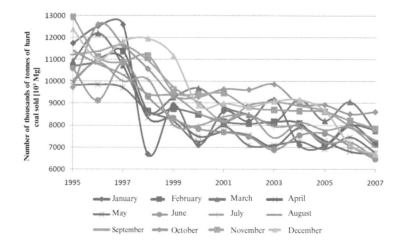

Figure 3.3.   The number of tonnes of hard coal sold on a monthly basis in 1995–2007 (Manowska, 2010).

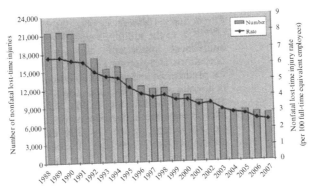

Figure 3.4. The lost time incident rate for the mining industry in the USA (Saporito and Self, 2012).

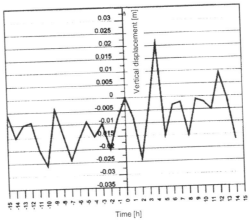

Figure 3.5. Vertical displacements of rocks over the time interval connected with tremor recorded in the 'Rydułtowy-Anna' coal mine (Szewioła-Sokoła, 2011).

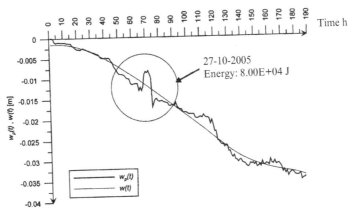

Figure 3.6. The theoretically determined subsidence $w(t)$ of point S and the observed vertical displacement $w_p(t)$ in an area of the 'Halemba' underground coal mine during the period 24-10-2005 to 31-10-2005 (Sokoła-Szewioła 2011).

Sometimes this information is alarming. By analysing it more carefully and in more depth one may get information that the existence of the object will very likely be jeopardised or even that its existence will be seriously threatened in the near future. The responsibility falls on the person performing the analysis if he/she does not generate running events.

In the regular operation of technical objects, non-stationarity seldom occurs unless the periods of running-in and intensive aging are neglected. However, there are some technical objects that have a non-stationary process of wear nearly from the beginning of their use (*vide*: hoist head ropes). There are also some objects for which the operation process can change drastically because of significant changes in their surroundings. It concerns powered support systems, for example. When a longwall is being constructed in an area where another longwall is in operation (this second excavation can be below or above the longwall), the stresses that exist in the rocks change and soon have a considerably different dispersion than in an intact rock. Drastically changed stresses in rocks can also be observed when the excavation comes into contact with geological faults. All of these situations cause a higher number of failures of powered support elements.

Look now at Figures 3.3–3.6. Information about the non-stationarity of the sequences noted in these figures 'gives' an analyser the ground to make some important decisions. Generally, in Figure 3.3 one can detect the process of changes in coal production in Poland. This occurred during the period of the restructuration of the Polish mining industry after the collapse of the previous political system. One can guess that the situation on the market stabilised over time and a certain sales level was reached. This information is useful for planning the further production of hard coal in Poland. Figure 3.4 in turn 'carries' different information. There is a positive tendency in mining in the USA that the lost time incident rate decreases over time. This is a clear indication that its policy in this regard is a correct one that should be maintained. However, it can be predicted that the general tendency tends to a certain asymptote if there are no significant changes in the mining methods being applied. Figures 3.5 and 3.6 contain essential information, which is a precursor of a coming tremor caused by running mining operations. If such an event is observed in a mine monitoring station, an appropriate action should be undertaken immediately. This problem will be discussed in the chapter on prediction.

Non-stationarity testing of sequences in mining engineering is of immense importance in some areas.

Let us conduct our consideration presuming, so far, that the sequence observed and later analysed is a stationary one in a sense of the mean value.

3.4   OUTCOME DISPERSION TESTING

If during the analysis being conducted there is no basis to reject the hypothesis that proclaims the stationarity of the sequence noted, i.e. that the values in the sequence do not change over time on average, then this does not mean that this sequence is free from any dependence on time. This relationship can be subtler, e.g. the dispersion of the sequence values is not constant but changes with time on average. As a rule, in mining practice, if a value is not constant, it increases over time.

A problem of this type can occur, for instance, when an examination of the data is being carried out and the point of interest is to find a *good* function that can satisfactorily describe the information that has been obtained. Such a situation can be observed when analysing records concerning the fatigue-wearing processes of a head rope of a hoist. Usually, a default presumption is that this function should describe the data that are in hand *well* and this *goodness* should be approximately constant over time. Unfortunately, very often the opposite regularity is noticed—the *goodness* of the description worsens with the amount of work executed by the rope. This goodness worsens because of the increasing dispersion of the values noted (total number of breaks of wires) around the expected value function.

■ **Example 3.7**

A plot of the function is shown in Figure 3.7: the total number of cracks in the wires of a hoist head rope versus the number q of winds executed by this hoist[12].

More than a half century ago, it was proposed (Kowalczyk 1957) that the empirical data in such case be approximated by a power function determined by the formula:

$$\Theta_t = \delta t^{\chi} c^{\zeta_t} \tag{3.24}$$

where: δ, χ—structural parameters of the function
Θ_t—total number of cracks of rope wires
t—time (or number q of winds)
c—constant
ζ_t—random component of the function.

Usually, linearisation of function (3.24) is being done by applying the natural logarithm and for this reason $c = e$ which makes that the random component ζ_t is in an additive way connected with the regression function.

It was presumed that the random component of the model comprises the whole stochastic nature of the process of the accumulation of cracks during the rope's utilization. It exclusively explains the fact that the theoretical function does not precisely cover the

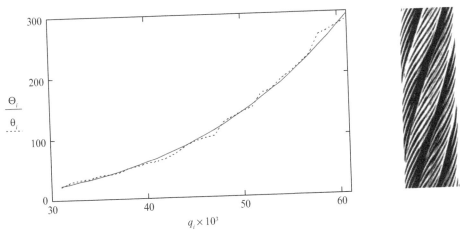

Figure 3.7. Total number of cracks in the wires of a hoist head rope vs. the number q of winds executed; θ_i empirical plot, Θ_i theoretical plot.

---

[12] Notice, that this plot begins at a certain distance from the inception point (0, 0) because after its installation the rope in the hoist works for a certain period of time without any break. It makes no sense to consider any approximation function to fit the data during this period.

empirical values. The total number of breaks at a given moment is a random variable (the left side of the equation). This means that the right-hand side of equation is also a random variable. The only stochastic element of this side is the random component. It is unobservable directly; however, it can be estimated by determining the sequence of residuals which measure how different the theoretical and empirical values are. The series of empirical values of θ_i are usually noted following the utilisation of a rope and create statistical data.

Considering function (3.24) the random component can be defined in two ways:

a. As a sequence of the differences between the empirical values of θ_i and its theoretical counterparts; it is the sequence for formula (3.24):

$$u_i = \theta_i - at_i^b \quad i = 1, 2, \ldots, \tag{3.25}$$

where a and b are estimates of the unknown structural parameters δ, χ

b. As the sequence of residuals defined by pattern:

$$\hat{u}_i = \ln \frac{\theta_i}{at_i^b} \quad i = 1, 2, \ldots, \tag{3.26}$$

which is the result of the appropriate conversion of formula (3.24). Notice, that the residual here is the index of power and $c = e$.

Both measures are correctly defined and they are two different measures of the random component for function (3.24).

The structural parameters are estimated by making a linearsation of the power function, that is:

$$\ln \Theta_t = \chi \ln t + \ln \delta + \zeta_t \tag{3.27}$$

and applying the method of least squares. The appropriateness of this method is ignored here (see Chapter 6 for more on this topic). Presuming that the process parameter is the number of winds executed, the results of the estimation of the function given in Figure 3.7 are:

$$\Theta_i = 5.28 \times 10^{-5} q_i^{3.78} e^{u_i}$$

This function is shown as a continuous one in Figure 3.7.

Incidentally, many engineers use this way of reasoning and they are usually convinced that everything is correct. But this is not true. For more on this topic, see Czaplicki 2010, Chapter 5.3.2.

Now observe what the course of the residuals looks like. Figure 3.8 illustrates the residuals u_i determined by function (3.25) for the sequence of the data observed and Figure 3.9 shows the residuals determined by function (3.26) against the number of winds executed by the rope. In both figures, the points of the residuals are connected by straight lines.

Both plots are different from a typical realisation of a pure stochastic process with a zero mean and a finite variance, as could be expected because of the least squares method that was applied. If this is so, this supposition should be verified. A suggestion can be formulated that there has to be a certain factor that generates this difference. Looking at both figures, a hypothesis can be postulated that the dispersion of the values is not uniformly distributed over time in a stochastic sense.

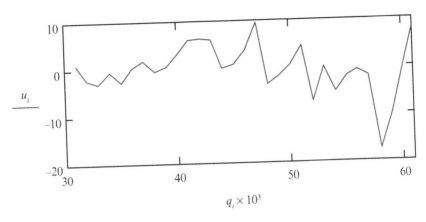

Figure 3.8. Residuals determined by function (3.25) for the sequence of data noted.

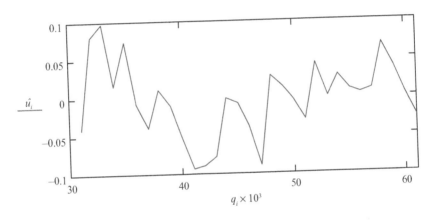

Figure 3.9. Residuals determined by function (3.26) vs. the number of winds executed by a rope.

To verify this supposition the data were divided in half and the standard error of estimation was calculated for both sequences giving $S_1 = 3.32$ for the first half and $S_2 = 6,75$ for the second. The problem now is to verify whether this difference is statistically significant or only random. An appropriate statistical test must be applied to clear up these doubts.

At least two tests can be considered:

a. The test applying the *F*-Snedecor's distribution
b. The Kruskal-Wallis one-way analysis of variance.

The reasoning based on (a) assumes a Gaussian distribution of the random variable being tested. This is the most frequently used method; however, this assumption must be verified before further analysis.

The reasoning based on (b) is a non-parametric method for testing the equality of the population medians among groups. It is identical to the one-way analysis of variance with the data being replaced by their ranks. Since it is a non-parametric method, this test does not assume the normal population (Kruskal and Wallis 1952, Siegel and Castellan 1988);

therefore, this approach is more general (more general than the test based on the *F*-Snedecor's random variable). There is a presumption in the test that the probability distribution describes a continuous random variable[13].

In the case being considered, an assumption can be made that the calculated sequence of differences has a normal probability distribution. Therefore, case (a) can be analysed.

Formulate a null hypothesis $H_0 : \sigma_1^2 = \sigma_2^2$ that the variance in the first half of the data does not differ significantly from the variance in the second half of the data. Looking at the data, an alternative hypothesis can be: $H_1 : \sigma_1^2 > \sigma_2^2$.

In order to verify the basic hypothesis, calculate the ratio of the empirical variances. This should be done in such way that the variance in the numerator must be not greater than the variance in the denominator. In our case we have the ratio S_2^2/S_1^2, which is the random variable that has the *F*-Snedecor's distribution with $(n_1 - 1, n_2 - 1)$ degrees of freedom if the hypothesis H_0 is the true one.

If the ratio is greater than the critical value $F_\alpha(n_1 - 1, n_2 - 1)$, which is taken from the appropriate statistical table presuming a level of significance α, the hypothesis H_0 must be rejected in favour of the alternative supposition. This means that the variances of residuals are statistically different, which is identical to the statement that the standard deviations are significantly different from each other.

In the case being analysed we have:

$$F = \hat{S}_2^2/\hat{S}_1^2 = 4.14 \quad \text{whereas} \quad F_{0.05}(14, 15) = 2.42$$

for the presumed level of significance $\alpha = 0.05$ (Table 9.6). The critical value should be determined in such way that the number of degrees of freedom associated with the greater variance is in the upper row of the table. The number of degrees of freedom associated with the smaller variance is placed in the side column of the table.

Due to the fact that the calculated *F* value is greater than that taken from Table 9.6: $F_{0.05}(14, 15) = 2.42$, we have the ground to reject the null hypothesis. Thus, it can be assumed that the dispersion of the residuals of the random variable being investigated increases with the number of winds that are executed by the hoist.

◄

By the way, when looking at Figure 3.3, we may suspect that the dispersion of the number of tonnes of hard coal sold on a monthly basis between the years 1995–1999 is significantly greater than in the next eight years.

Let us make some comments on an analysis of the outcomes of dispersion testing
The first piece of crucial information is that the data that are at hand is in fact the realisation of a certain stochastic process, and what is more, the dispersion of its values around the mean increases. If so, a further part of the analysis being conducted should be placed on an analysis of random processes.

Secondly, there must be a physical factor which generates the variability of the dispersion. Usually, this will be a certain process running in time. This process and its properties should be identified because much important information is associated with it.

Thirdly, the expected value function for the process is often estimated using a mathematical model that can be linearised and the least squares method is applied to estimate its unknown parameters. However, if the dispersion is not constant, one of the

---

[13] It is true that this test is more universal than Snedecor's *F* test but it is weaker in its power than the *F* test.

basic assumptions that must be fulfilled in order to allow the least squares method to be applied is not satisfied. Thus, if the least squares method is still applied, the estimates obtained can be less credible and in some cases false conclusions can be drawn based on them.

If the analysis concerns the wear process of a hoist head rope, this false inference can affect the assessment of the level of safety of the rope and this can be dangerous. Moreover, when an inference on the degree of rope wear in the future is conducted periodically, the formulated prognoses will be of a diminishing likelihood over time because of the increasing dispersion. This means that uncertainty increases with time where the accuracy of the prediction is concerned.

Because it is impossible to renew a rope that is in operation in a shaft, the only solution to keep safety at an appropriate level is to execute more precise and more comprehensive rope diagnostics and obviously to perform them more often.

When analysing the regular exploitation processes of pieces of equipment or processes of changing states, the variability of dispersion over time is rarely noted. Sometimes such a regularity can be observed when the object is in operation for a long time without any maintenance or planned prophylactic actions which should have been made.

3.5 CYCLIC COMPONENT TRACING

Many processes in mining have a cyclic character.

The cyclic character of processes is generated by the periodic character of the organisation of work. The cycle of a process can be connected with the calendar—season, day or production progression per day or shift. By looking more carefully at the operation of many machines, it is easy to notice that their exploitation process has a periodic nature, and that it can be understood in a different sense. Periods alone can have a more or less stochastic character. Many years ago, a hypothesis was formulated that proved that this cyclic nature of operation can periodically distort the processes of changes of states (Czaplicki 1974, 1975). The hypothesis proclaimed that in some periods of operation time—during the exploitation of a technical object (a single machine or a system of machines), the occurrence of some states are more probable than that of others. If this is so, the probability of the appearance of a given state is not constant in time but is rather a function of time and this function is a cyclic one. In addition, a stream of rock being extracted or hauled by transport means very often also has a periodic character. These two functions are frequently correlated with each other. In some cases, a stronger statement can be formulated: if a stream of rock being transported increases, and usually has greater dispersion in value, it means that the probability of the occurrence of a failure in the transporting units increases. Therefore, the output of a hauling unit that is calculated as the product of the probability of the work of the unit and the mean mass of the rock being transported gives an incorrect estimation because the higher the mass being transported, the lower the probability of its displacement. Thus, this estimation gives higher values than it should.

Consider the problem of the existence of a cyclic component from a formal point of view. There are two cases to consider:

a. the period of cycle is known
b. the period of cycle is unidentified.

Consider these cases in a sequence.

3.5.1   *The period of cycle is known*

If the period of a cycle is known, we can presume that during the cycle one observes consecutive stochastic copies of the same phenomenon. Denote this cycle by [0, *T*]. Presume that the object[14] can be in any of *k* mutually excluding states, *j* = 1, 2, ..., *k*. The object is observed *N* times—that is, records of what was going on with the object during *N* cycles are given. If we pay attention to one unit of cycle time (it can be any), we notice one out of the *k* events. If so, the construction of the probability distribution of a random variable that a given state is observed *k* times in *N* independent trials (*N* ≥ *k*) is now possible. This **distribution** is **multinomial** and is given by the formula:

$$P\{X_1 = b_1, X_2 = b_2, ..., X_k = b_k\} = \frac{N!}{b_1!b_2!...b_k!} p_1^{b_1} p_2^{b_2} ... p_k^{b_{k1}}, \qquad (3.28)$$

where:

$$\sum_{j=1}^{k} p_j = 1 \quad \sum_{j=1}^{k} b_j = N$$

and b_j is the multiplicity of the occurrence of the *j*-th state.

Consider one state of an object. Its realisation in time consists of the realisations in *N* periods of time. If all of these realisations are put together, the frequency of the occurrence of this state versus the time cycle will be obtained for *N* independent trials. The relative frequency, in turn, is the *j*-th estimator that is unbiased, consistent and most efficient for parameter *p* of the distribution (formula 3.28).

The following hypothesis can be formulated: the method of the exploitation of an object can generate a significant irregularity in the process of the changes of the states of the object. In other words, some states can occur more frequently in some periods of time during the cycle and some states will be observed less often.

If we have the diagrams of the relative frequency against time in the cycle for all states, the above hypothesis can be verified. It is obvious that a certain irregularity of the process of the appearance of a given state will be visible due to the stochastic character of the process. However, the problem arises of whether the changes that are observed are connected exclusively with the stochastic nature. By reversing the problem, a question can be formulated: how many times can a given state occur in a moment of cycle time in *N* trials that such an event can be assessed as very rare—so rare that a certain exploitation factor probably generated this irregularity?

Let us simplify our consideration and study only one state. Denote it by *s*. If it is the only one of interest, it can be specified in the following way:

$$P\{X_s = b_s\} = \frac{N!}{b_s!\prod_{j \neq s} b_j!} p_s^{b_s} \prod_{j \neq s} p_j^{b_j}$$

However

$$\prod_{j \neq s} b_j! = (N - b_s)! \quad i \quad \prod_{j \neq s} p_j^{b_j} = (1 - p_s)^{N - b_s},$$

---

[14] In some cases the term 'technical object' stands for a system of pieces of equipment. If this system is observed during a longer period of time, it may happen that some pieces are withdrawn or added. A cardinal feature of mine systems is their changeability because the lengths of the hauling distances change almost continuously. If only one piece of equipment is added or withdrawn from the system, the system is not the same. Its characteristics change.

thus

$$P\{\boldsymbol{X}_s = b_s\} = \binom{N}{b_s} p_s^{b_s}(1-p_s)^{N-b_s}. \tag{3.29}$$

The multinomial distribution is reduced to a binomial one. The problem of a significant irregularity in the occurrence of state s is reduced to finding the number b_s, which has a probability of appearance lower than that presumed, a small level of probability, say υ (where $\upsilon \ll 1$), that is:

$$\binom{N}{b_s} p_s^{b_s}(1-p_s)^{N-b_s} < \upsilon. \tag{3.30}$$

Due to the well-known properties of the binomial distribution, there will be two values b_s that fulfil this inequality. Denote them by $b_s^{(l)}$ and $b_s^{(u)}$, where $b_s^{(u)} > b_s^{(l)}$ (see Figure 3.10). From this figure it is easy to observe that the critical area is determined by level υ; all events that have a probability below this level should be comprehensively considered; we may suspect that their appearance was non-random.

The probability density function of this distribution has a maximum for:

$$b = (n+1)p \quad \text{if} \quad (n+1)p \notin \mathfrak{R}$$

$\mathfrak{R}$—set of natural numbers and has two maximum values:

$$b_1 = (n+1)p \quad \text{and} \quad b_2 = (n+1)p - 1,$$

if $(n+1)p \in \mathfrak{R}$.

Large reliability investigations comprising continuous mechanised systems operating in both underground and surface mining were carried out in Poland in the mid-seventies of the previous century. The point of interest was, among other things, the problem of whether the cyclic character of the work of these systems had an influence on the course of the operation process of these systems.

A histogram illustrating the frequency of the occurrence of a repair state in a certain series system operating in the underground coal mine in the Silesian District of Poland is shown in Figure 3.11. The system consisted of a coal shearer, two armoured flight conveyors and a certain number of belt conveyors that delivered the coal that was won to the shaft bin. The observation consisted of N elementary observations repeated during every morning shift excluding the first hour. Because of the properties of the series system, any repair of any piece of equipment of the system meant a repair state for the whole system. In the figure two probability levels are visible for which the probability υ was presumed to be 0.05.

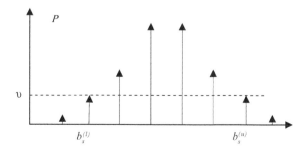

Figure 3.10.   Binomial probability density function.

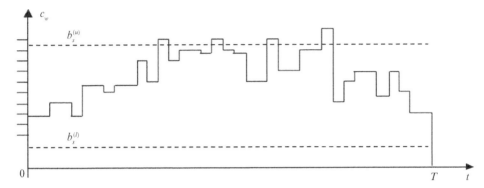

Figure 3.11.   Histogram of the relative frequency c_w of the occurrence of a repair state for a certain series system obtained by observing the system during N exploitation shifts.

Looking at Figure 3.11, four events should be considered—all of them connected with the fact that the frequency of the occurrence of a repair state was above the presumed level. Further analysis comprised an examination of the records to find out whether the reasons for the appearance of this state were repeated. If the reasons were repeated, they generated such a rare event that it can be assessed as not entirely random. Immediately, a recommendation can be formulated to eliminate them from the further operation of the system. However, if the reasons were different in each, they can be evaluated as purely random and therefore it does not matter how rare this event was[15].

It is very important to understand that *as a result of the application of a statistical procedure, information is obtained that indicates which events should be taken into further comprehensive consideration*. And that is all. Advanced analyses must proceed outside of the area of mathematics; physical aspects have to be taken into account before any final assessment can be made.

To complete these considerations it is necessary to construct an estimator of the unknown probability p_s that is given in formula 3.29. Following the relative frequency approach, the number of favourable events is represented by the area of the histogram, whereas all possible events are represented by an extraordinary event when only one state is observed in all N trials. Thus, an interesting estimator is determined by the function

---

[15] The elimination of every first hour of the operation was connected with the fact that the operation of the system was not a full one. For this reason the frequency c_w was below the lower critical level as a rule for obvious reasons.

$$\hat{p}_s = \frac{\sum_i t_{si}}{NT} \tag{3.31}$$

where t_{si} is the i-th time of state s. The denominator determines the total observation time whereas the numerator defines the total time of state s in the period of observation[16].

Consider now the second case.

3.5.2   *The period of the cycle is unknown*

In the theory of stochastic processes $X(t)$, it has been proved that each stochastic process can be **decomposed**[17] into three components:

a. The trend that is associated with the expected value function (the systematic component), $S(t)$
b. The cyclic component, $C(t)$
c. The pure random component, $\Xi(t)$.

The first two components are deterministic functions whereas the third one is a random one. This comprises the whole stochastic nature of the process. At the very beginning of the analysis of the process, the problem arises as to whether the composition of these three items should be an additive, multiplication and mixed one. It is suggested that if the process being analysed has an explosive character, the multiplication model should be applied. If the course of the process is rather smooth, the additive model would be better.

In mining engineering practice, the majority of the processes that are analysed have no explosive character even if they are non-stationary ones. For this reason, in our further consideration we presume that the model of the process is additive, multiplicative and mixed one. i.e.:

$$X(t) = S(t) + C(t) + \Xi(t) \tag{3.32}$$

If this is so, with data usually in the form of a time series, $x(t_1)$, $x(t_2)$, ..., $x(t_n)$ that is the discrete observation of the realisation of a certain stochastic process, the first step in the analysis of the process is the identification of the trend of the process. If this function is identified, $S(t)$, then the data should be transformed in order to obtain a new time series:

$$y(t_i) = x(t_i) - S(t_i); \quad i = 1, 2, ..., n \tag{3.33}$$

The above sequence has no trend and for this reason is stationary but still has both a cyclic and a pure component[18]. From a theoretical point of view we have the following situation. The sequence is a realisation of a mixture of two stochastic processes that are mutually uncorrelated and stationary with average values that equal zero. The properties of these processes are completely different. One process is strictly cyclic; the second one has no such property. Therefore, if these processes overlap, the final process can have a periodicity that is difficult to trace. The intensity of the obliteration of this periodicity depends on the autocorrelation of the cyclic process and increases when its variance increases (Granger and Hatanaka 1969). Nevertheless, it is necessary to identify these processes beginning with the cyclic component.

------

[16] For more on this subject from a mining engineering point of view, see Czaplicki 2010, p. 34–36.
[17] Decomposition generally means to express something in terms of a number of independent simpler components.
[18] Obviously, there may be a case in which the realisation has no cyclic component.

There are some methods in mathematical statistics that allow a periodic component in time series to be identified. These methods are: spectral analysis and harmonic analysis.

To describe the concept of the method of spectral analysis, we must introduce the term: function of the power spectrum, which is the first derivative of the spectrum distribution $F(\omega)$ of the stochastic process. This function can be expressed as:

$$dF(\omega) = f_2(\omega) + \sigma_i^2 \quad f_2(\omega) = dF_2(\omega)/d(\omega) \tag{3.34}$$

where $f_2(\omega)$ is the spectral density of the process $X_2(t)$ that 'hides' the cyclic component and σ_r^2 is the r-th variance of the process $X_1(t)$ of strict periodicity.

The relationship (3.34) can be used as a tool for an analysis of periodicity for two main reasons:

a. if the cyclic component equals zero ($X_1(t) = 0$), then the function $dF(\omega) = f_2(\omega)$, which means it covers the spectral density of the process $X_2(t)$, which is an absolutely continuous function
b. if there is a cyclic component ($X_1(t) \neq 0$), then the function $dF(\omega)$ is not absolutely continuous; in points $\omega = \omega_r$ the value of the function jumps up because to the value of the function $f_2(\omega)$ is added the variance σ_r^2 of the r-th component of the process $X_1(t)$.

In practice, when a graphical picture of the function $dF(\omega)$ is drawn, it is easy to notice such points (ω_r; $r = 1, 2, \ldots, s$) where the value of the function increases drastically. Therefore, one can say on the periodicity of the process in its points $2\pi/\omega_r$. A difficult problem arises when in some points the function increases only slightly. Unfortunately, estimation the function $dF(\omega)$ is usually done with a certain accuracy only and there is no clear indication whether the observed increment of the value of the function is significant or not. However, there are several methods that can be used to dispel any doubts. Different authors recommend different tests; however, many of them are complicated procedures.

Let us first consider the idea of harmonic analysis due to its simplicity.

If the observed time series $y(t_i)$; $i = 1, 2, \ldots, n$ has no trend, it can be expanded in a Fourier series. In mathematics, a Fourier series decomposes a periodic function into the sum of simple oscillating functions, namely sines and cosines. Following this line of reasoning, we can write:

$$y(t_i) = \frac{1}{2}a_0 + \sum_{i=1}^{n}\left(a_i \cos \omega_i t_i + b_i \sin \omega_i t_i\right) \tag{3.35}$$

where a_0, a_i, b_i are the Euler–Fourier coefficients.

The estimators of these coefficients are as follows:

• The expected value estimator

$$\hat{a}_0 = \frac{1}{n}\sum_{i=1}^{n} y(t_i) \tag{3.36}$$

• Further estimators of the Euler–Fourier coefficients

$$\hat{a}_j = \frac{2}{n}\sum_{i=1}^{n} y(t_i)\cos\frac{2\pi ij}{n} \qquad \hat{b}_j = \frac{2}{n}\sum_{i=1}^{n} y(t_i)\sin\frac{2\pi ij}{n} \tag{3.37}$$

for $j = 2, \ldots, n/2$.

The values of these statistics can be applied to verify the hypothesis that a cyclic component is significant in a given sample.

Consider the square amplitude of the process. It is given by the equation:

$$\hat{A}_j^2 = \hat{a}_j^2 + \hat{b}_j^2 \tag{3.38}$$

Its expected value is

$$E(\hat{A}_j^2) = A_E^2 = 4\frac{\sigma^2}{n} \tag{4.39}$$

where σ^2 is the variance of the process.

By replacing the unknown variance σ^2 with its unbiased estimator (the variance estimated from the sample), one obtains:

$$E\left(\hat{A}_j^2\right) \cong \frac{4}{n(j-1)}\sum_{i=1}^{n}\left(y(t_i)-\overline{y}\right)^2 \tag{3.40}$$

where $\overline{y}$ is the estimate of $\hat{a}_0$, i.e. the arithmetic mean calculated from the sample.

The probability that an event that $\hat{A}_j^2$ will be α times greater than A_E^2 is determined by the pattern:

$$P\left(\hat{A}_j^2 > \alpha A_E^2\right) = e^{-\alpha} \tag{3.41}$$

which—after rearrangement—gives:

$$P\left\{\left(\frac{\hat{A}_j}{A_E}\right)^2 > -\ln\alpha\right\} = \alpha \tag{3.42}$$

If the level of significance presumed equals α, one can verify a null hypothesis stating that the i-th jump that is the value of the quotient $\left(\frac{\hat{A}_j}{A_E}\right)^2$ is significant. If the inequality holds

$$\left(\frac{\hat{A}_j}{A_E}\right)^2 > -\ln\alpha$$

then the verified null hypothesis should be rejected. This means that we can presume that the time series shows important oscillations with the period that equals n/i. This regularity holds with the probability $1 - \alpha$.

In some cases, information that some periodic oscillations are significant is enough to take the proper decision in relation to the source of the observed data. But in some other cases it is not enough; for instance, we need to predict what the probable course of the process observed in the near future will be. Let us ignore for the time being what the prognosis means and let us conduct our consideration on a decomposition of a time series a little further. Here a more advanced approach will be presented that makes use of the fundamental monograph written by Box and Jenkins (1976).

If all of the significant fluctuations are identified, we are able to construct the cyclic component function $C(t)$[19]. Then, having specified two deterministic components, we are able to identify the third element, a purely stochastic one. Consider the following sequence:

---

[19] A cyclic component can consist of a few functions.

$$u(t_i) = x(t_i) - S(t_i) - C(t_i); \ i = 1, 2, \dots, n \tag{3.43}$$

The above relationship determines the time series of the residuals generated by the differences. It can be presumed that this sequence is a realisation of the unknown purely stochastic component ξ_t. Usually, it is also presumed that the sequence $u(t_i)$ has a zero expected value and the finite and constant standard deviation σ_ξ. In the majority of cases, a stronger presumption is formulated namely: $N_\xi(0, \sigma_\xi)$. Ignoring how strong the presumptions are, it is necessary to verify—by applying the appropriate statistical tests—whether all of these assumptions hold when confronted with the properties of data in hand.

■ **Example 3.8** (Based on Manowska's Ph.D. dissertation, 2010, Chapter 14)

In Figure 3.12 is a graph of a mass of hard coal sold in Poland versus time counted in months. The data comprise the period of the turn of the century: the late nineties—beginning of 21st century. A great restructuring of the mining industry (mainly coal) took place in Poland during this period.

It is easy to notice that the time series is a non-stationary one and that it decreases with time. Therefore, a linear function[20] was applied as the first approximation of the trend in the data, i.e.

$$\hat{y}_S(t) = \alpha t + \beta + u_t$$

The classical method of least squares was applied in order to estimate the unknown parameters a and b (see Chapter 6.2) obtaining the following equation:

$$\hat{y}_S(t) = -25.56t + 10947 + \zeta_t$$

where ζ_t is a random component. The linear regression function is visible in this Figure.

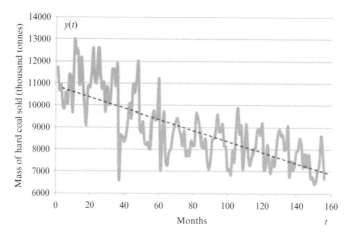

Figure 3.12.   Mass of hard coal sold versus time.

---

[20] This function makes sense in the interval observed and perhaps, for only a few months ahead. The more proper one should be a decreasing function that tends to a certain horizontal asymptote.

It was suspected that there was a cyclic component in the sequence, thus ζ_t consisted of the constituent and the purely random one.

As the next step, the differences were calculated: $y(t) - \hat{y}_S(t)$ for all of the months of the sample. All of these differences are shown in Figure 3.13.

The sequence visible in this figure was the income data for the Fourier analysis with its transform applied. The Matlab 5.3 program was used and the result of the Fourier study is presented in Figure 3.14.

It is important whether a dominating frequency is observed in terms of the module of the spectrum in such a graph. If so, a cyclic component very likely exists.

It was presumed that the general form of a cyclic function is described by the following pattern:

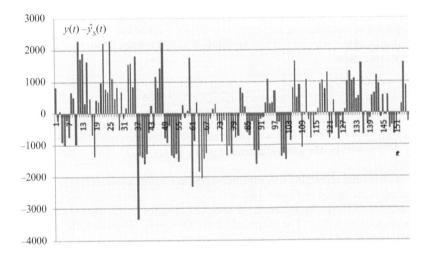

Figure 3.13.   Time series of the differences $y(t) = \hat{y}_S(t)$ (Manowska 2010).

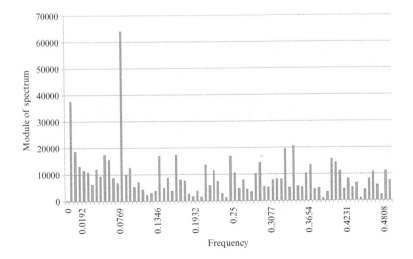

Figure 3.14.   Result of the application of the Fourier transform (Manowska 2010).

$$y_C(t) = \sin\left(2\frac{\pi}{T_0}t + \varphi\right)A \qquad\qquad (3.44)$$

where: T_0—period, number of months in one cycle,
　　　φ—the phase displacement,
　　　A—the amplitude.

Next, the harmonic analysis was applied that relies on a description of the residuals as the sum of the sinus functions for whatever period, amplitude and phase were selected using the appropriate algorithm. This algorithm relies on the analysis of the mean square error that is the result of the application of the sinus functions. The minimum of this error was a point of interest. In order to search for this minimum, the period was changed from 1 to 156 (sample size), the amplitude varied from 0 to 1000 tonnes and the phase displacement varied from 0 to 2π.

The minimum of the mean square error was 10,400 tonnes for the period equal to 12 and the phase displacement was 2.17. The amplitude was 820 tonnes. Therefore, the formula for the cyclic function was:

$$y_{C1}(t) = \sin\left(2\frac{\pi}{12}t + 2.17\right)820$$

If so, a further point of interest was the next time series, the second residuals. They were obviously determined by the general formula: $y(t) - \hat{y}_S(t) - y_{C1}(t)$. This sequence is shown in Figure 3.15.

This way of further reasoning was repeated and the Fourier analysis applied. The result of its application is presented in Figure 3.16.

The minimum of the mean square error was 8,272 tonnes for the period equal to 114 and the phase displacement 0.72. The amplitude was 506 tonnes. Therefore, the formula for the cyclic function was:

$$y_{C2}(t) = \sin\left(2\frac{\pi}{114}t + 0.72\right)506$$

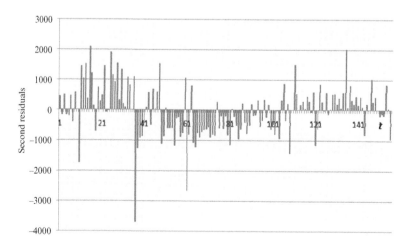

Figure 3.15.   Time series of the differences: $y(t) - \hat{y}_S(t) - y_{C1}(t)$ (Manowska 2010).

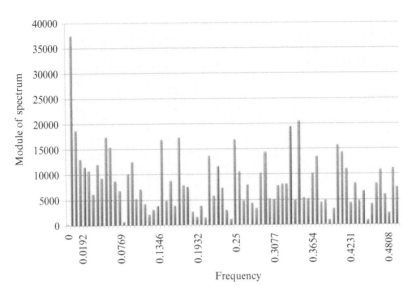

Figure 3.16. Result of the application of the Fourier transform for the second residuals of the time series (Manowska 2010).

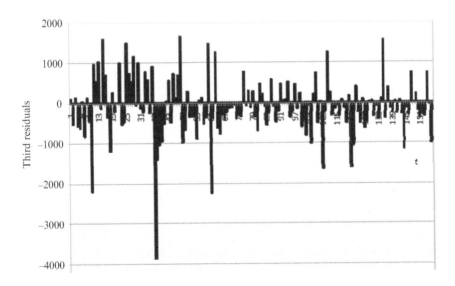

Figure 3.17. Time series of the differences: $y(t) - \hat{y}_S(t) - y_{C_1}(t) - y_{C_2}(t)$ (Manowska 2010).

A further point of interest was the next time series, the third residuals, given as the difference: $y(t) - \hat{y}_S(t) - y_{C_1}(t) - y_{C_2}(t)$, for all points noted. This sequence is shown in Figure 3.17.

Based on this data, a third plot was constructed that showed the result of the application of the Fourier analysis. Figure 3.18 illustrates again the relationship between the modules of spectrum against time.

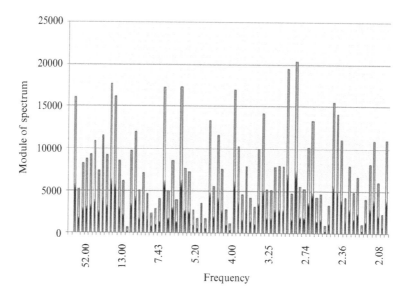

Figure 3.18.   Result of the application of the Fourier transform for the third residuals of the time series (Manowska 2010).

By analysing Figure 3.18 more carefully it is easy to conclude that there is no dominating frequency that is clearly greater than the others. Therefore, it can be suspected that no other cyclic functions should be constructed. However, this way of reasoning was repeated as in the previous cases and the mean square error was greater than before. It is obvious that the construction of the next cyclic function is not needed; a model with such a function will be worse than without it.

Thus, the final model to generate the observed time series is:

$$y(t) = \hat{y}_S(t) + y_{C_1}(t) + y_{C_2}(t) + \xi_t$$

□

3.6   AUTOCORRELATION ANALYSIS

There are some processes in mining engineering that depend on many factors that can be grouped together to create two specific sets: the properties of the object that is the point of the investigation and the main characteristic features of its operational process. They can comprise many elementary components such as: material fatigue, corrosion, friction wear, local weight loss, pitting and so on, all of which concern technical objects. However, there are some other processes running in the rocks surrounding a mine that also depend on many features and it is difficult to take all of them into account. In such cases, the course of an interesting variable can be described by the values of it that were noted in the past. This leads to the application of an autoregression model, which can be used if the values of the variable that were noted in sequent moments of time depend on each other.

There are also some other cases when the dependence of the actual values of the variable being investigated, for which the values were recorded in the past, are significant. This can concern a purely random component in time series.

Let us consider two examples.

■ **Example 3.9**

The object of consideration was a hoist head rope with a triangular shape of strands working in the main shaft of an underground mine. The point of interest was the course of the wear of the rope. Observations were made every $v = 10^3$ hoist cycles and the number of breaks in the wires were noted. The empirical data are shown in Figure 3.19. The plot of the theoretical function is also visible in this figure:

$$N_i = av_i^b$$

for which the estimates of the unknown structural parameters were: $a = 6.48 \times 10^{-5}$ and $b = 3.22$. These estimates were obtained after the linearisation of this power function and after the application of the least squares method.

The goodness of the estimation using this power function was investigated by analysing the residuals that were determined by the formula:

$$\varepsilon_i = n_i - N_i$$

The graph of these differences is presented in Figure 3.20.

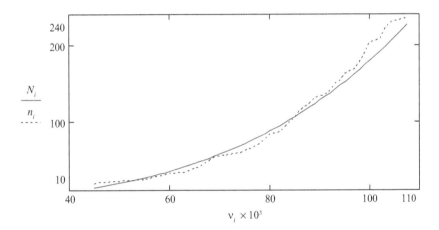

Figure 3.19.   Plot of the total number of cracks in the wires of the hoist head rope vs. the number of winds executed by the hoist; n_i empirical plot, N_i theoretical plot.

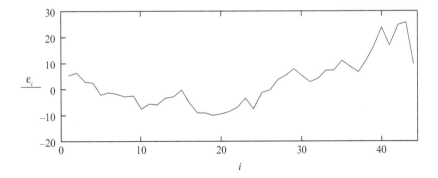

Figure 3.20.   Graph of sequent residuals ε_i.

This graph is not a typical realisation of a purely random process with a zero mean and a constant variance, which is what might be expected.

Let us investigate the residuals by searching for internal stochastic relationships.

One of assumptions of the classical method of least squares says that the residuals should not be correlated with each other, i.e.

$$E(\boldsymbol{\varepsilon}_i \boldsymbol{\varepsilon}_j) = 0 \quad i \neq j \tag{3.45}$$

The term correlation was introduced here in relation to the stationary testing of the realisation of the random variable that was observed. Recall that correlation is a certain type of stochastic relationship. It relies on such regularity that when the values of one variable increase (or decrease), on average the values of the second variable decrease or increase. Thus, it is a statistical relationship. If the consideration concerns only one variable, we say it is an autocorrelation.

Calculate the correlation coefficients of first, second and third order: $r_1^{(a)}$, $r_2^{(a)}$, $r_3^{(a)}$. This means that we investigate the interdependence between the two sequences that are noted; the original sequence and the sequence derived by the first, second and third beginning elements, respectively. Note, that the number of elements taken under consideration decreases one by one when one calculates the sequent autocorrelation coefficients. As the measure of correlation, the classic Pearson's linear correlation coefficient can be applied if a sample is given in the form $(x_i, y_i; i = 1, 2, ..., N)$. The coefficient can be expressed as:

$$R_{XY} = \frac{\sum_{i=1}^{N}(x_i - \bar{x})(y_i - \bar{y})}{\sqrt{\sum_{i=1}^{N}(x_i - \bar{x})^2 \sum_{i=1}^{N}(y_i - \bar{y})^2}} \tag{3.46}$$

In the case when the autocorrelation coefficient of the first order is being calculated, pattern (3.46) takes the form:

$$r_1^{(a)} = \frac{\sum_{i=1}^{N-1}(x_i - \bar{x}_1)(x_{i+1} - \bar{x}_2)}{\sqrt{\sum_{i=1}^{N-1}(x_i - \bar{x}_1)^2 \sum_{i=1}^{N-1}(x_{i+1} - \bar{x}_2)^2}} \tag{3.47}$$

where:

$$\bar{x}_1 = \frac{1}{N-1}\sum_{i=1}^{N-1} x_i \quad \bar{x}_2 = \frac{1}{N-1}\sum_{i=2}^{N} x_i \tag{3.47a}$$

Notice, that for an increasing sample size, the difference in the mean values becomes negligible. One can construct further autocorrelation coefficient formulas in a similar way.

The results of the calculation of the autocorrelation of the residuals in the sample were as follows:

$$r_1^{(a)} = 0.932 \quad r_2^{(a)} = 0.866 \quad r_3^{(a)} = 0.884$$

The values obtained are high, but we have no idea whether they are significant in a statistical sense. In order to answer this question, the Durbin-Watson test is usually the one that comes to mind (Durbin 1953, Durbin and Watson 1950, 1951). However, the Durbin-Watson statistic is only valid for stochastic regressors and first order autoregressive schemes (such as AR(1)). Furthermore, it is not relevant in many cases; for example, if the error distribution is not normal, or if it concerns the dependent variable in a lagged form as an independent variable. In these cases, it is not an appropriate test for autocorrelation. The tests that are suggested and which do not have these limitations are the Breusch-Godfrey test (Breusch 1979, Godfrey 1978, 1988) as well as the Pawłowski *J* test (Pawłowski 1973). Because the first of these tests is well-known and much simpler to apply when compared to the latter test, further reasoning will be performed using this kind of examination.

The Breusch-Godfrey statistic is determined by the following formula:

$$\chi^2(c) = (N-c)\left(r_c^{(a)}\right)^2 \tag{3.48}$$

where c is the autocorrelation order.

A verified hypothesis is H_0: $\rho_c = 0$, i.e. there is no autocorrelation of the order c in the random variable being tested. If the following inequality holds:

$$\chi^2(c) > (N-c)\left(r_c^{(a)}\right)^2$$

then there is no ground to reject the null hypothesis. Otherwise, one can presume that the autocorrelation of the order c is significant.

Making all of the necessary calculations and reading the critical values from the table of χ^2 distribution for a presumed level of significance $\alpha = 0.05$ (Table 9.4), we have:

$$37.35 \ (3.84) \quad 32.97 \ (5.99) \quad 32.04 \ (7.82)$$

where the first number is the empirical value and the corresponding critical one is in the brackets.

The Breusch-Godfrey test can also be supported by the *F*-Snedecor's statistic by making use of the well-known relationship between the χ^2 statistic and the *F*-Snedecor's statistic. It has been proven using a simulation technique for small samples that such an approach is better than that one based on the χ^2 statistic. In the case being analysed, it does not matter which statistic is applied (either χ^2 or *F*-Snedecor's), the result of verification is identical: all empirical values are significant.

By translating this result into engineering language, one can say that there is a significant dependence between the degree of rope wear in a given moment of time and the degree of rope wear a while ago and two whiles before. It also means that the wear process of the rope has a memory and—as investigations showed (Czaplicki 2010, Chapter 5)—this memory can be constant with time in a stochastic sense but it can also be variable depending on the number of winds that have been executed by the rope.   ◄

Some important remarks can be formulated in connection with autocorrelation testing when random variables are the objects of engineering interest.

- If during a statistical investigation, the autocorrelation of the random variable being tested was traced, then one can be almost certain that there is a physical reason generating this statistical regularity.
- Autocorrelation means that there is a 'memory' in the process that is observed and the future state depends—as a rule—on the state just before, sometimes on some states that happened earlier.
- If the autocorrelation was traced, it gives ground to formulate the supposition that the adequate model describing the course of the random variable being tested is an autoregression function and *vice versa*.
- If autocorrelation was found, the further investigation should be focused on recognising the physical grounds that are the source of this autocorrelation. This can provide important knowledge on the nature of the process that is being observed which can allow for some

counteraction if this phenomenon is disadvantageous (if possible) or can allow engineers to make use of it if it is useful.

3.7   HOMOGENEITY OF DATA

The statement that something is **homogeneous** means that it is alike, similar or uniform from a certain point of view. In the statistical sense, homogeneity is a property of a data set and relates to the validity of the very advantageous assumption that the statistical properties of the samples that were taken are identical to the whole population.

There are some areas of analysis in mining engineering in which we cannot complain that the data are poor or small in number. This often concerns analyses from ore dressing areas. A similar situation applies to gathering data on the regular vibrations of surrounding rocks of a mine. Moreover, there are also some different fields of mining engineering interest where the information that is collected is usually poor. In reality, the machines that are in operation are often of a high quality and reliability and failures do not occur very often. Therefore, in order to gather an appropriately large enough sample of information on how a given machine fulfils its duties from a reliability point of view is difficult; such a technical object should operate for a long time—frequently too long in comparison to mining reality. In some other cases, when research concerns destructive tests, we cannot permit so many items to be destroyed. However, in many cases, it is possible to examine a certain number of similar objects operating in similar conditions and it can be expected that data that are obtained will be homogeneous; all of the observations can be gathered together in order to create a large sample so that statistical inference will have a strong foundation. In other words—we have observed a certain number of stochastic copies of the same phenomenon and these data create the entirety.

The homogeneity of random variables is also of interest in comparative studies. One has two slightly different technical objects or slightly different processes and the point of interest is to answer the question of whether the difference that exists is significant or not.

In engineering studies of a probabilistic nature, homogeneity is usually understood in two ways, namely:

a. As the equality of distributions, i.e. the probability distributions of random variables that are the subject of interest are identical ones
b. As the equality of parameters; the parameters of a statistical nature characterising the selected properties of the object of investigation, e.g. average values, standard deviations, probabilities of occurrence of determined events or states etc., differ from each other only negligibly from a statistical point of view.

Consider case (a).

There are a number of statistical tests that allow the hypothesis that the data are homogeneous to be verified from the distribution point of view. Divide this problem into two separate cases:

i. There are two samples
ii. There are three or more samples

and we are interested whether we can assume that they come from the same population. This division is connected with the properties of the tests that are applied.

When the data in hand comprise two samples, using the Smirnov test is recommended[21]. The variable tested should be a continuous one. A model considered for this test is as follows.

---

[21] There is also the Smirnov test, which allows three samples to be compared (Birnbaum and Hall 1960).

Let $(X_1, X_2, ..., X_n)$ and $(Y_1, Y_2, ..., Y_n)$ be two simple samples. A verified null hypothesis states that both samples come from the same population.

Construct two empirical distribution functions for both random variables according to the following patterns:

$$F_n(x) = \begin{cases} 0 & \text{for} & x \leq x_{(1)} \\ i/n & \text{for} & x_{(i)} < x \leq x_{(i+1)} & i = 1, 2, ..., n-1 \\ 1 & \text{for} & x > x_{(n)} \end{cases} \tag{3.49}$$

and

$$G_m(y) = \begin{cases} 0 & \text{for} & y \leq y_{(1)} \\ j/m & \text{for} & y_{(j)} < y \leq y_{(j+1)} & j = 1, 2, ..., m-1 \\ 1 & \text{for} & y > y_{(m)} \end{cases} \tag{3.50}$$

where $(x_{(1)}, x_{(2)}, ..., x_{(n)})$ and $(y_{(1)}, y_{(2)}, ..., y_{(m)})$ are the samples arranged monotonically into non-decreasing sequences.

The statistic that measures the distance between distributions $F_n(x)$ and $G_m(y)$ is determined in one of the following ways:

$$\begin{aligned} D_{n,m}^+ &= \sup_x [G_m(x) - F_{mn}(x)] \\ D_{n,m}^- &= -\inf_x [G_m(x) - F_{mn}(x)] \\ D_{n,m} &= \sup_x [G_m(x) - F_{mn}(x)] \end{aligned} \tag{3.51}$$

which means that the point of interest is the maximum of the mismatch. In other words, if you plot the sorted values of sample x against the sorted values of sample y as a series of increasing steps then the test statistic is the maximum vertical gap between these two plots.

Obviously, $D_{n,m} = D_{m,n}$. Both statistics $D_{n,m}^+$ and $D_{n,m}^-$ have the same distribution. Let us devote our attention to only one of them.

Denote by $D_{n,m}^+(\alpha)$ and $D_{n,m}(\alpha)$ the critical values for both statistics if the level of significance in the test is α. Due to the discontinuity of the statistic distribution $D_{n,m}$, the corresponding critical value is determined by the formula:

$$D_{n,m}(\alpha) = \inf\{d : P(D_{n,m} \geq d) \leq \alpha\}$$

The critical value $D_{n,m}^+(\alpha)$ is determined analogically.

In practice statistics are calculated by applying one of the formulas:

$$\begin{aligned} D_{n,m}^+ &= \max_{1 \leq j \leq m} \left(\frac{j}{m} - F_n(y_{(j)}) \right) = \max_{1 \leq i \leq n} \left(G_m(x_{(i)}) - \frac{i-1}{n} \right) \\ D_{n,m}^- &= \max_{1 \leq j \leq m} \left(F_n(y_{(j)}) - \frac{j-1}{m} \right) = \max_{1 \leq i \leq n} \left(\frac{i}{n} - G_m(x_{(i)}) \right) \\ D_{n,m} &= \max\left(D_{n,m}^+, D_{n,m}^- \right) \end{aligned} \tag{3.51a}$$

To apply the above test it is necessary to make use of Table 9.16a, which gives the critical values $D_{n,m}(\alpha)$. These values are valid for $n = 3(1)20$, $m = 2(1)n$ and $\alpha = 0.01; 0.02; 0.05; 0.10$. The intersection of the line that corresponds with data n and m and the column that corresponds with the probability α determines two numbers: the integer $d_{n,m}(\alpha)$ and the fractional number α^*. At the same intersection of the line with the column denoted by k read the integer $k_{n,m}$. The critical value is given by the formula:

$$D_{n,m}(\alpha) = \frac{d_{n,m}(\alpha)}{k_{n,m}} \tag{3.52}$$

The number α^* is the real level of significance of the test, in which $\alpha^* \le \alpha$. The differences between α^* and α come from the discontinuity of the distribution of statistic $D_{n,m}$.

The distribution of statistic $D_{n,m}$ for $n = m = 1(1)40$ is presented in Table 9.16b. For a given n and $k = 1(1)12$, the probability $P\{D_{n,n} \le k/n\}$ can be read off. Due to the fact that for $\alpha \le 0.10$, the following approximate equality holds:

$$D_{n,m}^+(\alpha) \approx D_{n,m}(2\alpha)$$

There is no table for the statistic $D_{n,m}^+(\alpha)$.

▪ Example 3.10

A durability investigation of a certain mechanical part of an articulated dump truck was carried out. The point of interest was the number of load cycles but not as related to the failure occurrence but the number of load cycles that were the difference between the assumed level and the number achieved. Two parts were tested and for this reason two samples were obtained. They were as follows:

$(0.46 \quad 0.14 \quad 2.45 \quad -0.32 \quad -0.07 \quad 0.30) \times 10^3$ cycles

$(0.06 \quad -2.53 \quad -0.53 \quad -0.19 \quad 0.54 \quad -1.56 \quad 0.19 \quad -1.19 \quad 0.02) \times 10^3$ cycles

A hypothesis was formulated stating that these two samples came from the same population. The alternative supposition rejects this.

The calculation procedure is as follows.

1. Construction of the empirical distribution $F_n(x)$ for the first sample:

i	X	$F_n(x)$
1	−0.32	0
2	−0.07	1/6
3	0.14	2/6
4	0.30	3/6
5	0.46	4/6
6	2.45	5/6

2. Construction of the empirical distribution $G_m(y)$ for the second sample:

j	y	$G_m(y)$
1	−2.53	0
2	−1.56	1/9
3	−1.19	2/9
4	−0.53	3/9
5	−0.19	4/9
6	0.02	5/9
7	0.06	6/9
8	0.19	7/9
9	0.54	8/9

3. Sort values of both samples to get distribution functions $F_n(x)$ and $G_m(y)$ as k/r where r is the minimum common multiple for numbers n and m
4. Further calculation procedures are as follows:

u	$F_n(u)$	$G_m(u)$	$F_n(u) - G_m(u)$
−2.53	0	0	0
−1.56	0	2/18	−2/18
−1.19	0	3/18	−3/18
−0.53	0	4/18	−4/18
−0.32	0	8/18	−8/18
−0.19	3/18	8/18	−5/18
−0.07	3/18	10/18	−7/18
0.02	6/18	10/18	−4/18
0.06	6/18	12/18	−6/18
0.14	6/18	14/18	−8/18
0.19	9/18	14/18	−5/18
0.30	9/18	16/18	−7/18
0.46	12/18	16/18	−4/18
0.54	15/18	16/18	−1/18
2.45	15/18	1	−3/18

5. Look for the maximum inconsistency in the last column. Here we have:

$$D_{6,9} = \max|F_6(u) - G_9(u)| = 8/18$$

From Table 9.16a one gets the critical value:

$$D_{6,9}(0.05) = 13/18$$

By looking at both values, it is easy to conclude that there is no ground to reject the verified hypothesis proclaiming that both samples come from the same population. ◄

Remark. If $n, m \to \infty$ then the statistic:

$$D_{n,m}^+ \sqrt{\frac{nm}{n+m}}$$

has χ^2 distribution with 2 degrees of freedom.

The statistic:

$$D_{n,m}\sqrt{\frac{nm}{n+m}}$$

has a Kolmogorov $K(y)$ distribution[22].

We continue to consider case (a) but the number of samples is two, three or more.

In statistics there are a few tests that can be applied in such a case but the most popular in engineering practice seems to be the **Kruskal-Wallis test** based on the sum of ranks (Kruskal-Wallis 1952). It is a non-parametric test and is used to compare more than two samples that are independent, or not related. The model considered for this test is as follows.

Presume that k objects are observed with regard to a certain feature and therefore k samples are obtained. A convenient feature of the test is that the samples can have different sizes. Denote them by n_i; $i = 1, 2, ..., k$. Assume that the random values of a measure of the feature can be described by a certain probability distribution $F(x)$ and a statistical hypothesis H_0 is formulated that all probability distributions are identical, which is:

$$H_0 : F_1(x) = F_2(x) = \cdots = F_k(x)$$

The alternative hypothesis H_1 rejects the null supposition.

3.7.1 *The test procedure*

All elements of all of the samples are gathered together and ranks are assigned for the monotonically ordered set—from 1 to N, where N is the total number of elements in all samples, $\sum_{i=1}^{k} n_i = N$. If tied values exist, the average of ranks must be assigned to tied values. Next, the value of the following statistic is calculated:

$$K_N = \frac{12}{N(N+1)} \sum_{i=1}^{k} \frac{T_i^2}{n_i} - 3(N+1) \tag{3.53}$$

where T_i is the sum of ranks in i-th sample.

Looking at formula (3.53), it is easy to notice that if there are more differences between the average sample ranks and the general mean rank, statistic K_N is larger. A low dispersion in this regard, in turn, will be favourable for the hypothesis H_0—providing that there is no ground to reject it.

Kruskal and Wallis observed that if k grows and if the sizes n_i increase, the random variable K_N has the asymptotic probability distribution of χ^2 with $k - 1$ degrees of freedom.

Therefore, if the following inequality holds:

$$K_N \geq \chi_\alpha^2(k-1) \tag{3.54}$$

where $\chi_\alpha^2(k-1)$ is the critical value for the assumed level of significance α, the verified null hypothesis should be rejected.

For a large k, the random variable $\sqrt{2\chi_{k-1}^2} - \sqrt{2(k-1)-1}$ has approximately the standardised normal distribution $N(0, 1)$. Accordingly, for a large k, the following approximations can be applied:

---

[22] The **Kolmogorov distribution** has a cumulative function: $P(K \leq y) = \frac{\sqrt{2\pi}}{y} \sum_{i=1}^{\infty} \exp\left[-\frac{(2i-1)^2\pi^2}{8y^2}\right]$.

$$\chi_\alpha^2(k-1) \cong (k-1)\left(1 - \frac{2}{9(k-1)} + u_{1-\alpha}\sqrt{\frac{2}{9(k-1)}}\right)^3 \tag{3.55a}$$

or

$$\chi_\alpha^2(k-1) \cong \frac{1}{2}\left(\sqrt{2(k-1)} + u_{1-\alpha}\right)^2 \tag{3.55b}$$

Large amounts of computing resources are required to calculate the exact probabilities for the Kruskal-Wallis test. Existing software only provides exact probabilities for sample sizes of less than about 30 participants. These software programs rely on asymptotic approximation for larger sample sizes. Exact probability values for larger sample sizes are actually available. Spurrier (2003) published exact probability tables for samples with as many as 45 participants. Meyer and Seaman (2006) made precise probability distributions for samples as large as 105 participants.

If some of n_i values are small (that is, less than 5), the probability distribution of K_N can be quite different from this Chi-square distribution.

In order to obtain a more precise reasoning when tied ranks are in samples, a correction should be done by first calculating the following measure:

$$\upsilon = \left(1 - \frac{\sum_{j=1}^g (t_j^S - t_j)}{N^S - N}\right)^{-1} \tag{3.56}$$

where g is the number of groups of tied ranks, and t_j is the number of tied ranks in j-th group.

Then multiply number υ by the estimate K_N. It can be proved that $\upsilon > 1$ always and for this reason the new value of statistic (3.53) will be greater than the one calculated without correction. This means that by taking the correction into account the chance of the rejection of the verified null hypothesis increases.

■ **Example 3.11**

The investigation concerned *four* scrapers with the same parameters, made by the same producer. Repair times were noted and the following data were gathered:

Machine I: 85, 150, 430, 30, 170, 600, 210
Machine II: 50, 80, 750, 140, 320, 260, 360, 180
Machine III: 135, 90, 490, 110, 145, 190
Machine IV: 580, 120, 330, 100, 160, 240.

All times are given in minutes.

Here, we have small size samples. In order to create a large sample, a hypothesis was formulated that all of these data are homogeneous.

The Kruskal-Wallis was applied to verify this supposition. The calculation procedure was as follows.

No.	Sample I Time min	Rank	Sample II Time min	Rank	Sample III Time min	Rank	Sample IV Time min	Rank
1	85	4	50	2	135	9	580	25
2	150	12	80	3	90	5	120	8
3	430	23	750	27	490	24	330	21
4	30	1	140	10	110	7	100	6
5	170	14	320	20	145	11	160	13
6	600	26	260	19	190	16	240	18
7	210	17	360	22				
8			180	15				
	Σ	97	Σ	118	Σ	72	Σ	91

Calculate the estimate for K_N statistic. We have:

$$K_N = \frac{12}{27 \times 28}\left(\frac{97^2}{7} + \frac{118^2}{8} + \frac{72^2}{6} + \frac{91^2}{6}\right) - 3 \times 28 = 0.6$$

Compare this value with the critical one. From the table of χ^2 critical values (Table 9.4), we have $\chi_\alpha^2(3) = 7.8$ for the presumed level of significance $\alpha = 0.05$. Because the critical value is greater than the empirical value, there is no ground to reject the verified hypothesis. This means that all of the data can be treated as one sample. If so, this new sample has 27 elements. Now, we can try to find a theoretical probability distribution that will satisfactorily describe the empirical data.   ◀

Our previous considerations on homogeneity of data are important for two reasons at least.

- If there is no ground to reject the hypothesis that the investigated data are homogeneous, there is a possibility to gather all of the data and to create a large sample. This is important because the unit samples in some cases can be small. Working with a large sample, there is a greater likelihood that stronger statistical inferences can be made.
- If there is a basis to reject the hypothesis on homogeneity in the data, a further investigation should be done to find the reason why the data are inhomogeneous and which object has 'made' it so. Discovering the reason that is generating this 'unfitness' can give valuable information from an operational point of view.

Let us now consider the second of the cases listed—(b); some parameters are the points of our interest. It was stated that our interest in homogeneity is not always as strong as the equality of distributions. One can be interested in the identity of certain parameters of a statistical nature that characterise specific features of the object of interest. In reliability investigations of mine equipment, we can be interested, for instance, in whether some probabilities of failure occurrence are identical from a stochastic point of view.

Let us study the following probabilistic model.

We investigate k technical objects and we are interested in the number of work cycles that are executed by these objects. Let $n_1, n_2, \ldots, n_k$ denote these numbers till the moment of the occurrence of m-th failure. Denote by Q the probability of the appearance of one failure in one work cycle. We would like to check whether the following hypothesis holds:

$$H_0\colon Q_1 = Q_2 = \cdots = Q_k$$

which states that the probability of the occurrence of a failure in a work cycle for all of the objects is the same. An alternative hypothesis rejects this.

To verify the basic supposition, one can apply **the Cochran's test**[23].

A measure in this test is the statistic:

$$\Theta = \frac{\max(n_1, n_2, ..., n_k) - 1}{n_1 + n_2 + \cdots n_{k-m}} \tag{3.57}$$

The verified hypothesis should be rejected if the following inequality holds:

$$\Theta \geq q_{k,2m}(1-\alpha) \tag{3.58}$$

where $q_{k,2m}(1-\alpha)$ is the quantile of order $(1-\alpha)$ of the Cochran's statistic (Table 9.9).

In the literature on the subject it is recommended (see for instance Migdalski 1992) that the Hartley's test should be applied when the number of objects being observed is small ($k \leq 12$).

A measure in this study is:

$$\Lambda = \frac{\max(n_1, n_2, ..., n_k) - 1}{\min(n_1, n_2, ..., n_k) - 1} \tag{3.59}$$

The verified hypothesis should be discarded if the following inequality:

$$\Lambda \geq \eta_{k,2m}(1-\alpha) \tag{3.60}$$

where $\eta_{k,2m}(1-\alpha)$ is the quantile of order $(1-\alpha)$ of the Hartley's statistic is the true one.

- **Example 3.12**

In a certain quarry *seven* wheel loaders operated that loaded blasted rock into the crushers and onto dumpers; they were also used in some auxiliary works. The operating machines came from two different producers. However, their reliability was similar. In a different quarry, owned by the same contractor, it was planned to replace some machines of this type but the problem was which producer has better machines.

Let us ignore a problem of negotiations, the possible discounts offered by producers, conditions of payment, realisation of the purveyance and assurance of spare parts—all of which are connected with the potential transaction, and let us devote our attention to the reliability of the equipment that will be purchased.

---

[23] There are a few tests in mathematical statistics that are connected with the name of William Cochran.

A reliability study of machines comprising maintenance problems and especially repairs was done. The applied statistical test has no ground to reject the hypothesis stating that all of the repair times could be satisfactorily described by one probability distribution. It was presumed that a satisfactory deciding criterion would be satisfactory frequency of satisfactory occurrence of failures. A new reliability investigation was performed with this criterion in mind and a day was presumed as a basic elementary period of operation.

A decision was made to observe machines up to the moment of the 10-th failure occurrence. For machines from the first producers, the following sequence was noted:

$$22.6 \quad 18.0 \quad 26.0 \quad 19.8 \text{ days}$$

and

$$10.9 \quad 18.6 \quad 16.7 \text{ days}$$

was noted for machines from the second producer.

At first glance, the reliability of the machines of the first producer looks better than those of the second one. However, formulating the problem from a statistical point of view, we should answer the question of whether this 'difference' is significant statistically or not.

Because there are only *seven* machines in operation, we should apply the Hartley's test. Calculate an estimate of the statistic:

$$\frac{\max(n_1, n_2, \ldots, n_k) - 1}{\min(n_1, n_2, \ldots, n_k) - 1} = \frac{25}{9.9} = 2.52$$

This value should be compared with the critical value. Presume a level of significance $\alpha = 0.05$ as usual. The corresponding critical value (Table 9.10) is:

$$\eta_{k,2m}(1-\alpha) = \eta_{7,20}(1-\alpha) = 3.94$$

Comparing these two values, we have no doubts that we have no basis to reject the verified hypothesis. The observed differences in values are not statistically significant. ◄

Let us notice that the above test rather carelessly used the information that was in hand. It only takes into consideration the maximum and minimum values. The rest of the information is 'useless'. It looks more proper if the hypothesis will be formulated stating that the times of repair can be described by one probability distribution and to apply an appropriate statistical procedure to verify this supposition. Another approach can be used to check whether our guess that the data can be treated as homogeneous one is correct.

We can also investigate, using the Wilcoxon-Mann-Whitney's test (see for instance Lehmann & D'Abrera 2006), whether the supposition stating that the data come from two different populations having the same expected values is a true one. The condition of the application of the test is in the form of a probability distribution that should be similar to the probability distribution of the random variables tested.

■ **Example 3.13**

In the seismic station of a certain underground coal mine, tremors of the rock surrounding the mine were noted. A point of interest was the rock vibrations whose energy exceeded 10^5 J. In the selected period of observation, 33 events were recorded:

$$(7, 15, 280, 190, 900, 8, 8, 100, 10, 2, 800, 2000, 1000, 900, 850, 95,$$
$$25, 100, 950, 9, 320, 210, 20, 20, 40, 6, 600, 40, 5, 7, 105, 2, 80) \times 10^5 \, J$$

The randomness of the sample was tested first. The sample has 33 elements and for this reason the 17-th element is the sample median for the sample arranged monotonically. This element is recorded as the last one: $80 \times 10^5 \, J$.

Converting the sample into a sequence of signs we have:

$$- - + + + - - + - - + + + + + + + - + + - + + - - - - - + - - - + -$$

The number of series in this sequence is 15. The number of n_+ signs $= n_+ = 16$. Presuming a level of significance $\alpha = 0.05$ and using Table 9.8, we have two critical values:

$$K_{\alpha/2}(16, 16) = 11 \quad \text{and} \quad K_{1-\alpha/2}(16, 16) = 22$$

The empirical number of the series is above the critical numbers and for this reason we have no ground to reject the hypothesis stating randomness. We can presume that the sample has a random property.

Figure 3.21 is an illustration of the sequence of tremors noted taking into account their energy.

Let us now investigate whether this sequence is a stationary one. Apply the test based on the Spearman's rank correlation coefficient. Using the procedure described in Chapter 3.3, we can construct a table which facilitates further reasoning (Table 3.3). It contains the sum of squares of differences that equal 6975.5, the number that is important for further analysis.

Calculate the Spearman's rank correlation coefficient—formula (3.18) supported by (3.19). We have

$$r_S = 1 - \frac{6 \times 6975.5}{33(33^2 - 1)} = -0.166$$

Formulate hypothesis H_0 stating there is no dependence between the values with respect to time. This hypothesis is set against a hypothesis $H_1: \rho \neq 0$, stating that the values of the variable depend on time.

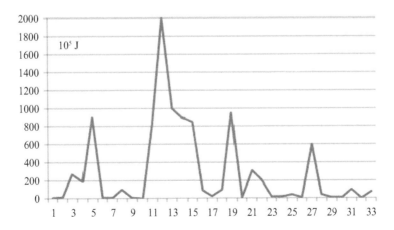

Figure 3.21.   The sequence of the energy of several seismic tremors ($\times 10^5 \, J$).

Table 3.3.   Auxiliary calculations.

No.	Value	Rank	$(v_i - i)^2$
1	7	5,5	20,25
2	15	11	81
3	280	24	441
4	190	22	324
5	900	29,5	600,25
6	8	7,5	2,25
7	8	7,5	0,25
8	100	19,5	132,25
9	10	10	1
10	2	1,5	72,25
11	800	27	256
12	2000	33	441
13	1000	32	361
14	900	29,5	240,25
15	850	28	169
16	95	18	4
17	25	14	9
18	100	19,5	2,25
19	950	31	144
20	9	9	121
21	320	25	16
22	210	23	1
23	20	12,5	110,25
24	20	12,5	132,25
25	40	15,5	90,25
26	6	4	484
27	600	26	1
28	40	15,5	156,25
29	5	3	676
30	7	5,5	600,25
31	105	21	100
32	2	1,5	930,25
33	80	17	256
		Σ	6975,5

To verify the null supposition the critical value $r_\alpha(n)$ should be taken from Table 9.14 for a given level of significance α, and the sample size $n = 33$. However, the sample size is large and in such case the approximation can be applied, i.e.

$$r_S(\alpha, n) \approx \frac{u_{1-\alpha}}{\sqrt{n-1}}$$

where $u_{1-\alpha}$ is the quantile of order $(1 - \alpha)$ of the standardized normal distribution $N(0, 1)$.
   In our case we have:

$$r_S(\alpha, n) \approx \frac{1.96}{\sqrt{32}} = 0.346$$

Looking at the empirical value and the corresponding critical one we have no ground to reject the null hypothesis. We can assume that the sequence noted is free from dependence of time.

Let us conduct our consideration further. Check whether a memory exists in the realisation of the random variable. Thus, we formulate a hypothesis stating that the mutual correlation between the values of the random variable does not exist versus an alternative supposition which rejects this.

Let us check the autocorrelation of the first order. Firstly, using formula (3.47), we have estimations of the means:

$$\bar{x}_1 = 300.8 \times 10^5 \, J \quad \bar{x}_2 = 303 \times 10^5 \, J$$

which allow the correlation coefficient to be calculated

$$r_1^{(a)} = \frac{\sum_{i=1}^{n-1} (x_i - \bar{x}_1)(x_{i+1} - \bar{x}_2)}{\sqrt{\sum_{i=1}^{n-1}(x_i - \bar{x}_1)^2 \sum_{i=1}^{n-1}(x_{i+1} - \bar{x}_2)^2}} = 0.434$$

In order to check whether this value is significant or not, calculate the Breusch-Godfrey statistic:

$$\chi^2(c=1) = (N-1)(r_1^{(a)})^2 = 5.842$$

Now, we can compare this value with the one from theory. Let us read the critical value from the Table of χ^2 distribution for the presumed level of significance $\alpha = 0.05$ (Table 9.4). Here we have:

$$\chi^2_{\alpha=0.05}(r=1) = 3.841$$

The empirical value clearly exceeds the critical one. We have the ground to discard the null hypothesis. There is a memory in the realisation of the random variable of interest. It concerns the neighbouring values. This is important information for the researcher.

Let us now check whether the memory can be extended into the next step. Formally, verify the hypothesis stating that there is no autocorrelation of the second order. We calculate the appropriate average values:

$$\bar{x}_3 = 310.4 \times 10^5 \, J \quad \bar{x}_4 = 312.2 \times 10^5 \, J$$

and we calculate the linear correlation coefficient for every second value. Here we have:

$$r_2^{(a)} = \frac{\sum_{i=1}^{n-2} (x_i - \bar{x}_1)(x_{i+2} - \bar{x}_2)}{\sqrt{\sum_{i=1}^{n-2}(x_i - \bar{x}_1)^2 \sum_{i=1}^{n-2}(x_{i+2} - \bar{x}_2)^2}} = 0.186$$

Using this estimate, calculate the Breusch-Godfrey statistic:

$$\chi^2(c=2) = (N-2)(r_2^{(a)})^2 = 5.654$$

Compare this value with the one from theory. Read the critical value from the table of χ^2 distribution for presumed level of significance $\alpha = 0.05$ (Table 9.4). Here we have:

$$\chi^2_{\alpha=0.05}(r=2) = 5.991$$

The empirical value does not exceed the critical one. We have no ground to discard the null hypothesis. There is no memory of the second type in the realisation of random variable tested. ◄

There is no doubt that the problem of sudden rock displacement of great energy is extremely important in underground mining. This problem is a multi-dimensional one. One dimension is time; more precisely, a random variable which is the time from one tremor to the next tremor. Notice that in example 3.10, time was excluded from the analysis; the data comprised only the sequence of events. The second dimension is the energy that is connected with a given tremor. Again, it is a random variable. The dispersion of this variable is sometimes very high and the range of values is very broad. For these reasons, probability distributions serving as theoretical models often use logarithms (log-normal, log-gamma or log-Weibull).

There is also a subtle problem connected with the effect of a given tremor. Some tremors produce displacement of rock masses that are not serious for mining production even when the energy level is very high. In some other cases, the energy involved in a given tremor is not so great but the result is very serious—e.g. a roof collapse in some underground openings. Thus, one considers two random variables:

a. The energy of a rock tremor with 'safe' repercussions—the variable characterised by a certain density function $f(x)$
b. The energy of a rock tremor with serious repercussions—the variable characterised by a certain density function $g(x)$.

The probability density functions of these two random variables overlap (see Figure 3.22). Generally, the location of these probability distributions is as follows. The density function $f(x)$ is located on the left because the values that take the random variable X, which express energy are low, on average. In Figure 3.22 the level of energy x_0 is such that mine detectors record tremors only above a certain value. The density function $g(x)$ in turn is on the right compared to the location of the density $f(x)$. Therefore, the range of the variation of energy is divided into three intervals. In the first interval (I) the energy involved in the occurring tremor is low and the repercussion is not serious. The second is interval (II) in which there is an increasing probability of serious repercussions—the probability that the tremor from this interval will cause severe damage increases. The last interval (III) is characterised by a certain event. A tremor from this interval has drastically serious consequences.

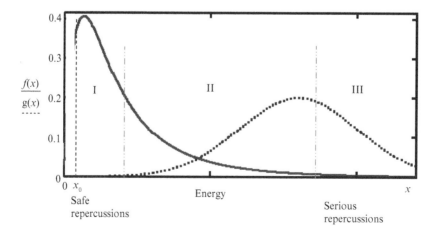

Figure 3.22. The probability density functions of the energy of rock tremors and their repercussions.

CHAPTER 4

Synthesis of data

Having behind us all of the investigations that are connected with statistical diagnostics (previous chapter), we now presume that the data in hand are a classical random sample.

If so, we can make an evaluation of this information in the sense of an estimation of the selected numerical characteristics as well as in the sense of certain functions which will describe the data that were obtained *well*. The parameters of first choice in the engineering world are: the expected value and the standard deviation. These parameters are well-known, communicative and contain significant, useful information.

Some further parameters can be enumerated, but what will be taken into account depends on the scope of the consideration.

If the points of interest are econometrics problems (mineral sales in time, market demand for a given mineral commodity, the prognosis of its price and so on), these supplementary parameters are usually taken from descriptive statistics, e.g. quantiles (especially—median) and typical functions connected with the description of random variables.

In a reliability investigation, parameters such as the intensity of the failures for an object that can be repaired and the mean times of the states are commonly applied. But the most important and most frequently used parameter is the steady-state availability, which is a function of the repair rate. Where functions are concerned, the probability density function is very important. In addition, the survival function, the hazard function and the renewal function can be joined to the list but they are concerned with objects that cannot be repaired. We neglect here functions that are of less importance.

In safety studies, the basic parameters are mainly probabilities such as: the probability of correct performance of a technical object during a given period of time and the probability that a load will exceed the strength of the machine. Some further parameters are: the factor of safety[1], the confidence factor and the mass of the probability common to the distribution of the load and strength. Functions are first of all the probability distributions of the operation load and operation strength.

All of these parameters are well defined on the basis of their theories. However, a problem comes into play here, how to adequately estimate these characteristics when the data (sample) are in our hands. Therefore, a further consideration that will be conducted here concerns the theory and practice of estimation.

A concept of the function that permits estimating an unknown value of a parameter was introduced in Chapter 1.3 and the desired properties of a *good* estimator were described. Let us now enlarge the scope of our consideration of estimation.

If we would like to estimate an unknown value of one or more parameters of the distribution of a random variable, we say it is a **parametric estimation**.

If our inference does not concern the parameters and if information about the distribution function is not available, which means that we have no idea about the class of the distribution, we say that it is a **non-parametric estimation**.

Let us now consider the problem of the estimation of the unknown parameters of random variables.

---

[1] Remember, the factor of safety has a loose connection with safety. Probabilistic measures of safety are much better.

4.1   ESTIMATION OF THE PARAMETERS OF A RANDOM VARIABLE

There is a scope of consideration in the theory of estimation in which the point of interest is to obtain an estimate of an unknown parameter in the form of a number. This is **point estimation**. It involves the use of a sample in order to calculate a single value that is to serve as the 'best guess' or 'best estimate' of an unknown (fixed or random) population parameter. It is the application of a point estimator to the data.

Consider a problem of the methods of the construction of functions that allow estimates of the unknown values of parameters to be obtained.

There are a number of methods of construction of estimators that have desirable properties. The oldest general method proposed for this purpose is the **method of moments** introduced by K. Pearson (1894, 1898). This method consists in equating a convenient number of the sample moments to the corresponding moments of the distribution, which are functions of the unknown parameters. By considering as many as there are parameters to be estimated, and solving the resulting equations with respect to the parameters, estimates of the latter are obtained. This method usually leads to comparatively simple calculations. Estimators obtained by this method are consistent but frequently biased and of a low efficiency.

Consider, for instance, the application of this method to get estimators of the unknown structural parameters of gamma distribution given by formula (1.48). The relationships between the expected value and the variance and these structural parameters were defined by patterns (1.49). Having a sample, we replace the expected value with its estimate ($\bar{x}$—the mean of the sample) and the variance by its estimate (S_x^2—the variance of the sample). At this moment, we have to replace the structural parameters with their evaluating functions. Therefore, a new pair of patterns is as follows:

$$\frac{\hat{\xi}}{\hat{v}} = \bar{x} \quad \text{and} \quad \frac{\hat{\xi}}{\hat{v}^2} = S_x^2 \tag{4.1}$$

By rearranging them, we have:

$$\hat{v} = \frac{\bar{x}}{S_x^2} \quad \text{and} \quad \hat{\xi} = \frac{\bar{x}^2}{S_x^2} \tag{4.2}$$

Notice the difference by comparing them to the formula (1.49). The above patterns are estimators and these functions are random variables because one can get different estimates for different samples that are taken.

Similarly, one can obtain, for example, the estimators for the structural parameters of the beta distribution that is determined by formula (1.61). Using patterns for the expected values and the variance (1.63), we can construct the following set of equations:

$$\bar{x} = \frac{\hat{c}}{\hat{c}+\hat{d}} \quad \text{and} \quad S_x^2 = \frac{\hat{c}\hat{d}}{(\hat{c}+\hat{d})^2(\hat{c}+\hat{d}+1)} \tag{4.3}$$

Finding estimators for the unknown structural parameters for this distribution is now not a problem.

A different method that is used to obtain estimators is the **method of quantiles**. There are several methods for estimating the quantiles (Serfling 1980). The idea is to compare a theoretical quantile to the corresponding one that is constructed from the sample, similar to that in the method of moments. If several parameters are evaluated, then several equations should be made using the appropriate number of quantiles based on the sample taken.

It can be proved (Gnyedenko et al. 1968) that if some general assumptions are fulfilled, the quantile of order p based on a sample (recall, $F_n(z_p) = p$) has an asymptotically Gaussian distribution with the expected value being the quantile of order p of the theoretical distribution ($F(\zeta) = p$) and the variance as determined by the pattern: $[p(1-p)]/nf_n^2(\zeta)$.

This method does not give estimators with good statistical properties and has actually almost been abandoned.

The method that is recommended is the **method of maximum likelihood**. From a theoretical point of view it is the most important general method. In specific cases, this method was already applied by Gauss (1880), although it was introduced as a general method of estimation by R.A. Fisher (1912) and later was developed in a series of works by the same author.

A cardinal term of this method is the sample likelihood that is expressed by the likelihood function. The sample likelihood is the joint density of the continuous probability distribution of the outcomes $(x_1, x_2, ..., x_n)$ given by the sample where this likelihood depends on the real value of the estimated parameter. For discrete distribution, one can say on the joint probability that corresponds with the density. Denote by $f(x, \xi)$ the probability density function. The function:

$$L(x_1, x_2, ..., x_n; \xi) = \prod_{i=1}^{n} f(x_i; \xi)$$ (4.4)

is called a **likelihood function**.

When the values of a sample are given, the likelihood function L becomes the function of the single variable ξ. The method of maximum likelihood relies on the selection of such an estimator of ξ that assures the maximum of the likelihood function. It is known that a function attains its maximum at a certain point if the first derivative of the function equals zero and the second derivative in this point is less than zero. Thus, it is necessary to solve the equation:

$$\frac{\partial L(x_i; \xi)}{\partial \xi} = 0$$ (4.5)

with respect to ξ. It is very often much more useful to use the equation:

$$\frac{\partial \ln L(x_i; \xi)}{\partial \xi} = 0$$ (4.5a)

instead of (4.5) because its logarithmic function is more convenient for differentiation. Both functions $L(x_i; \xi)$ and $\ln L(x_i; \xi)$ have the extreme at the same point.

A significant limitation of this method is the necessity to have information on the probability distribution function.

Remark. Estimators obtained by this method are sometimes biased.

Consider, as an example, a population that is characterised by the exponential distribution with parameter θ. A sample was taken of elements x_i; $i = 1, 2, ..., n$ and the point of interest is the construction of an estimator of the distribution using the method of likelihood.

The likelihood function in this case is as follows:

$$L = \theta e^{-\theta x_1} \theta e^{-\theta x_2} ... \theta e^{-\theta x_n} = \theta^n e^{-\theta \sum_{i=1}^{n} x_i}$$

Calculate the logarithm of both sides of the equation

$$\ln L = n\ln\theta - \theta\sum_{i=1}^{n} x_i$$

Therefore

$$\frac{d\ln L}{d\theta} = n\frac{1}{\theta} - \sum_{i=1}^{n} x_i$$

and

$$n\frac{1}{\theta} - \sum_{i=1}^{n} x_i = 0$$

Now, the estimator that takes the following form can be constructed:

$$\hat{\theta} = \frac{n}{\sum_{i=1}^{n} x_i} = \frac{1}{\bar{x}} \tag{4.6}$$

The investigation plan has significant influence on the final form of the estimator in reliability investigations of technical objects as well as in investigations of their operation processes.

A **plan of investigation** in reliability should include this trio:

$$< n, \mathfrak{A}, \mathfrak{U} >$$

where: n—sample size, number of items investigated,
  $\mathfrak{A}$—sampling principle,
  $\mathfrak{U}$—criterion for the termination of the investigation.

The most common sample principles are:

- Sampling without return ($\mathfrak{B}$); a failed element is not replaced with a new one
- Sampling with return ($\mathfrak{Z}$); a failed element is replaced with new one.

The criteria of an investigation of a termination can be different. Those most frequently applied are:

- Moment of the r-th failure occurrence
- Amount of work executed
- Research is carried out over time T
- Research is carried out either over time T or up to the r-th failure occurrence depending on which criterion will be achieved earlier.

Graphical illustrations of the realisations of a reliability investigation are presented in Figure 4.1 are shown.

Let us illustrate, for instance, formulas that are used to determine the estimators of the parameter of the exponential probability distribution in relation to the tests performed.

– For plan $[n, \mathfrak{Z}, T]$

$$\hat{\lambda} = \frac{d(T)}{nT} \tag{4.7}$$

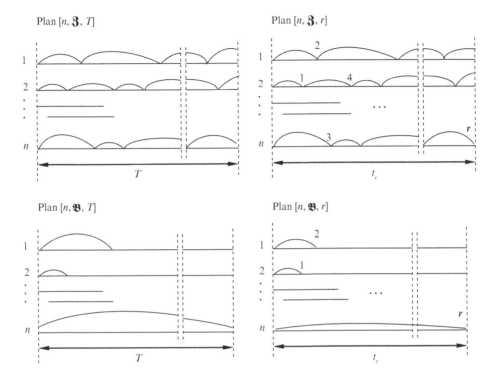

Figure 4.1.   Graphical illustration of the mode of reliability investigations for a given criterion of their termination.

where $d(T)$ is the number of failures noted during the investigation

– For plan $[n, \mathbf{3}, r]$

$$\hat{\lambda} = \frac{r-1}{nt_r} \qquad (4.8)$$

where t_r is time up to the r-th failure occurrence

– For plan $[n, \mathbf{B}, T]$

$$\hat{\lambda} = \frac{d(T)}{S(T)} \qquad (4.9)$$

where:

$$S(T) = \sum_{i=1}^{d(T)} t_i + [n - d(T)] \, T \qquad (4.9a)$$

t_i—the i-th time to failure occurrence

– For plan $[n, \mathbf{B}, r]$

$$\hat{\lambda} = \frac{r-1}{S(t_r)} \qquad (4.10)$$

where:

$$S(t_r) = \sum_{i=1}^{r} t_i + [n - r] t_r \qquad\qquad (4.10a)$$

Remark. Estimator (4.9) is a biased one.

Our previous considerations were devoted to the construction of estimators in order to assess the unknown values of the structural parameters of probability distribution functions; 'structural' means that they are in formulas of these distribution functions. However, there is also an area in mathematical statistics in which estimation is oriented on a different problem. We are still interested in building formulas that will permit unknown structural parameters of functions to be estimated but these relationships are not connected with the distribution functions. These multi-element relations express a variety of dependences between random variables. In order to estimate the structural parameters in this case, the **least squares method** is applied as a rule. This method originated in the fields of astronomy and geodesy as scientists and mathematicians sought to provide solutions to the challenges of navigating the Earth's oceans during the Age of Exploration. It was created by Carl Gauss in 1795 (*vide*: Bretscher 1995). The most important application of the least squares method is in data-fitting. The best fit in the least squares sense minimises the sum of squared residuals; residual here is understood as the difference between the observed value and the corresponding value that is provided by the model. A description of this method along with applications will be given in Chapter 6.

When data have been obtained and the estimator selected one calculates an appraisal in the form of a number that is a **point estimation**. However, it is easy to perceive the probability that the estimate will be identical to the unknown value of the parameter, which is practically zero, when the population is a continuous one. This means that we made an error during such estimation with the probability that is near one. Bearing this in mind, statisticians developed a different method of estimation—**interval estimation.** Its idea is to construct an interval instead of calculating a single value. For the time being, we will presume that this is an interval of possible or probable values of the unknown population parameter. Spława-Neyman (Neyman 1937), who introduced this idea, identified interval estimation as distinct from point estimation. In doing so, he recognised that the then-recent work quoting results in the form of an estimate plus-or-minus a standard deviation indicated that interval estimation was actually the problem that statisticians really had in mind.

The most widespread forms of interval estimation are:

- Confidence intervals
- Credible intervals applied using the Bayesian method.

Other common approaches to interval estimation that are included in statistical theory are:

- Tolerance intervals
- Prediction intervals
- Likelihood intervals[2].

For our further analyses, we will presume that the constructed numerical interval **covers** the unknown value of population parameter being estimated with a certain presumed probability. The idea of a confidence interval is as follows.

Presume that the random variable X has in its population the distribution function determined by the unknown parameter θ. Presume also that a sample was taken x_i; $i = 1, 2, ..., n$

---

[2] Non-statistical methods that can lead to interval estimates include fuzzy logic.

from this population and the estimator $\hat{\theta}_n$ of the parameter was constructed. The estimator, as a random variable has its own probability density function $h(\hat{\theta}_n; \theta)$. Define two such functions $c_1(\theta)$ and $c_2(\theta)$ that fulfil two equations:

$$P\{\hat{\theta}_n < c_1(\theta)\} = \int_{-\infty}^{c_1(\theta)} h(\hat{\theta}_n; \theta)\, d\hat{\theta}_n = \frac{1}{2}\alpha \qquad (4.11a)$$

$$P\{\hat{\theta}_n > c_2(\theta)\} = \int_{c_2(\theta)}^{\infty} h(\hat{\theta}_n; \theta)\, d\hat{\theta}_n = \frac{1}{2}\alpha \qquad (4.11b)$$

where α is any number from (0, 1) interval.

Bearing the above relationships in mind and making any necessary transformations, the following equation can be obtained:

$$P\{g_1(\hat{\theta}_n) < \theta < g_2(\hat{\theta}_n)\} = \int_{c_1(\theta)}^{c_2(\theta)} h(\hat{\theta}_n; \theta)\, d\hat{\theta}_n = 1 - \alpha = \gamma \qquad (4.12)$$

The interval obtained in this way $[g_1(\hat{\theta}_n), g_2(\hat{\theta}_n)]$ is called a **confidence interval** and its boundary values **lower and upper confidence limits**, respectively. The number γ is called the **confidence coefficient** or **level of confidence**.

Observe the following regularities and pay attention to their correct interpretation:

a. An interval obtained in this way is random
b. For a given γ an infinite number of intervals can be found that fulfil relationship (4.12) and depending on sample size the point of interest will be those intervals with the narrowest range
c. The statement that for a given parameter θ the confidence interval was found, keeping the confidence coefficient γ in mind, means that in $100\gamma\%$ of the cases the interval *covers* the unknown value of parameter θ.

Figure below is an illustration of the above statement. The lengths of intervals are random, many of them cover the unknown value of parameter θ but sometimes constructed interval does not cover it.

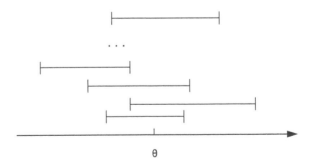

The construction of a confidence interval can concern an inference on the one side (either (4.11a) or (4.11b)) and here we have one side interval where attention should be paid to the limits of the determination of the variable.

In practice, one is obviously confined to one sample and to the determination of only one confidence interval. It is worth noticing that presuming a confidence interval near one is not very convenient. The greater the value of γ, the wider the confidence interval and for this reason the accuracy of the estimation lower.

In the majority of statistics books, many examples of a Gaussian distribution are given. In engineering analyses, the application of this distribution is not so wide. Moreover, the frequently set up presumption (in many statistics books) that one parameter of this distribution is known does not hold in practice. Therefore, consider the construction of a confidence interval for the mean of a random variable of an unknown distribution.

In such a case, we are in an inconvenient situation. If we would like to make a credible study, we should have a large sample size.

It is a well-known fact in mathematical statistics that the arithmetic mean based on a sample of size n taken from any distribution has the following limited distribution:

$$N\left(m, \frac{\sigma}{\sqrt{n}}\right) \quad \text{for } n \to \infty$$

where m is the expected value and σ is the variance. Observe that if the distribution is normal the above pattern holds for any sample size n.

It can be assumed that for a large n (greater than 30) the mean has such a distribution in which the statistic:

$$U = \frac{\bar{x} - m}{s}\sqrt{n} : N(0,1)$$

Thus, we can construct the following equation:

$$P\left(-u_\gamma < \frac{\bar{x} - m}{s}\sqrt{n} < u_\gamma\right) = \gamma \tag{4.13}$$

where u_γ is the multiplication factor of the estimation error of the unknown average value. This factor can be taken from the table of standardised normal distributions for the level of confidence γ and for the value $(1 + \gamma)/2$. This factor is in fact the quantile of the standardised distribution $N(0, 1)$.

Transforming equation (4.13) correctly, we get:

$$P\left(\bar{x} - u_\gamma \frac{s}{\sqrt{n}} < m < \bar{x} + u_\gamma \frac{s}{\sqrt{n}}\right) = \gamma \tag{4.14}$$

Remark. The confidence interval for the unknown expected value m that is defined by the above equation is the same as the confidence interval for the mean in the normal distribution when the standard deviation is σ except that in the formula (4.14) is presumed that $\sigma \cong s$.

In engineering practice, we often use exponential distribution. Therefore, let us look at the confidence intervals for a structural parameter of this distribution in cases of different investigation plans being used (the estimators for this parameter were determined by formulas (4.7)–(4.10)).

– For plan $[n, \mathbf{3}, T]$ the interval is:

$$\left[\frac{\Delta_{1-\alpha/2}[d(T)-1]}{nT}, \quad \frac{\Delta_{\alpha/2}[d(T)]}{nT}\right] \tag{4.15a}$$

where: $\Delta_\beta[d(T)]$ is the quantile of order β of the Poisson distribution for the number of failures noted $d(T)$

− For plan $[n, \mathbf{3}, r]$ the interval is:

$$\left[\frac{\Delta_{1-\alpha/2}[r-1]}{nt_r}, \frac{\Delta_{\alpha/2}[r-1]}{nt_r}\right] \tag{4.15b}$$

− For plan $[n, \mathbf{B}, T]$ the interval is:

$$\left[\frac{1}{T}\ln\frac{1}{1-\underline{p}[\boldsymbol{d}(T)]}, \frac{1}{T}\ln\frac{1}{1-\overline{p}[\boldsymbol{d}(T)]}\right] \tag{4.15c}$$

where $\underline{p}[\boldsymbol{d}(T)]$ and $\overline{p}[\boldsymbol{d}(T)]$ are the lower and upper limits of the confidence interval for parameter p in the binomial distribution.

− For plan $[n, \mathbf{B}, r]$ the interval is:

$$\left[\frac{\Delta_{1-\alpha/2}[r-1]}{S(t_r)}, \frac{\Delta_{\alpha/2}[r-1]}{S(t_r)}\right] \tag{4.15d}$$

where α is the presumed probability near one.

The quantiles of the Poisson distribution are in the construction of confidence intervals (4.15a), (4.15b) and (4.15d). This is connected with the fact that the most important information of a statistical nature that is obtained from the investigation is statistic $\boldsymbol{d}(T)$—the number of failures noted and the distribution of its random variable is simply the Poisson distribution. The exception to this rule is pattern (5.15c) because the value of the statistic $\boldsymbol{d}(T)$ in this plan of investigation does not depend on the parameter λ for technical objects of high reliability in practice (more on this topic see Gnyedenko et al. 1969, Chapter 3.4). Hence, the transformation was made by replacing the Poisson distribution with the binomial one.

Now, let us make some generalisations about the consideration being conducted by looking at interval estimations for a technical object whose operational process is of the work-repair type and both distributions of states are exponential.

Consider the process cycle $t_{wi} + t_{ni}$; $i = 1, 2, \ldots$; t_w—work time, t_r—repair time. Let us devote our attention firstly on the work state. A sequence of times is associated with this state. Consider the sum $\sum_{i=1}^{n} t_{wi}$. This sum is a random variable and it can be presumed that the component variables are independent and obviously exponentially distributed. If so, this sum has the Erlang distribution of order n or, what is the same thing, the product $2\lambda\sum_{i=1}^{n} t_{wi}$ has χ^2 distribution with $2n$ degrees of freedom (*vide* formula (3.3)). This statement allows the interval estimation for the unknown parameter λ to be obtained because the following equation holds:

$$P\left\{\chi_1^2(2n) < 2\lambda\sum_{i=1}^{n} t_{wi} < \chi_2^2(2n)\right\} = 1-\alpha$$

which yields

$$P\left\{\frac{\chi_1^2(2n)}{2\sum_{i=1}^{n} t_{wi}} < \lambda < \frac{\chi_2^2(2n)}{2\sum_{i=1}^{n} t_{wi}}\right\} = 1-\alpha \tag{4.16}$$

The above formula determines the interval estimation for the intensity of failures of the object whose operation process is being observed.

For $\chi_1^2 = 0$ and $\chi_2^2 = \chi_\alpha^2(2n)$ we have the left side confidence interval, for $\chi_2^2 \to \infty$ and $\chi_1^2 = \chi_{1-\alpha}^2(2n)$ we have the right side confidence interval and for $\chi_1^2 = \chi_{1-(\alpha/2)}^2(2n)$, $\chi_2^2 = \chi_{\alpha/2}^2(2n)$ we have sided interval.

Analogically, the interval estimation can be obtained for the intensity of repair β.

Based on the previous study, an interval estimation for the unknown expected values of the times of states can be constructed. It is enough to notice that the random variable $2\Sigma_{i=1}^{n} t_{wi}/E(t_w)$ has an χ^2 distribution with $2n$ degrees of freedom. Thus

$$P\left\{\chi_1^2(2n) < \frac{2\sum_{i=1}^{n} t_{wi}}{E(t_w)} < \chi_2^2(2n)\right\} = 1 - \alpha$$

or

$$P\left\{\frac{2\sum_{i=1}^{n} t_{wi}}{\chi_2^2(2n)} < E(t_w) < \frac{2\sum_{i=1}^{n} t_{wi}}{\chi_1^2(2n)}\right\} = 1 - \alpha \qquad (4.17a)$$

Similarly

$$P\left\{\frac{2\sum_{i=1}^{n} t_{ri}}{\chi_2^2(2n)} < E(t_r) < \frac{2\sum_{i=1}^{n} t_{ri}}{\chi_1^2(2n)}\right\} = 1 - \alpha \qquad (4.17b)$$

When the repair rate κ is analysed[3] it is enough to perceive that if $\chi^2(r_1)$ and $\chi^2(r_2)$ are the independent random variables of the Chi-squared distribution with r_1 and r_2 degrees of freedom, respectively, then the random variable

$$\frac{\chi^2(r_1)}{r_1} : \frac{\chi^2(r_2)}{r_2} = F_{r_1, r_2}$$

has the Snedecor's distribution with (r_1, r_2) degrees of freedom (compare formula (1.120)). Because we presume that our observation comprised n cycles of the process, thus $r_1 = r_2 = 2n$ and then

$$\frac{E(t_r)\sum_{i=1}^{n} t_{wi})}{E(t_w)\sum_{i=1}^{n} t_{ri})} = F_{2n, 2n}$$

Hence, it is true that

$$P\left\{F_1(2n, 2n) < \kappa \frac{\sum_{i=1}^{n} t_{wi}}{\sum_{i=1}^{n} t_{ri}} < F_2(2n, 2n)\right\} = 1 - \alpha$$

---

[3] Recall, $\kappa = \lambda/\beta$.

By transforming the above we have:

$$P\left\{\frac{\sum_{i=1}^{n}t_{ri}}{\sum_{i=1}^{n}t_{wi}}F_1(2n,2n)<\kappa<\frac{\sum_{i=1}^{n}t_{ri}}{\sum_{i=1}^{n}t_{wi}}F_2(2n,2n)\right\}=1-\alpha \tag{4.18}$$

This equation determines the confidence interval for the unknown repair rate from the population.

It is worth noticing that

$$\frac{\sum_{i=1}^{n}t_{ri}}{\sum_{i=1}^{n}t_{wi}}=\hat{\kappa}_n$$

is the estimator of the repair rate based on the sample.

Determination of the confidence interval for the steady-state availability A is not a problem now and the following equation holds:

$$P\left\{\frac{1}{1+\hat{\kappa}_n F_2(2n,2n)}<A<\frac{1}{1+\hat{\kappa}_n F_1(2n,2n)}\right\}=1-\alpha \tag{4.19}$$

The reasoning above can be successfully enlarged on series systems that consist of identical elements (which almost always holds true in practice).

Bearing in mind the first principle of reduction for series systems (see for instance Czaplicki 2010 p. 130), we can state that the intensity of failures λ_S of a series system consisting of k identical elements is the sum of all of the intensities $\lambda^{(j)}$; $j=1, 2, \ldots, k$ that is $\lambda_S=k\lambda$ and for this reason the random variable:

$$2\lambda_S\sum_{i=1}^{N}t_{w_i}$$

where N is the number of the process cycles of the system observed has the Chi-square distribution with $2Nk$ degrees of freedom.

The interval estimation for the intensity of failures in a series system λ_S can be obtained from the equation:

$$P\left\{\frac{\chi_1^2(2Nk)}{2\sum_{i=1}^{N}t_{w_i}}<\lambda_S<\frac{\chi_2^2(2Nk)}{2\sum_{i=1}^{N}t_{w_i}}\right\}=1-\alpha \tag{4.20}$$

while the interval estimation for the unknown expected value of work time can be obtained from the pattern:

$$P\left\{\frac{2\sum_{i=1}^{N}t_{w_i}}{\chi_2^2(2Nk)}<E[t_w]_S<\frac{2\sum_{i=1}^{N}t_{w_i}}{\chi_1^2(2Nk)}\right\}=1-\alpha \tag{4.21}$$

In accordance with the principles of reduction for series systems, the intensity of repair in the system is:

$$\beta_S=\beta$$

and this means that the confidence interval is the same for any element of the system and for the whole system. Analogically, the confidence interval for the expected value of the repair of an element is the same as for the whole series system. The only difference is that for the system, we have N instead of n.

Consider now the following random variable:

$$\frac{\mathbf{\chi}^2(2Nk)}{2Nk} : \frac{\mathbf{\chi}^2(2N)}{2N} = F_{2Nk,2N}$$

This variable is useful to obtain the interval estimation for the unknown repair rate κ_s of the considered system because:

$$P\left\{F_1(2Nk, 2N)\frac{\sum_{i=1}^{N}t_{r_i}}{\sum_{i=1}^{N}t_{w_i}}k < \kappa_S < F_2(2Nk, 2N)\frac{\sum_{i=1}^{N}t_{r_i}}{\sum_{i=1}^{N}t_{w_i}}k\right\} = 1 - \alpha \qquad (4.22)$$

The above equation is advantageous to get the interval estimation for the unknown steady-state availability A_S of the system

$$P\left\{\left(1 + kF_2(2Nk, 2N)\frac{\sum_{i=1}^{N}t_{r_i}}{\sum_{i=1}^{N}t_{w_i}}\right)^{-1} < A_S < \left(1 + kF_1(2Nk, 2N)\frac{\sum_{i=1}^{N}t_{r_i}}{\sum_{i=1}^{N}t_{w_i}}\right)^{-1}\right\} = 1 - \alpha$$

$$(4.23)$$

After certain modifications, the interval estimations can be obtained for a series system that is composed of non-identical elements (Czaplicki 1977).

In reliability analyses of **technical objects working to first failure occurrence**, it is important to assess the lower limited value for the unknown mean time of the population if **no failure** was recorded during investigation.

Generally, it should be distinguished whether a random variable associated with the population is described by a function that has one or more structural parameters. If only one parameter characterises the distribution function, no further information is required.

Presume that n objects have been tested in time T and no failure was recorded, then by presuming an appropriately high level of probability δ, we will search for a formula that defines the lower limited value L_L for the unknown expected value expressed in time units for the population.

Thus, if the probability distribution is, for example, exponential (see pattern (1.53)) then the lower limit L_L for the expected value of life time can be obtained by solving the equation:

$$\exp\left(-\frac{nT}{L_e}\right) = 1 - \delta \qquad (4.24)$$

with respect to the limit L_e, which yields to the formula:

$$L_e = \frac{-nT}{\ln(1-\delta)} \qquad (4.25)$$

If the probability distribution has two structural parameters, it is necessary to have some information on at least one of the parameters.

If we postulate that the probability distribution is the Weibull one (1.56), then the lower limited value L_W for the expected value can be obtained by solving the equation

$$\exp(\lambda T^\alpha) = \sqrt[n]{1-\delta} = \gamma \tag{4.26}$$

Believe that information on the shape parameter α can be available. Bearing in mind the equation to determine the expected value for the Weibull distribution (1.58) and making all of the necessary mathematical transformations, one obtains:

$$L_W = \frac{T\Gamma\left(1+\dfrac{1}{\alpha}\right)}{(-\ln\gamma)^{1/\delta}} \tag{4.27}$$

In some cases the Gaussian distribution is applied but we already have knowledge on the limitations of this distribution (see Chapter 1). For this reason, we presume that the variation coefficient v is low ($v < \frac{1}{3}$) and that we know this value. If so, the equation to be solved is:

$$F_N\left(\frac{L_N - T}{vL_n}\right) = \sqrt[n]{1-\delta} = \gamma \tag{4.28}$$

The above equation allows the limited value L_N to be found, which is in this case is determined by the equation:

$$L_N = \frac{T}{1 - vu_\gamma} \tag{4.29}$$

where u_γ is the quantile of order γ of the standardised normal distribution (Table 9.2).

If the probability distribution is a log-normal one and information on the standard deviation σ is in hand, the equation to determine the limited value is:

$$lgL_L = lgT + \sigma u_\gamma + 1.1513\sigma^2 \tag{4.30}$$

■ **Example 5.1**

Suppose that $n = 100$ identical elements were tested over time $T = 1000$ h. We compare the limited values for different distributions presuming that the probability $\delta = 0.95$.

Assume that the shape parameter in the Weibull distribution $\alpha = 2$, the coefficient of variation in the normal distribution $v = 0.10$ and the standard deviation $\sigma = 0.10$ in the log-normal distribution.

We have:

$-\ L_e = \dfrac{-nT}{\ln(1-\delta)} = 3338\ \text{h}$

$-\ L_W = \dfrac{T\Gamma\left(1+\frac{1}{\alpha}\right)}{(-\ln\gamma)^{1/\delta}} = 3558\ \text{h}$

$-\ L_N = \dfrac{T}{(1-vu_\gamma)} = 1232\ \text{h}$

$-\ L_L = 1583\text{h}$

By changing the probability δ as well as changing the values of the supplementary parameters (α, v, σ), we obtain different numbers of the limited values for the means.   ◄

4.2   PROBABILITY DISTRIBUTION DESCRIPTION

When a sample is taken and it is the representative one, i.e. it is such that it represents the population satisfactorily in a statistical sense, and the random variable tested can then be characterised in a synthetic way by means of selected parameters as was shown in the previous chapter, we can try to find a theoretical distribution that will describe the data in a satisfactory way; satisfactory in a statistical sense.

Sometimes we have some information *a priori* on the distribution of the random variable of interest. Basically, there are two sources of such information, namely:

– Theoretical; a theoretical study indicates what kind of distribution we are dealing with
– Empirical; there is information about what kind of statistical model can be applied from research that had previously been carried out.

However, if there is a lack of any information in this regard, we should follow the scheme presented below.

A **histogram** should be constructed based on the sample. This is a graphical representation of the distribution of the amount or frequency of a given set of elements that is classified with regard to the feature that is being investigated. Because a histogram is constructed based on the sample, the graph shows the empirical distribution of the feature. In the case of a frequency diagram, the chart illustrates the empirical density distribution function. This method of presentation was introduced by Pearson (1895). A histogram consists of tabular frequencies, which are shown as adjacent rectangles, erected over discrete intervals (bins) with an area equal to the frequency of the observations in the interval. The height of a rectangle is also equal to the frequency density of the interval, i.e., the frequency divided by the width of the interval. The total area of the histogram is equal to the number of data. A histogram may also be normalised to display relative frequencies. It then shows the proportion of cases that fall into each of several categories with the total area equalling 1. The categories are usually specified as consecutive, non-overlapping intervals of a variable. The categories (intervals) must be adjacent, and often are chosen to be of the same size. The rectangles of a histogram are drawn so that they touch each other in order to indicate that the original variable is continuous.

Practically, the number of intervals should be at least *five*; otherwise, the histogram will be difficult to read. In order to inference appropriately, a few elements to each interval should be ascribed at least. It follows that the sample should have at least 30 elements. If, however, there had been a previous investigation concerning the population of the specified and known distribution, then the sample size could be a little smaller.

In a mathematical sense, a histogram is a function m_i that counts the number of observations that fall into each of the disjoint categories, whereas the graph of a histogram is only one way to represent a histogram. Thus, if we let n be the total number of observations and k be the total number of bins, the histogram m_i meets the following conditions:

$$n = \sum_{i=1}^{k} m_i$$

A conclusion can be formulated from the consideration presented above that the way to find the theoretical distribution that will describe the data *well* relies on the construction of the empirical probability density function and searching for such a theoretical counterpart to which the graph is similar to the empirical one.

Selection of such a method for proceeding results from the fact that plots of density functions for different classes of distributions usually differ significantly from one another. If we would like to find the theoretical distribution using, for instance, the distribution function, then it will be very difficult to discover it. Plots of distribution functions are very similar to each other. However, a certain different way of finding a theoretical model in this regard and in this mode is sometimes used in a reliability investigation when technical objects working to the first failure occurrence are considered. The point of interest here is the function of the conditional intensity of failures (the so-called hazard function). Graphs of this function differ distinctly from each other in many cases.

Notice a certain subtleness. Often a few theoretical distributions can be ascribed to a given histogram. Principally, the decision of which model will be chosen depends on the researcher. However, it is worth considering some additional problems or using some hints. Some indications in this regard can be obtained by considering the goal of the investigation to be conducted. If the distribution function will be used in a further part of the investigation, its properties should be convenient to use in planned analysis. Sometimes the point of interest in further study is associated with the tail of the probability density function. Many probability distributions differ in the properties of their tails. Generally, if two density functions can be applied, attention should be paid to the distribution of the mass along the values of the variable.

☐ **Example 3.5 (cont.)**

In example 3.5 stationary testing was done in connection with the sample taken. There was no basis to reject the hypothesis stating the stationarity of the sequence of the numbers noted. Now, we try to find a theoretical probability distribution that will satisfactorily describe the data.

Let us divide the range variation of the variable into five separable 50-minute-long intervals. Taking into consideration that 0 does not belong to the set of data, the right side limited values of the bins will belong to their bins. The histogram is presented in Figure 4.2. It gives information that the theoretical density function should be asymmetric supported on the positive real values.

The probability distributions that are most frequently applied in such a case are the gamma or Weibull function. Let us try to describe our data using the gamma distribution.

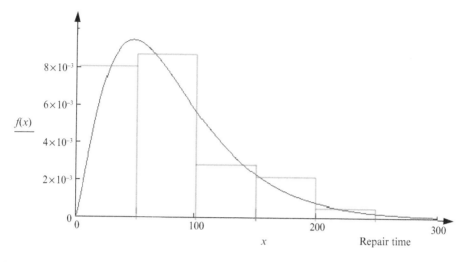

Figure 4.2.   The histogram and the theoretical probability density function for the repair times from Example 3.5.

Recall that the probability density function for this distribution is given by formula (1.48)

$$f(x) = \frac{v^{\xi}}{\Gamma(\xi)} x^{\xi-1} e^{-vx}, \qquad x \geq 0, \quad v > 0, \quad \xi > 0$$

Estimate the unknown structural parameters for this function.

Let us apply the method of moments—patterns (4.2). The estimates of the expected value and the standard deviation are as follows:

$$\bar{x} = 83.0 \, \text{min} \qquad S_x = 53.1 \, \text{min}$$

which give these estimations of the structural parameters:

$$\hat{\xi} = 2.44 \qquad \hat{v} = 0.03$$

Making use of the formulas taken from the method of maximum likelihood, we calculate the logarithm of the function L for a sample of size n

$$\ln L = n\xi \ln v - n \ln \Gamma(\xi) + (\xi - 1) \sum_{i=1}^{n} \ln x_i - v \sum_{i=1}^{n} x_i \qquad (4.31)$$

Now, it is necessary to solve the set of equations that allow the estimates of structural parameters to be found, namely

$$\frac{\partial \ln L}{\partial \xi} = 0 \qquad \frac{\partial \ln L}{\partial v} = 0$$

Combining the above equations we get:

$$\frac{\partial \ln \Gamma(\xi)}{\partial \xi} - \ln\left(\frac{\xi}{\bar{x}}\right) = \frac{1}{n} \sum_{i=1}^{n} \ln x_i \qquad v = \frac{\xi}{\bar{x}}$$

This set can only be solved applying an appropriate computer program.

It turns out that the estimates obtained in this way differ insignificantly from the estimates obtained using the method of moments.

Now, we have the theoretical model that should describe the distribution of data *well*. The next step is to verify this supposition, which has the following statistical form: the theoretical function describing the distribution of the empirical data differs insignificantly from the corresponding empirical function. The alternative hypothesis rejects this.   □

The research situation that has been just described touches the sphere of consideration in mathematical statistics that belongs to non-parametric[4] tests of significance and especially tests of the goodness-of-fit.

Non-parametric tests can generally be divided into three groups:

- Tests of the goodness-of-fit
- Randomness tests
- Independence tests.

These tests characterise a lower power than parametric tests; however, they have the advantage of simple construction and uncomplicated calculations. An example of an application of randomness analysis was presented in Chapter 3.1. Now the point of interest will be the application of tests of the goodness-of-fit. Independence tests will be considered in the next chapter.

Tests of the goodness-of-fit investigate the compatibility (accordance) of the empirical distribution with the theoretical one. It is also possible to verify the hypothesis that few distributions are consistent using some of these tests, which means that they can be described *well* by means of one probability function.

In the case just considered the point of interest is a problem of whether the theoretical model that has just been selected describes the data *well*. In order to answer this question, a test of the goodness-of-fit should be applied. Two tests are used most frequently—the Pearson's Chi-squared test and the Kolmogorov test. They are based on two different approaches to the issue of compatibility. In the Pearson's test attention is paid to a comparison of the frequencies; empirical and theoretical. In the Kolmogorov test attention is focused on the maximum of the inconsistency between the distributions, again, a theoretical one with an empirical distribution[5].

Pearson's Chi-squared test. The test has one disadvantage—the data to be analysed should be large. This is due to a requirement that is part of the test procedure. The range of the variation of the variables is divided into a certain number of separate intervals that are in contact with a neighbouring interval. It is suggested that approximately 8 to 10 sample outcomes, on average, should belong to each interval. The test procedure is as follows.

Presume that the population of interest is characterised by some distribution. A sample is taken of size n. The outcomes are divided into r separate bins. Denote the number of outcomes in a given bin by n_i; $i = 1, 2, ..., r$. Obviously

$$\sum_{i=1}^{r} n_i = n$$

The empirical probability density function is obtained in this way.

---

[4]There is a large amount of information in mathematical statistics dealing with non-parametric problems (see for instance Corder and Foreman (2009), Gibson et al. (2003) or Wasserman (2007)).
[5]There are also some tests that are focused on one family of probability distributions, e.g. the Shapiro-Wilk test deals exclusively with the normal distribution.

A verified hypothesis states that the population can be characterised by a certain specified probability function $F(x)$. Using the information contained in the empirical distribution, one can calculate the probability p_i; the probability that the random variable tested will take a value from the bin. Multiplying these probabilities by the sample size, one gets a theoretical number that should occur in a given bin if the verified hypothesis is a true one. Using all of the data, the following statistic should be calculated:

$$\chi^2 = \sum_{i=1}^{r} \frac{(n_i - np_i)^2}{np_i} \tag{4.32}$$

which has—if the verified hypothesis is the true one—the asymptotical χ^2 distribution with $r - k - 1$ degrees of freedom where k denotes the number of parameters of the distribution $F(x)$.

The critical region in this test is the right-side one. The critical value $\chi_\alpha^2 (r-k-1)$ should be read from the Table (9.4) of χ^2 distribution for $r - k - 1$ degrees of freedom and a presumed level of significance α.

If the following inequality holds

$$P\left\{\chi^2 \geq \chi_\alpha^2(r-k-1)\right\}$$

then the verified hypothesis should be rejected, i.e. the difference between the empirical distribution and the theoretical one is statistically significant. Otherwise, there is no basis to reject the null hypothesis. It is said that if the value given by formula (4.32) is closer to zero, then the verified hypothesis is more reliable.

■ **Example 4.2** (Tumidajski 1993; Example 9.2)

In a mine dressing plant 263 portions of the material that was fed mechanically for dressing were collected. The data were used to verify a hypothesis that states that the useful component (copper) in this material can be described by a Gaussian distribution. The samples had an identical mass and a chemical analysis was carried out. Outcomes were grouped in bins, the necessary calculations done and a table to facilitate further calculations was constructed.

Table 4.1. Auxiliary calculations.

Bin intervals Cu content	n_i	u_i	$\Phi(u_i)$	p_i	np_i	$\dfrac{(n_i - np_i)^2}{np_i}$
<1.78	30	−1.063	0.144572	0.144572	38.02	1.6918
1.78 ÷ 1.89	28	−0.719	0.235762	0.091190	23.98	0.6739
1.89 ÷ 2.00	35	−0.375	0.351973	0.116211	30.56	0.6451
2.00 ÷ 2.11	44	−0.031	0.488034	0.136061	35.78	1.8884
2.11 ÷ 2.22	42	0.312	0.621720	0.133686	35.16	1.3306
2.22 ÷ 2.33	29	0.656	0.745373	0.123653	32.52	0.3810
2.33 ÷ 2.44	26	1.000	0.841345	0.095972	15.24	0.2290
>2.44	29	–	1.000000	0.158655	41.73	3.8834

where $p_i = \Phi_{i+1} - \Phi_i$ for $i \neq 1$.

Information was obtained from a previous investigation that the distribution of the data could be satisfactorily described by a Gaussian distribution. Therefore, a null hypothesis was constructed repeating that way of reasoning. The arithmetic mean from the samples that had just been taken was calculated obtaining $\bar{x} = 2.1181$, and the standard deviation $s = 0.3228$.

Upper limits of bins were standardised as $u_i = \frac{x_i - \bar{x}}{s}$, whereas values $\Phi(u_i)$ were read from the Table of the standardised normal distribution $N(0, 1)$.

By calculating the Chi-squared statistic, the following result was obtained $\chi^2 = 10.72$.

Presume a level of significance $\alpha = 0.05$. The number of degrees of freedom is $r - 2 - 1 = 5$. From Table 9.4, we have the critical value: 11.07.

The empirical value is below the critical one—conclusion: there is no ground to reject the verified hypothesis. We can assume that the distribution $N(2.12; 0.32)$ describes the distribution of copper in the broken rock delivered to the dressing plant *well*. ◄

■ Example 4.3

In the seismic station of a certain underground coal mine tremors of rock surrounding the mine were noted between 14-09-2001 and 21-10-2010. The point of interest was the time (counted in days) between two neighbouring tremors whose energy was above the defined level. Within the selected period of observation, 32 events were recorded, namely:

> 10, 131, 81, 116, 17, 196, 29, 178, 20, 147, 23, 134, 20, 222, 129, 207, 95,
> 89, 80, 75, 145, 112, 15, 85, 219, 53, 99, 49, 71, 318, 381, 211 days   Figure 4.3.

Let us perform a statistical analysis of the sample.

First of all, the randomness of the sample should be tested. The sample has 32 elements and for this reason the median is the arithmetic mean of two neighbouring elements—number 16 and 17—for the sample ordered monotonically. Thus, the median is $\frac{1}{2}(95 + 99) = 97$ days.

The original sample is now converted into a sequence of signs: all of the observations in the sample larger than the median value are given a + sign and those below the median are given a – sign. Thus, the sequence is as follows:

$$- + - + - + - + - + - + + + - - - - + + - - + - + - - + + +$$

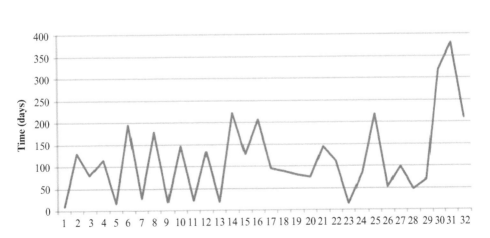

Figure 4.3.   Time between two neighbouring tremors for the period 14-09-2001 and 21-10-2010.

The number of series in this sequence is 22. The number of signs $n_- = n_+ = 16$. Presuming a level of significance $\alpha = 0.05$ and using Table 9.8, we have two critical values for our case:

$$K_{\alpha/2}(16, 16) = 11 \quad \text{and} \quad K_{1-\alpha/2}(16, 16) = 22$$

The empirical number of the series touches the right side critical value. The outcome is ambiguous. Formally we can reject the hypothesis proclaiming randomness. But we may try to conduct our analysis a bit further.

Usually, in such a case, there are two ways to conduct further analysis. One way is to collect a new sample, but in the case being considered this was not possible. A second way is to apply a different test. Unfortunately, the author of this book is not familiar with any different test for the randomness of a sample than that presented in Chapter 3.1.

Thus, let us conduct a further study assuming with a certain carefulness that the sample has a randomness property.

There is no outlier in the sample.

Let us check whether the sample is a stationary one by applying the test based on the Spearman's rank correlation coefficient. Calculations connected with this test are as follows:

Order i	1	2	3	4	5	6	7	8	9	10	11	12	13	14	15	16	17
Value	10	131	81	116	17	196	29	178	20	147	23	134	20	222	129	207	95
Rank v_i	1	21	13	19	3	26	7	25	4, 5	24	6	22	4, 5	30	20	27	16

Order i	18	19	20	21	22	23	24	25	26	27	28	29	30	31	32
Value	89	80	75	145	112	15	85	219	53	99	49	71	318	381	221
Rank v_i	15	12	11	23	18	2	14	28	9	17	8	10	31	32	29

Calculate the coefficient (formula 3.18). We have:

$$r_S = 1 - \frac{6\sum_{i=1}^{n}(v_i - i)^2}{n(n^2 - 1)} = 0.255$$

Let us check whether this value is significant. The null hypothesis states that the stationarity of the sequence versus an alternative supposition rejecting it. The critical value can be calculated by applying formula (3.23). Here we have:

$$r_S(\alpha, n) = \frac{1.645}{\sqrt{n-1}} = 0.295$$

Comparing these two values, we can state that there is no basis to reject the null hypothesis that states the stationarity of the sample.

If so, we can try to find a theoretical model which will describe the distribution of the data *well*. Let us divide the range of observations into five 80-day-long intervals. By ascribing to each interval the corresponding values, a histogram is obtained which is presented in Figure 4.4. The theoretical models that could be considered here are: exponential, gamma or Weibull distributions. Let us reject the exponential model because this is a particular case for gamma distribution as well as for the Weibull distribution.

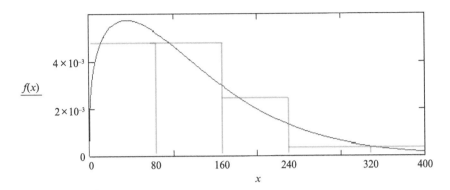

Figure 4.4. The histogram and the theoretical probability density function for the times from Example 5.3.

Let us try to use the Weibull distribution. It was defined by pattern (1.54). Recall

$$f(x) = \alpha\lambda x^{\alpha-1} e^{-\lambda x^{\alpha}} \quad x \geq 0, \alpha > 0, \lambda > 0$$

In order to estimate unknown values of the structural parameters, we can apply the method of moments first. This means that we need to solve the following set of equations:

$$\Gamma(1+\hat{\alpha}^{-1})\,\hat{\lambda}^{-1/\hat{\alpha}} = \bar{x}$$

$$\frac{\Gamma\left(1+\dfrac{2}{\hat{\alpha}}\right) - \Gamma^2\left(1+\dfrac{1}{\hat{\alpha}}\right)}{\hat{\lambda}^{\frac{2}{\hat{\alpha}}}} = S_x^2 \qquad (4.33)$$

where $\bar{x}$ is the average value of the observations and S_x^2 is the corresponding variance.

This set can easily be solved using an appropriate computer program.

For the data in hand

$$\bar{x} = 117.7 \text{ days} \quad \text{and} \quad S_x^2 = 7995 \text{ days}^2$$

and therefore

$$\hat{\alpha} = 1.33 \quad \text{and} \quad \hat{\lambda} = 1.58 \times 10^{-3}$$

The probability density function with these parameters is depicted in Figure 4.4. Notice that two pieces of information here indicate that the exponential distribution has no validity. Firstly, the standard deviation of the random variable investigated is distinctly different than the mean. Secondly, the shape parameter α is clearly different than unity.

Apply now the method of most likelihood. We have the set of equations:

$$\hat{\lambda} = \frac{n}{\sum_{i=1}^{n}(x_i)^{\hat{\alpha}}}$$

$$\hat{\alpha}\left[\hat{\lambda}\sum_{i=1}^{n}(x_i)^{\alpha}\ln x_i - \sum_{i=1}^{n}\ln x_i\right] = n \qquad (4.34)$$

Inserting the estimates of parameters that were obtained using the method of moments, we come to the conclusion that the estimates are good enough. Changes in the values of the parameters because of the application of the method of most likelihood are negligible.

Now we have the theoretical model that describes the distribution of data. Let us check whether this model describes the distribution of data *well* in a statistical sense. It is necessary to apply the test of the goodness-of-fit. Our data are not so rich and for this reason, it seems advisable to use the Kolmogorov test.

□

In the **Kolmogorov test** two distribution functions are compared—the empirical one with the theoretical one. A null hypothesis H_0 that is verified states that these distribution functions are identical, i.e. $H_0: F^{(e)}(x) = F^{(t)}(x)$, where $F^{(e)}(x)$ is an empirical function and $F^{(t)}(x)$ is the corresponding theoretical one. If the population has a distribution that is in accordance with the theoretical one—this means that the null hypothesis is true—then the values of both distribution functions should be close to one another. In the Kolmogorov test, the absolute differences between these functions are investigated and attention is paid to the largest gap between these values. This difference is the basis for the construction of the Kolmogorov statistic whose distribution is independent of the theoretical distribution function. The critical values were calculated based on this statistic and they are given in Table 9.12. If the maximum difference between the theoretical and empirical functions is greater than the critical value, the verified hypothesis should be rejected.

The test procedure is as follows.

Observations are ordered monotonically and they are grouped in separate intervals (bins). In this way information on a number $n_j; j = 1, 2, \ldots, k, \ldots, r$ of the outcomes belonging to each interval is obtained. Obviously, $\sum_{j=1}^{r} n_j = n$.

The empirical distribution function is determined as:

$$F^{(e)}(x) = \frac{1}{n} \sum_{j \leq k} n_j \qquad (4.35)$$

where $\sum_{j \leq k} n_j$ is the total number of observations up to the k-th interval, inclusively.

Next, based on the theoretical distribution function for each sample value (if the size of the sample is small) or for each interval (bin), one calculates the value of the function. Thus, we have two sequences of values of these distribution functions. Afterward, one calculates the absolute difference between the two value functions and the greatest difference is noted. This value is compared with the corresponding critical value for a given number of observations and a presumed level of significance. If the largest gap is below the critical value, we have no ground to discard the verified hypothesis.

□ **Example 4.3 (cont.)**

Following the reasoning presented above, calculate the values of the distribution functions—empirical and theoretical—in the points that correspond with the empirical values. The results of the calculation are presented in Table 4.2.

Presume a level of significance $\alpha = 0.05$. For such a level of significance and the sample size $n = 32$, the critical value in the Kolmogorov test is 0.234 (see Table 9.12). The maximum difference between these functions is 0.090. Thus, there is no ground to reject the verified hypothesis. This means that there will be no objection to the statement that the Weibull distribution with the parameters $(1.33; 1.58 \times 10^{-3})$ describes the empirical distribution of data *well*.

Table 4.2.   The empirical and theoretical distribution functions for
Example 4.3.

j	x_j	$F^{(e)}(x)$	$F^{(t)}(x)$	$\max|F^{(e)}(x)-F^{(t)}(x)|$
	days			
1	10	0.031	0.033	
2	15	0.062	0.056	
3	17	0.094	0.066	
4	20	0.125	0.081	
5	20	0.156	0.097	
6	23	0.187	0.097	0.090
7	29	0.219	0.13	
8	49	0.250	0.244	
9	53	0.281	0.267	
10	71	0.313	0.367	
11	75	0.344	0.389	
12	80	0.375	0.415	
13	81	0.406	0.421	
14	85	0.438	0.441	
15	89	0.469	0.461	
16	95	0.5	0.491	
17	99	0.531	0.51	
18	112	0.563	0.568	
19	116	0.594	0.585	
20	129	0.625	0.637	
21	131	0.656	0.645	
22	134	0.688	0.656	
23	145	0.719	0.694	
24	147	0.750	0.7	
25	178	0.781	0.789	
26	196	0.813	0.829	
27	207	0.844	0.851	
28	219	0.875	0.871	
29	221	0.906	0.874	
30	222	0.937	0.876	
31	318	0.969	0.965	
32	381	1	0.986	

$$F^{(e)}(x)=\frac{j}{n}$$

$$F^{(t)}(x)=\int_0^{x_j} f(x)dx$$

◀

□ Example 3.5 (cont.)

As a result of the preliminary analysis and bearing in mind the histogram shown in Figure 4.2, we came to the conclusion that the distribution of data could be described by the gamma function (1.49) with parameters (2.44; 0.03).

Now it is time to check whether this description is *good* in a statistical sense. The sample is not so rich, and therefore the Kolmogorov test can be applied.

Therefore, formulate the hypothesis $H_0:F^{(e)}(x) = F^{(t)}(x)$ versus the hypothesis H_1 which denies it.

In order to check whether the basic supposition is not false, the sample was ordered monotonically in an increasing way. Both distribution functions were calculated at the points that corresponded to the empirical values. The results of the calculation are presented in Table 4.3.

Presume a level of significance $\alpha = 0.05$. For such a level of significance and the sample size $n = 35$, the critical value in the Kolmogorov test is 0.224 (see Table 9.12). The maximum difference between these functions is 0.064. Thus, there is no ground to reject the verified

Table 4.3.   The empirical and theoretical distribution functions for example 3.5.

j	x_j [min]	$F^{(e)}(x_j)$	$F^{(t)}(x_j)$ gamma	$\max\|F^{(e)}(x_j) - F_0(x_j)\|$	$F^{(t)}(x_j)$ Weibull
1	10	0.029	0.014		0.028
2	15	0.057	0.033		0.053
3	20	0.086	0.061		0.083
4	25	0.114	0.095		0.116
5	30	0.143	0.134		0.152
6	30	0.171	0.134		0.152
7	35	0.200	0.177		0.191
8	35	0.229	0.177		0.191
9	40	0.257	0.222		0.230
10	40	0.286	0.222	0.064	0.230
11	45	0.314	0.268		0.271
12	50	0.343	0.315		0.312
13	60	0.371	0.408		0.394
14	60	0.400	0.408		0.394
15	65	0.429	0.453		0.434
16	65	0.457	0.453		0.434
17	70	0.485	0.496		0.473
18	70	0.514	0.496		0.473
19	80	0.543	0.575		0.548
20	80	0.571	0.575		0.548
21	85	0.600	0.612		0.583
22	90	0.629	0.646		0.616
23	95	0.657	0.678		0.648
24	100	0.686	0.707		0.678
25	100	0.714	0.707		0.678
26	105	0.743	0.735		0.706
27	110	0.771	0.76		0.733
28	120	0.800	0.804		0.781

(Continued)

Table 4.3. (Continued).

j	x_j [min]	$F^{(e)}(x_j)$	$F^{(t)}(x_j)$ gamma	$\max\|F^{(e)}(x_j) - F_0(x_j)\|$	$F^{(t)}(x_j)$ Weibull
29	130	0.829	0.842		0.822
30	140	0.857	0.872		0.857
31	155	0.886	0.908		0.898
32	170	0.914	0.935		0.929
33	170	0.943	0.935		0.929
34	180	0.971	0.948		0.945
35	230	1	0.984		0.986

hypothesis. This means that there will be no objection to the statement that the gamma distribution with the parameters (2.44; 0.03) describes the empirical distribution of data *well*. ◀

If there is no indication of what kind of distribution should be applied, one can choose any of them provided that the selected distribution describes—in a statistical sense—the distribution of empirical data *well*. In the case just considered, the Weibull function has a similar shape as the distribution. By repeating the calculation for the Weibull distribution, one obtains the theoretical distribution function values, which are shown in the last column of this Table. It is easy to notice that the Weibull distribution of the calculated structural parameters can also be used because the maximum difference between the cumulative functions does not exceed the critical value in the Kolmogorov test.

4.3 AN EXAMPLE OF EMPIRICAL–THEORETICAL INFERENCE ABOUT THE DISTRIBUTION OF A RANDOM VARIABLE

Many systems in mining are of the series type in a reliability sense, especially the mechanised ones, e.g., BWE—belt conveyors—stacker in surface mining; shearer—AFC—belt conveyors in underground coal longwall mining; crusher (sizer, Australian)—belt conveyor—dumping machine; screen—receiving conveyor. These systems consist of pieces of equipment that can be repaired (are renewable). There are also series systems in mining that consist of non-renewable objects.

In reliability theory, a system has a series structure if, and only if, the failure of any element of the system means the failure of the system.

This type of structure is the worst of all of those that are possible, i.e. such a system has the most inconvenient reliability parameters and characteristics.

If, for instance, any piece of equipment that is engaged in the excavation and haulage of the coal won in an underground mine fails, then the whole system will not have any stream of coal quite soon. An exception to this rule is when such a system has a bin in its structure and a certain amount of the mineral can be stored in it.

The reliability parameters of renewable series systems are calculated by applying the Markov processes theory if the times of the states (work and repair) can be satisfactorily described by exponential distributions. If, however, the probability distributions of the times of the states are not exponential but are still stochastically independent, then the theory of semi-Markov processes should be applied.

If a **system** is **of a series structure** and its components **work to the first failure occurrence**, which means that the whole system works to the first failure occurrence, then order statistics should be applied. A system of this type can be found, for instance, in electronics.

Before a definition of order statistics is given, let us consider a random variable that describes the time of work of such a system that consists of identical elements. It is defined by the equation:

$$X_{1,n} = \min(X^{(1)}, X^{(2)}, ..., X^{(n)}) \tag{4.36}$$

where: $X^{(i)}$; $i = 1, 2, ..., n$ denote the work times of the elements
$\qquad X_{1,n}$ denotes the first (shortest) time amongst n possible.

Denote by $F^{(i)}(x)$ the probability distribution of the random variables $X^{(i)}$.

Assuming that the work times of the elements are stochastically independent, the probability distribution of the work time for a series system is determined by the formula:

$$F_{1,n}(x) = 1 - \prod_{i=1}^{n}\left(1 - F^{(i)}(x)\right) \tag{4.37}$$

which means that the reliability (survival) function is given by the pattern:

$$R_{1,n}(x) = \prod_{i=1}^{n} R^{(i)}(x) \tag{4.38}$$

Recall

$$R^{(i)}(x) = 1 - F^{(i)}(x) \tag{4.38a}$$

If the probability distributions of the work time of the elements are exponential, i.e.

$$F^{(i)}(x) = 1 - \exp(-\lambda^{(i)}x) \qquad x \geq 0 \qquad \lambda^{(i)} > 0 \tag{4.39}$$

where $\lambda^{(i)}$ is the intensity of the failures of the i-th element then the probability distribution of the work time of the system is also exponential and given by the formula:

$$F_{1,n}(x) = 1 - \exp\left(-x\sum_{i=1}^{n}\lambda^{(i)}\right) \qquad x \geq 0 \tag{4.40}$$

The formula above contains information that the intensity of the failures of the system is the sum of all of the intensities of the failures of all of its elements.

The mean time of work of the system is determined by the equation:

$$E(X_{1,n}) = \left[\sum_{i=1}^{n}\frac{1}{E(X^{(i)})}\right]^{-1} \tag{4.41}$$

Particularly, if all of the elements are the same in a reliability sense then

$$R_{1,n}(x) = [R(x)]^n$$

and for the exponential law

$$\lambda = n\lambda^{(1)} \qquad E(X_{1,n}) = \frac{1}{n}E(X^{(1)})$$

■ **Example 4.4**

Often, durability tests of chains, which are not used only in mining, rely on the following scheme. In testing a device, a three-link or five-link segment of the chain is fixed and an alternating load is applied periodically. Such tests are usually normalised. The test is finished the moment that any link breaks off. The number of load cycles until the moment of the occurrence of the break is noted[6].

When the data collected from strength tests are gathered, it is usually presumed that the number of cycles noted contains information about the durability of the links in the chain and because this number is a random variable, the probability distribution is usually estimated and—as a rule—a log-normal distribution is applied.

The presented reasoning is flawed.

The random variable that is observed in testing—the number of load cycles until the moment of the failure of one out of three chain links or one out of five chain links—is the first order statistic $X_{1,n}$, $n = 3$ or 5. From a reliability point of view, a chain segment is a system consisting of three or five identical elements, stochastically taking. All of the elements work until the moment of the first failure occurrence. Thus, these systems are non-repairable ones.

Recall now the definition of an **order statistic**.

Let $X_1, X_2, ..., X_n$ be a random vector of size n and let $x_1, x_2, ..., x_n$ be the realisation of this vector. An order statistic $X_{k,n}$ is a function of the random variables $X_1, X_2, ..., X_n$ taking k-th largest value in every sequence $x_1, x_2, ..., x_n$.

If $f(x)$ denotes the probability density function of the random variable x and $F(x)$ is the cumulative function, then the probability density function $f_{k,n}(x)$ is determined by the pattern:

$$f_{k,n}(x) = \frac{n!}{(k-1)!(n-k)!} f(x)[F(x)]^{k-1}[1 - F(x)]^{n-k} \tag{4.42}$$

As you can see the probability distribution functions of order statistics are different.

If the estimation of the probability distribution of the random variable—the number of load cycles until the moment of the failure occurrence—is done, this distribution is $F_{1,n}(x)$. Therefore, if there is interest in the probability distribution of the durability of the links in a chain $F(x)$ then, bearing in mind formula (4.37) and rearranging it appropriately, we have:

$$F(x) = 1 - (1 - F_{1,n}(x))^n \tag{4.43}$$

---

[6]Some other investigations use a slightly different scheme. A given segment of a chain is tested for a given number of load cycles. If there is no failure during this test, the investigation is finished and it is assumed that the chain fulfils the stipulated requirements.

As the investigations show, the data gathered during durability tests can be satisfactorily described by a log-normal distribution and this distribution is $F_{1,n}(x)$. Now, the problem is what can be stated about the probability distribution $F(x)$.

Let us analyse this problem in more depth.

Presume that as a result of testing, a sequence of numbers was obtained that created a sample of size n; x_1, x_2, ..., x_n. Knowing *a priori* that the theoretical model is log-normal, it is necessary to convert the sample into a sequence of logarithms $\log x_1$, $\log x_2$, ..., $\log x_n$. The next step is the estimation of the structural parameters, i.e. the estimation of the average value and the standard deviation.

Presume that the estimation gives the following result:

− The estimate of the average value of the number of load cycles until the moment of the first failure occurrence out of n links tested, equals 3×10^4 ($\mu = \log 3 \times 10^4 = 4.477$)
− The estimate of the standard deviation equals 1.5×10^4 ($\sigma = \log 1.5 \times 10^4 = 4.176$).

Having the estimations of these parameters of the probability function, a test of the goodness-of-fit should now be applied in order to verify the hypothesis which states that the empirical and the theoretical functions are identical statistically. It should be possible to describe the distribution of logarithms by the Gaussian distribution.

Presume that the test that was used gave no ground to reject the verified supposition.

If so, the probability density function is given by the pattern:

$$f_{1,n}(x) = \frac{c}{x\sigma\sqrt{2\pi}}\exp\left(-\frac{(\log x - \mu)^2}{2\sigma^2}\right) \qquad x > 0 \qquad c = 0.4343$$

and this is shown in Figure 4.5.

Let us now start a statistical inference about the probability density function $f(x)$ that already has the probability density function $f_{1,n}(x)$ determined. Unfortunately, it is easy to come to the conclusion—looking at formula (4.43)—that there is no way to present the function that is being searched for in an explicit form.

We can only:

− state that the distribution that is being searched for is not a log-normal one
− find a plot of the probability density function $f(x)$ by applying an appropriate computer program.

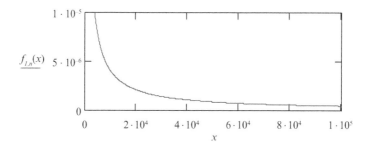

Figure 4.5.   The probability density function of a log-normal distribution of the first order statistic of a chain link.

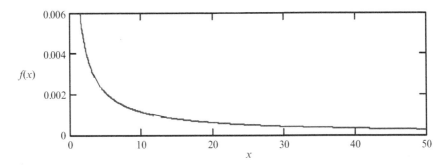

Figure 4.6. The probability density function of chain link durability for $n = 3$.

A graph of the probability density function of the number of load cycles of a chain link until the moment it fails for the data being considered and only the three chain links that created the segment (i.e. the diameter of the links of the chain is large) is presented in Figure 4.6.

Although the course of the function is similar to the function from Figure 4.5, it does not have a log-normal distribution. You can only try to find the theoretical distribution that will describe this function satisfactorily from a statistical point of view. Obviously the expected values and the standard deviations should be the same. ◄

CHAPTER 5

Relationships between random variables

We have reached the point in our considerations at which we have the statistical diagnostics (a preliminary analysis of data) and the sample that is characterised by synthetic numerical characteristics—the parameters of the random variable being investigated. Moreover, a theoretical distribution was found that describes the data collected *well*.

However, this is obviously not enough. Often there is such a situation in which we suspect that the random variable that is being investigated has an influence on a certain different random variable or even several random variables. It is therefore necessary to check this supposition. A similar problem can arise from different reasoning. We would like to theoretically describe the functioning of a certain technical object. A theoretical model was found but every model has some assumptions that have to be fulfilled in order to validate its application. Often the requirements of the model are such that some random variables have to be independent (often). And again, the problem of verifying whether this requirement has been fulfilled comes into play—the independence or dependence of some random variables. Sometimes, the following situation appears during the research that is being carried out: the independence of some random variables was expected but—as the result of statistical testing—something opposite has been found. Such a situation is very interesting from a research point of view. We should find the reason why such an irregularity exists.

The points of interest in this chapter are such problems. We start our study from an investigation of the independence of random variables.

5.1   THE CHI-SQUARE TEST OF INDEPENDENCE

We will presume for the time being that our investigation concerns only two random variables and that they are of a categorical nature. Generally, variables can be classified as categorical (qualitative) or quantitative (numerical).

- **Categorical variables** take on values that are names or labels. The colour of a ball (e.g., red, green, blue) or the type of hoist conveyance (e.g., skip, cage, bucket) would be examples of categorical variables.
- **Quantitative variables** are numerical. They represent a measurable quantity. For example, when we speak about the population of a country, we are talking about the number of people in the country—a measurable attribute of the country. Therefore, population would be a quantitative variable.

Let us formulate a question as to whether changes in the values of one variable are accompanied by changes in the values of another variable.

The simplest case is when the investigation concerns independence and when this freedom is confirmed by an appropriate statistical investigation, the investigation is finished. But, when information is obtained during investigation that the random variables are dependent, some questions immediately arise such as: how dependent—strongly or weakly? What is the character of this dependence—linear? Nonlinear? Etc.

A basic test for an investigation into the independence of two categorical random variables is the Pearson's Chi-squared test of independence. The idea behind the test is as follows.

A sample of size n from the population that is being investigated is taken; **remark**: the sample should be large. The outcomes are classified to create a so-called contingency table (independence table) with w rows and k columns. The interior of the table creates the matrix n_{ij} and n_{ij}, which means how many times the i-th realisation of the category (feature) X occurs and simultaneously the j-th realisation of the category Y. It is recommended that $\Lambda_{i,j} \, n_{ij} \geq 5$.

By summing up in rows and columns, we obtain the marginal numbers $n_{i\cdot}$ and $n_{\cdot j}$ that is:

$$n_{i\cdot} = \sum_{j=1}^{k} n_{ij} \quad n_{j\cdot} = \sum_{i=1}^{w} n_{ij} \tag{5.1}$$

but

$$n = \sum_{i=1}^{w} \sum_{j=1}^{k} n_{ij} = \sum_{i=1}^{w} n_{i\cdot} = \sum_{j=1}^{k} n_{\cdot j} \tag{5.2}$$

Table 5.1 shows the concept of an independence table.

A hypothesis that proclaims the independence of the categories being investigated as expressed by the random variables is formulated as:

$$H_0 : P\{X = x_i, Y = y_j\} = P\{X = x_i\} P\{Y = y_j\} \tag{5.3}$$

in accordance with the formula (1.10).

In the test the statistic criterion that is being considered is determined as:

$$\chi^2 = \sum_{i=1}^{w} \sum_{j=1}^{k} \frac{\left(n_{ij} - n_{ij}^*\right)^2}{n_{ij}^*} \tag{5.4}$$

where

$$n_{ij}^* = \frac{n_{i\cdot} n_{\cdot j}}{n} \tag{5.5}$$

The statistic (5.4) has the asymptotic Chi-squared distribution with $(w-1)(k-1)$ degrees of freedom if the verified hypothesis is a true one. The critical region in this test is the right-side test. Presuming the level of significance α and knowing the number of degrees of freedom, the critical value should be taken from Table 9.4. If the empirical value given by the

Table 5.1. Contingency table.

		Feature Y						
		y_1	y_2	...	y_j	...	y_k	$n_{i\cdot}$
Feature X	x_1	n_{11}	n_{12}		n_{1j}		n_{1k}	$n_{1\cdot}$
	x_2	n_{21}	n_{22}		n_{2j}		n_{2k}	$n_{2\cdot}$
	...							
	x_i	n_{i1}	n_{i2}		n_{ij}		n_{ik}	$n_{i\cdot}$
	...							
	x_w	n_{w1}	n_{w2}		n_{wj}		n_{wk}	$n_{w\cdot}$
	$n_{\cdot j}$	$n_{\cdot 1}$	$n_{\cdot 2}$		$n_{\cdot j}$		$n_{\cdot k}$	n

formula (5.4) is greater than the critical one, the null hypothesis should be rejected. Otherwise, there is no ground to reject the verified hypothesis.

Analysing the merits and demerits of this test more carefully, it is worth noticing that the numbers in the table have a discrete character. This suggests that the probability of the first type of error is underestimated—it is easier to reject the verified hypothesis. Therefore, this is a certain bias of the test. This takes on a special meaning if there is a case when the features have only two categories. Removal of the bias can be done by the introduction of a corrective amendment[1], the so-called **Yates** amendment and the formula (5.4) is:

$$\chi^2 = \sum_{i=1}^{2} \sum_{j=1}^{2} \frac{\left(\left|n_{ij} - n_{ij}^*\right| - 0.5\right)^2}{n_{ij}^*} \tag{5.6}$$

At the beginning of the description of the test, the statement was given that the sample should be large. Generally, in mathematical statistics a truism is the statement that it is advantageous if a sample is large. In some national standards, there are recommendations that sample size should have at least 100 elements and marginal numbers—8 elements at least.

Martin (1972) suggested that the Yates amendment should be applied to a four-fold table and also when the sample size does not exceed 60 elements. Siegel and Castellan (1988) made a different suggestion. They recommended not using the test when the sample size is below 20. They also made more suggestions related to the conditions of the application of the test.

In some areas of study in mining engineering, we have no possibility to gather a large sample. In such a case, a different approach should be used. We will now consider the outlook for tables of sizes 2×2 and 2×3 that are based on Bennet and Nakamura (1963) and Bolshev and Smirnov (1965).

The tables that are the point of interest are presented as:

(A)

m_1	m_2	m
$n_1 - m_1$	$n_2 - m_2$	$n - m$
n_1	n_2	n

or

(B)

m_1	m_2	m_3	m
$n_1 - m_1$	$n_2 - m_2$	$n_3 - m_3$	$n - m$
n_1	n_2	n_3	n

Table (A) is constructed in the following way: a sample of size n is taken from the population being investigated. A property of the sample is that m ($m < n$) elements of the sample possess a certain feature while the others do not. The sample is divided into two parts of size n_1 and n_2; $n_1 + n_2 = n$ in a random way. Among the items in the first part (n_1), m_1 elements have the feature that is of interest. Similarly, among the items of the second part (n_2), m_2 elements have the feature that is of interest.

Table (B) is constructed based on a similar experiment, where it is necessary to separate the features into three groups.

In order to fulfil the requirements of the test, a 2×2 table was constructed in such way that: $n_1 \geq n_2$ and $(m_1/n_1) \geq (m_2/n_2)$. A 2×3 table was constructed in such a manner that: $m_1 \leq m_2 \leq m_3$ and $m \leq n - m$. A limitation of this test is the requirement: $n_1 = n_2 = n_3$.

---

[1] This amendment is recommended for smaller samples.

Denote by M_i; $i = 1, 2, 3$ the random variables whose realisations are observed in the form of the numbers m_i and denote them by:

$$p_i = E\left\{\frac{M_i}{n_i}\right\}$$

the corresponding probability. If the probability of the classification of an element into the i-th part does not depend on whether this element has the feature of interest, then $p_1 = p_2$ in case (A) and $p_1 = p_2 = p_3$ in case (B).

The hypothesis that is verified—considering the 2×2 table—proclaims the identity of the probabilities, H_0: $p_1 = p_2$ versus an alternative supposition H_1: $p_1 \neq p_2$. There is also the possibility to verify the null hypothesis when the alternative hypothesis is: H_1: $p_1 > p_2$. If the table is 2×3, the verified hypothesis proclaims: H_0: $p_1 = p_2 = p_3$ versus the alternative hypothesis that rejects it.

The set of tables with the critical values for the test comprise almost 60 pages and for this reason it is not included in this book. Besides, many books that consist of sets of statistical tables include this set.

▪ Example 5.1

The durability of the pinions used in some mine machinery was investigated. To improve the reliability of toothed gear transition in which the pinions operate a high level of performance was established, which expressed by the number of hours of work that the pinions should operate without any failure.

During the reliability investigation 21 pinions from the first producer and 19 pinions produced by a second producer were tested. As a result of the investigation, it was observed that 11 pinions from the first producer and only 3 pinions from the second producer fulfilled the requirement.

However, a statistical hypothesis was formulated, which stated that the durability of pinions manufactured by the second producer was the same as the durability of pinions manufactured by the first producer.

The outcomes of the investigation are presented in the table below.

		Producer		
		first	second	Σ
Result	positive	11	3	14
	negative	10	16	26
	Σ	21	19	40

Presume the level of significance $\alpha = 0.025$. Using the appropriate statistical table (see for instance Zieliński 1972), we get the critical value $m_2(21; 19; 11; 0.025) = 3$. Due to the fact that the empirical value $m_2 = 3$ is not greater than the critical value, there is the ground to reject the verified hypothesis provided that the random variables are arranged monotonically in the same way. However, both values are the same, and this fact may generate some doubts where the reasoning is concerned.

Note some further problems connected with this investigation:

1. By presuming a slightly different level of significance, a different result of the statistical reasoning can be obtained; e.g. for $\alpha = 0.05$ the critical value is $m_2 = 4$
2. The alternative hypothesis can be formulated otherwise suspecting that the durability of the pinions from the second producer is worse. Readers can check such an approach for themselves. ◄

5.2   THE PEARSON'S LINEAR CORRELATION COEFFICIENT

If information is obtained that some random variables are mutually statistically dependent during an investigation, further research should be directed towards the analysis of this interdependence.

Generally, the first area of consideration as to whether statistical interdependence was stated is correlation analysis.

Correlation is the certain stochastic relation between two or more random variables. It relies on such a regularity that changes in the values of one variable are accompanied by systematic stochastic changes in the values of the second variable or variables. Sometimes, correlation is described as a certain degree of stochastic 'brotherhood' of the changes in random variables.

There are a few different methods that can be used to check whether or not there is a correlation between variables.

One of most commonly applied methods is the construction of a statistical table like the one that was presented in the previous chapter (table of independence). If the numbers lie principally on the main diagonal or the largest numbers lie on the main diagonal, then it can be expected that the random variables are correlated linearly. If the numbers are located in a certain characteristic way in the table but not linearly, a supposition can be formulated that a nonlinear correlation exists between random variables.

A different method, which is often applied in practice, is the construction of a diagram of the dispersion of the random variables in a rectangular coordinate system X, Y in which pairs of observations (x_i, y_i) are used. If these points are accumulated in such a way that they can be closed by an ellipse, then we can suspect that a linear correlation exists between random variables. If the points are closer to each other, a stronger relationship can be expected. If the points are scattered, one can expect a lack of correlation.

Example diagrams are shown in Figure 6.2.

A precise statement about whether the correlation between random variables exists can be obtained by analytical reasoning.

There are many correlation measures in mathematical statistics, such as: linear correlation coefficients, rank correlation coefficients, nonlinear correlation coefficients, partial correlation coefficients, multiple correlation coefficients and so on. Many of these measures have found applications in mining engineering and they have been used for years.

Let us discuss the problem of correlation between two random variables.

There is no doubt that the measure of correlation that has been most commonly applied for years is the Pearson's linear correlation coefficient. It is defined by formula (1.79). Recall that this is a normalised measure and is determined over a $[-1, +1]$ interval. In the case of functional relationship between variables X and Y, this coefficient becomes 1 when the increment in the values of one variable is accompanied by an increment in the values of a second variable. If the relationship is reversed but is still strict, then the coefficient reaches –1. Generally, the greater the value in the modulus of the correlation coefficient, the stronger the interdependence between the random variables. Remember, we are only considering the **linear** relationship between the variables. If the random variables are mutually independent, then the value of the correlation coefficient is zero.

Let us now turn from theory to practice.

Having some information about the general population, we can use estimator (3.46); however, some varieties of this measure can be found in statistical books.

Similarly, as in many previous research situations, the point of interest is whether the estimate that is obtained is statistically significant or not. Obviously, a test is needed to resolve the problem. Usually, the procedure is as follows.

A statistical hypothesis H_0 is formulated that proclaims that there is no correlation between the random variables that are being investigated which is noted as: H_0: $\rho = 0$; ρ means the correlation coefficient in the whole population. An alternative hypothesis rejects it.

There are at least two ways to verify the null hypothesis.

One way is to apply the Student's statistic, namely: if $n \geq 3$ and the verified hypothesis is true one then the statistic

$$t = \frac{R_{X,Y}}{\sqrt{1 - R_{X,Y}^2}} \sqrt{n-2} \tag{5.7}$$

has the Student's distribution with $n - 2$ degrees of freedom[2]. From the table with the critical values of the Student's distribution, we read off the value $t(\alpha, n - 2)$ for the presumed level of significance α and the number $n - 2$ degrees of freedom. If the estimate (5.7) is not lower in the modulus than the critical value, then the verified hypothesis should be rejected. Otherwise, there is no basis to reject the null hypothesis.

The second way is to use the critical values table for the Pearson's correlation coefficient. They are given in Table 9.13; parameter v here is $n - k$ and k is the number of random variables ($k = 2$) that are being investigated.

■ **Example 5.2**

At the beginning of seventies of the 20th century a reliability investigation of mine hoist head ropes comprising 35 mines was carried out in Poland.

The information that was gathered contained: the technical data of the hoists and operational parameters such as, for instance, the average number I of winds per day. The speed of the increment of wire breaks was assumed as the measure of rope wear.

The number of days T of work was calculated until the moment at which the speed attained a certain value presuming that this value was the maximum for all of the ropes that were investigated. All of the ropes had the same construction (triangle shape of strands) and for this reason it was presumed that the wear measure was selected well.

---

[2] The number two is connected with the fact that two random variables are considered.

After having defined these two parameters, an interesting problem was whether these two values were correlated, i.e. when the increment in the intensity of the hoist work was noticed, it should have been accompanied by a decrement in the rope durability.

The empirical pairs (T, I) are shown in Figure 5.1.

A statistical hypothesis was formulated which stated that there was no correlation between random variables, i.e. H_0: $\rho = 0$ versus an alternative hypothesis that rejected it. The level of significance was presumed $\alpha = 0.05$.

The Pearson's correlation coefficient was calculated obtaining $R_{T,I} = -0.333$.

For the given sample size and the presumed level of significance the critical value was calculated by interpolating values because the critical values are given for $n = 30$ and for 35 in Table 9.13. In the case that was considered $v = 33$. The formula (5.7) that gives can also be applied:

Table 5.2.   Auxiliary calculations.

	T Days	I Av number of winds/day
1	316	680
2	205	720
3	960	640
4	440	680
5	331	680
6	693	680
7	782	640
8	360	640
9	331	640
10	472	640
11	305	680
12	522	400
13	146	620
14	226	560
15	321	520
16	525	480
17	409	520
18	239	560
19	479	520
20	333	880
21	363	880
22	525	840
23	414	880
24	405	840
25	345	880
26	290	880
27	462	880
28	554	680
29	643	680
30	560	520
31	276	520
32	729	400
33	617	480
34	913	480
35	940	460

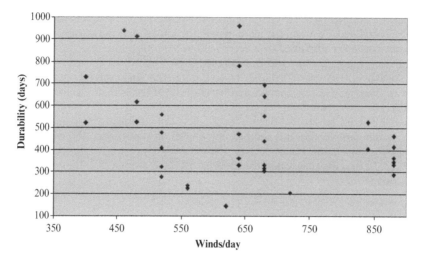

Figure 5.1. Empirical points for the hoist head ropes: durability of the rope versus the average number of winds that were executed.

$$t = \frac{-0.333}{\sqrt{1 - (-0.333)^2}} \sqrt{33} = -2.03$$

Making use of Table 9.3 we get the critical value:

$$t(\alpha = 0.05, n - 2 = 33) \cong 2.04$$

The critical value is greater than the empirical value and thus we have no ground to reject the null hypothesis.

Nevertheless, both values are very close to each other and this fact creates a sensitive point. If the level of significance is presumed to be slightly higher, the result of the reasoning is different and the final conclusion as well.

As there are some doubts about the result of inference, let us check it using the critical values for the Pearson's correlation coefficient. For the presumed level of significance and the known sample size (Table 9.13), we have the critical value 0.334, which is also obtained from the interpolation. Thus, because this way of reasoning gives the same result, there is no ground to reject the verified hypothesis, but—again—both values, the empirical and critical ones, are very close to each other. Notice that from an engineering point of view we suspect the exist-ence of a certain relationship between the random variables that are being investigated. This was the reason that we paid attention to the very small difference between the critical and the empirical values. At this that stage of our analysis, we accept the outcome of that part of the statistical inference, although it looks as though further research should be conducted.

Our conclusion stating that there is no correlation between variables being investigated can be strengthened by analysing Figure 5.1 and Figure 6.2. It is easy to notice that the empirical points scattering in Figure 5.1 does not correspond with any sketch visible in Figure 6.2. □

Our previous considerations comprised problems related to the stochastic interdependence between two random variables. Now it is time to enlarge the scope of interest for a greater number of random variables.

If the number of random variables of interest is k and the relationship between them is lin-ear, we can then estimate the linear correlation coefficients for the whole set of pairs of vari-ables. These coefficients can be arranged in matrix $\mathbb{R}$, which is called a correlation matrix:

$$\mathbb{R} = \begin{bmatrix} 1 & R_{12} & \ldots & R_{1k} \\ R_{21} & 1 & \ldots & R_{2k} \\ \ldots & & & \\ R_{k1} & R_{k2} & & 1 \end{bmatrix}, \tag{5.8}$$

where $R_{ij} = R_{ji}$.

Formally, if X_1, X_2, ..., X_k are random variables with the non-zero variances $\sigma_1^2 > 0$, $\sigma_2^2 > 0$, ... $\sigma_k^2 > 0$ then the entries R_{ij}; $i \neq j$ are equal to the correlation coefficients; for $i = j$ the element is defined as 1. The properties of the correlation matrix $\mathbb{R}$ are determined by the properties of the covariance matrix Σ, according to the relation:

$$\Sigma = \mathbb{D} \mathbb{R} \mathbb{D}$$

where $\mathbb{D}$ is the diagonal matrix with entries σ_1, σ_2, ... σ_k.

5.3 PARTIAL CORRELATION COEFFICIENT AND MULTIPLE CORRELATION COEFFICIENT

We are often interested in the stochastic interdependence between two random variables ignoring the influence of a third variable or other variables.

A measure that allows such a relationship to be investigated is the partial correlation coefficient. In a general case, it is defined as:

$$K_{ij} = - \mathbb{R}_{ij} / (\mathbb{R}_{ii} \mathbb{R}_{jj})^{1/2}, \tag{5.9}$$

where $\mathbb{R}_{ij}$, $\mathbb{R}_{ii}$ and $\mathbb{R}_{jj}$ are the algebraic components that correspond to the elements of the matrix $\mathbb{R}$.

In the case of three random variables that are marked by the numbers 1, 2 and 3, we have:

$$K_{12;3} = \frac{R_{12} - R_{13}R_{23}}{\sqrt{\left(1 - R_{13}^2\right)\left(1 - R_{23}^2\right)}} \tag{5.10}$$

Let us return to Example 5.2.

□ **Example 5.2 (cont.)**

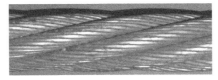

As a part of the research, a calculation of the maximum stress σ^2 daN/mm² (daN stands for 0.1 × Newton) for each rope was done[3]. An interesting problem that had been formulated during the investigation was to get the answer to the question of whether a statistical

---

[3] In engineering nomenclature, the symbol σ denotes stress and it has nothing to do with standard deviation as in the mathematical statistics that were just noted. Similarly, the symbol E in statistics means the expected value while in mechanics it is the Young's modulus.

relationship existed between the intensity of the hoist work and the durability of the rope ignoring the influence of the total rope stress. These data are presented in Table 5.3.

The following correlation coefficients were calculated:

$$R_{T\sigma} = -0.224 \qquad R_{I\sigma} = -0.061$$

The partial correlation coefficient between the random variables *T* and *I* was calculated using the formula (see pattern 5.10) ignoring the influence of the random variable σ:

$$K_{TI:\sigma} = \frac{R_{TI} - R_{T\sigma}R_{I\sigma}}{\sqrt{\left(1-R_{T\sigma}^2\right)\left(1-R_{I\sigma}^2\right)}} = \frac{-0.333-(-0.224)(-0.061)}{\sqrt{(1-(-0.224^2))(1-(-0.061)^2)}} = -0.356$$

Table 5.3.   Auxiliary calculations.

	T Days	*I* Av number of winds/day	σ daN/mm^2
1	316	680	30.0
2	205	720	30.0
3	960	640	30.
4	440	680	30.0
5	331	680	30.0
6	693	680	30.0
7	782	640	30.0
8	360	640	25.7
9	331	640	25.7
10	472	640	25.7
11	305	680	25.7
12	522	400	37.1
13	146	620	37.1
14	226	560	37.1
15	321	520	37.1
16	525	480	37.1
17	409	520	37.1
18	239	560	37.1
19	479	520	37.1
20	333	880	34.4
21	363	880	34.4
22	525	840	34.4
23	414	880	34.4
24	405	840	34.4
25	345	880	34.4
26	290	880	34.4
27	462	880	34.4
28	554	680	27.3
29	643	680	27.3
30	560	520	37.1
31	276	520	31.6
32	729	400	31.6
33	617	480	31.6
34	913	480	31.6
35	940	460	31.6

A statistical hypothesis was formulated stating that there is no correlation between the random variables of interest ignoring the influence related to the total rope stress. An alternative hypothesis rejected this. The level of significance was presumed $\alpha = 0.05$.

The critical value for the given sample size and the degrees of freedom equal to $(35 - 3)$ is 0.340 (Table 9.13). Taking into account that $|-0.356| > 0.340$, the null hypothesis should be rejected at the presumed level of significance. This means that there is a statistically significant negative relationship between the average number of winds that are executed by the rope and its durability if the influence of the total rope stress is ignored. However, this dependence is rather weak.

Notice that the sign before the coefficient value is very important.

The outcome that was obtained meets engineering expectations. ◄

Three significant remarks in connection with the correlation coefficient analysis are worth noting, namely:

a. If the correlation coefficient from the sample is near 1, it does not mean that a cause-effect relation exists: between the random variables; we can say on a strong stochastic dependence (it may be totally random *similarity* of two samples).
b. The small correlation coefficient does not mean that a statistical dependence exists between the random variables being investigated; there may be a dependence but it can have a character that is different than a linear one (more precisely, rectilinear).
c. The correlation coefficient can be underestimated due to measurement errors that may be connected with one or more variables; however, there are statistical measures that take such faults into account.

Let us make some generalisations about our considerations. The subject of interest will now be an investigation on correlation in a case where several random variables are concerned. Presume that there is given a random variable Y in the population and the point of interest is the influence of the set of different variables $(X_1, X_2, ..., X_k)$ on that variable. In econometrics, variable Y is treated as the variable that is being explained and variables X_i are treated as the variables that provide the explanation[4]. The measure R_m determines the degree of dependence of the correlation relationship between variables X_i; $i = 1, 2, ..., k$ and the variable Y is the multiple correlation coefficient that is determined by the formula:

$$R_m = [1 - (|\mathbb{R}|/\mathbb{R}_{YY})]^{1/2}, \tag{5.11}$$

where: $|\mathbb{R}|$—determinant of the matrix $\mathbb{R}$ of correlation coefficients,
$\mathbb{R}_{YY}$—algebraic component of the element YY of this matrix.

A multiple correlation coefficient is a normalised measure supported on the closed interval $[0, 1]$. The higher its value and the closer to unity, the stronger the correlation dependence is. There is a possibility to verify its significance by applying the appropriate statistical test. The critical values of the multiple correlation coefficient are given in Table 9.15.

Often several additional significant problems occur during correlation analysis especially those connected with the proper interpretation of results obtained. Every so often some supplementary statistical measures must be applied in order to obtain a final result. An example of such an extensive analysis was given in Czaplicki's book (2010, Chapter 3.9, which was entitled 'Mutual dependence of random variables'). The example concerned the results of the reliability investigations of the main hoists that were operating in the Polish mining industry in the late seventies of the 20th century.

---

[4] In many countries students are taught 'mathematical determinism', i.e. if x increases two times y increases four times strictly because $y = 2x^2$ and x is the independent variable and y is the dependent variable. In practice a strict relationship seldom exists. Moreover, in some research works x is the variable giving explanation and in some different analyses x is explained.

5.4 NON-LINEAR CORRELATION MEASURES

Let us make some generalisations about the concept of the linear correlation coefficient.

Assume that a sample (x_i, y_i); $i = 1, 2, \ldots, n$ was taken and a model in the form of the function $y = f(x)$ was found. Consider the following measure:

$$\varphi^2 = \frac{\sum_{i=1}^{n} [y_i - f(x_i)]^2}{\sum_{i=1}^{n} (y_i - \bar{y})^2} \tag{5.12}$$

The above measure, which is called the **goodness-of-fit factor** here, is normalised and it takes its values from closed interval [0, 1]. When the value of this measure is known, the multiple correlation coefficient R_m can be calculated because:

$$R_m = \sqrt{1 - \varphi^2} \tag{5.13}$$

This measure determines the conformity between the function that was found (called the regression function) and the empirical values that are contained in the sample. Accordance is greater if the correlation coefficient is closer to unity and obviously φ^2 is closer to zero.

There is a certain inconvenience in the application of the correlation coefficient. It can be estimated provided that the regression function is known. If this model is unidentified, a different measure must be applied instead of the correlation coefficient R (5.13)—**correlation ratio**. This ratio can be used if the dependence between the variables is non-linear. It can be calculated if:

a. the sample that was taken is rich
b. values of the variables are divided into categories.

Categories have a physical validation in an independent analysis. Categories can be constructed without being in touch with the physical background and can be based on a purely mathematical treatment. Note that such a sample allows for the construction of a **table of correlation** similar to what was done in the case of the contingency table. An example of a correlation table is given below.

For the data given in the correlation table, the goodness-of-fit factor is determined by the formula:

$$\varphi^2 = \frac{\sum_{i=1}^{w} \sum_{j=1}^{k} [y_j - f(x_i)]^2 n_{ij}}{\sum_{i=1}^{w} (y_j - \bar{y})^2 n_{\cdot j}} \tag{5.14}$$

where w and k denote the number of categories of the features X and Y.

The correlation ratio of feature Y with respect to feature X is the statistic:

$$\eta_{Y|X} = \sqrt{\frac{\sum_{i=1}^{w} (\Theta_i - Y_s)^2 n_{i\cdot}}{\sum_{j=1}^{w} (y_j - Y_s) n_{\cdot j}}} \tag{5.15}$$

where:

$$Y_s = \frac{1}{n} \sum_j y_j n_{\cdot j} \tag{5.16}$$

is the arithmetic mean in the marginal distribution of variable Y

$$\Theta_i = \frac{\sum_{j=1}^{k} n_{ij} y_j \mid x_i}{\sum_{j=1}^{k} n_{ij}} \tag{5.17}$$

is the average value in the marginal distribution of variable $X = x_i$; $i = 1, 2, ..., w$.

When the positions of both variables are replaced, the correlation ratio $\eta_{X|Y}$ of the feature X with respect to the feature Y will be obtained.

■ **Example 5.3**

Calculate both correlation ratios for data given in Table 5.4.

Look at pattern (5.15). It can be expressed as:

$$\eta_{Y|X} = \sqrt{\frac{\sum_{i=1}^{w}(\Theta_i - \bar{Y}_1)^2 n_{i\cdot}}{\sum_{j=1}^{k}(y_j - \bar{Y}_1)n_{\cdot j}}} = \sqrt{\frac{S^2(Y|X)}{S^2(Y)}} = \frac{S(Y|X)}{S(Y)}$$

where: $S^2(Y|X)$ and $S(Y|X)$ are the conditional variance and the conditional standard deviation among the categories for random variable Y

$S^2(X|Y)$ and $S(X|Y)$ are the conditional variance and the conditional standard deviation among the categories for random variable X

$S^2(Y)$ $S^2(X)$ and $S(Y)$ $S(X)$ are the variances and the standard deviations for random variables Y and X respectively in the marginal distributions.

Similarly, the formula for $\eta_{X|Y}$ can be developed.

Now, calculation is needed to obtain evaluation of correlation ratios of interest. It is presented in Table 5.5.

Table 5.4. An example of a correlation table.

		X							
	Y	1	2	3	4	5	6	7	
No of category k	No of category i	31–33	33–35	35–37	37–39	39–41	41–43	43–45	$n_{\cdot j}$
1	3.25–3.75	1	1						2
2	3.75–4.25		2	3	4	1			10
3	4.25–4.75	1	3	6	3	3	1		17
4	4.75–5.25	1	2	4	5	2	1		15
5	5.25–5.75			1	3	4	2		10
6	5.75–6.25					2	1	1	4
7	6.25–6.75						1	1	2
	$n_{i\cdot}$	3	8	14	15	12	6	2	60

Table 5.5. Supplementary calculations—the characteristics of the marginal distribution.

X_i	$n_{i\cdot}$	$X_i n_{i\cdot}$	$(X_i - X_s)^2 n_{i\cdot}$	$Y_s|x_i$	$(Y_s|x_i - Y_s)^2 n_{i\cdot}$	Y_j	$n_{\cdot j}$	$Y_j n_{\cdot j}$	$(Y_j - Y_s)^2 n_{\cdot j}$	$X_s|y_j$	$(X_s|y_j - X_s)^2 n_{\cdot j}$
32	3	96	97.47	4.33	0.78	3.5	2	7	3.59	33	44.18
34	8	272	109.52	4.37	1.77	4.0	10	40	7.06	36.8	8.1
36	14	504	40.46	4.61	0.74	4.5	17	76.5	1.96	36.82	13.16
38	15	570	1.35	3.87	14.11	5.0	15	75	0.38	37.07	5.95
40	12	480	63.48	5.12	0.94	5.5	10	55	4.36	39.4	28.9
42	6	252	110.94	5.5	2.61	6.0	4	24	6.73	41.5	57.76
44	2	88	79.38	6.25	3.98	6.5	2	13	5.51	43	56.18
	60	2262	502.6		24.93		60	290.5	29.59		214.24

where:

$$X_s = \frac{2262}{60} = 37.7$$

$$Y_s = \frac{290.5}{60} = 4.84$$

Calculate now the unconditional and conditional standard deviations:

$$S(X) = \sqrt{\frac{502.6}{60}} = 2.89 \qquad S(Y) = \sqrt{\frac{29.59}{60}} = 0.70$$

$$S(X_s|y) = \sqrt{\frac{214.24}{60}} = 1.89 \quad S(Y_s|x) = \sqrt{\frac{24.93}{60}} = 0.645$$

Thus, the correlation ratios are as follows:

$$\eta_{X|Y} = \frac{0.645}{0.70} = 0.921 \qquad \eta_{Y|X} = \frac{1.89}{2.89} = 0.654$$

Let us make some short comments.

Because both values are significantly different so the relationship between investigated random variables is nonlinear.

Greater cognitive sense has correlation ratio $\eta_{X|Y}$ than $\eta_{Y|X}$. ◄

The correlation ratio is a unitless measure, normalised and takes values between 0 and 1. Theoretically, the limit $\eta = 0$ represents the special case of no dispersion among the means of the different categories. The limit $\eta = 1$ refers to no dispersion within the respective categories. There is also a third special case when all of the data take the same value; in this case the correlation ratio is undefined. All of these cases are interesting from a theoretical point of view only. If the value of the correlation ratio increases, the relationship between the random variables will be assessed as becoming stronger. This measure does not allow the direction of changes to be stated but it is obvious taking into account that we have a non-linear relationship here. This measure is unsymmetrical, thus

$$\eta_{Y|X} \neq \eta_{X|Y} \qquad\qquad (5.18)$$

However, in the special case in which the relationship is rectilinear, the following formula holds:

$$\eta_{Y|X}^2 = \eta_{X|Y}^2 = R_{XY}^2 \qquad\qquad (5.18a)$$

In other cases the above relationship does not hold and the following relationships are true:

$$\eta_{Y|X}^2 \geq R_{XY}^2 \qquad \eta_{X|Y}^2 \geq R_{XY}^2 \qquad\qquad (5.18b)$$

Thus, if at least one of the correlation ratios is zero then $R_{XY} = 0$.

No relationship exists between the correlation coefficients $\eta_{Y|X}^2$ and $\eta_{X|Y}^2$ and they may radically differ from each other.

There is the possibility to test the significance of the correlation ratio.

Assume that the investigated random variables X and Y have a certain two-dimensional distribution with unknown correlation ratios:

$$H_{Y|X}^2 = \frac{E\{E(Y|X) - E(Y)\}^2}{\sigma^2(Y)} \tag{5.19a}$$

$$H_{X|Y}^2 = \frac{E\{E(X|Y) - E(YX)\}^2}{\sigma^2(X)} \tag{5.19b}$$

and a null hypothesis is formulated that proclaims a lack of the correlation $H_0: H_{Y|X}^2 = 0$ against an alternative hypothesis $H_1: H_{Y|X}^2 \neq 0$. If the null hypothesis is true then the statistic:

$$F = \frac{\eta_{Y|X}^2}{1 - \eta_{Y|X}^2} \frac{n-1}{k-1} \tag{5.20}$$

has the Snedecor's F distribution with $(w - 1, k - 1)$ degrees of freedom.

Presuming the level of significance α, the critical region is above the interval $(F_{1-\alpha}(w - 1, n - 1), +\infty)$. If the empirical value falls into that region, the verified hypothesis should be rejected. Otherwise, there is no basis for rejection.

Presume now that the point of interest is the verification of the hypothesis $H_0: H_{X|Y}^2 = 0$. If so, the following statistic should be considered:

$$F = \frac{\eta_{X|Y}^2}{1 - \eta_{X|Y}^2} \frac{n-w}{w-1} \tag{5.21}$$

It has the Snedecor's F distribution with $(w - 1, n - w)$ degrees of freedom. The further part of statistical inference is well known.

CHAPTER 6

Synthesis of data—regression analysis

6.1 PRELIMINARY REMARKS

If two or more random variables in a general population are of interest, they are interdependent stochastically and if an appropriately rich sample is taken, we can try to find an analytical formula of a functional nature that describes this interdependence *well*. This problem is extremely important in many engineering areas and mining engineering is no exception to this rule. This problem has had a very rich literature for many years and is still under development. In this chapter some elementary information as well as some connections with mine practice will be provided.

Presume for the time being, that only two random variables are of interest.

Usually, a research situation is of such a nature that the values of one variable Y are obtained by being read off at some regular intervals of time or after the execution of given piece of work (counted in the amount of the mass of extracted, hauled or dumped rock etc.) A second variable X is more or less determined on the values of the first variable. Let us call variable X an **explanatory variable** and variable Y the **variable being explained**[1].

The relation between variables can be of different nature, namely:

i. Variables are connected by a **cause-and-effect relationship**, i.e. because the explanatory variable took a certain value, the variable being explained took a corresponding value, e.g. for a greater intensity of machine usage, its durability is reduced by a certain number of work cycles, on average;

   a. A mathematical model has a physical sense and compatible dimensions.

ii. Variables are connected by a **symptomatic relationship**, i.e. we do not have the possibility to observe the variable that directly causes the value that will be taken by variable Y but we have the possibility to observe a different variable, which is significantly and strongly correlated with the causal variable;

   a. A mathematical model constructed in this case has no physical relationship, but there is often the possibility to make the dimensions compatible by assigning the appropriate dimensions to some model parameters.

   b. The possibility to apply symptomatic model seldom occurs in engineering practice.

iii. Variables are connected by neither a cause-and-effect relationship nor by a symptomatic relationship; the model describes only the process of how the values of variable Y are formed as a function of a certain parameter, e.g. the compressive strength of the soil as a function of the depth of the soil layer being investigated;

---

[1] The terms 'independent variable' and 'dependent' variable' are commonly used. However, in econometrics or technometrics these terms are not well-suited. In some cases the variables that are treated as independent—during a statistical analysis—are dependent on each other. The terms proposed here seem to be more appropriate ones and they are in common use in some European languages.

    a. The constructed model has no compatible dimensions and is a **model of the tendency of the development of the trend**[2].

iv. The explanatory variable and the variable being explained are the same variables but values of the explanatory variable are displaced over a certain parameter; the **model is of an autoregressive** character and is usually applied when the future values of the variable depends on the previous values.

    a. The model has compatible dimensions and contains information on a **memory in the process** that is being investigated.

v. Variables are not connected by any physical relationship; the **model has an adaptive character** and is only constructed in such a way that will describe the evolution of the values of variable *Y* well.

    a. Such a model has no physical logic and its only sense is to describe how the values of the variable being described change following certain parameter *well*.

Let us assume that a model whose analytical formula is being investigated is described by the formula:

$$Y = f(X) + \xi \tag{6.1}$$

where: *X*—the explanatory variable,
      ξ—the stochastic component of the model, the random variable.

The explanatory variable *X* can have a different nature. It can be a deterministic variable; in some cases it can be treated as random variable.

Notice (this is important) that the variable being explained is a random variable. In the case when the explanatory variable is not a random one, then the only variable that ensures the compliance of both sides of equation (6.1) is the stochastic component. It comprises the whole stochastic nature of the right side of equation (6.1).

Before the construction of a trial model, i.e. before the construction of the function *f*, a researcher involved in the investigation tries to become familiar with the literature concerning the subject under consideration and tries to find any model that has already been established. If such a model exists, then *basically* the problem of the identification of the model is resolved. However, there is a problem as to whether the data that have just been gathered conform to the model from the literature. The term that was just used—*basically*—makes sense. It does not matter how serious an investigation has been made and how magnificent the researcher who was involved in it is, validation of the model must be performed.

However, if there is a lack of publications in this regard, then it is necessary to choose an analytical model that will describe the data. In some cases, the results of the investigation can 'suggest' what kind of model should be applied. If there is an autocorrelation in the data gathered, it means that there is a memory in the process that is being investigated. For this reason, the first choice of a model should be an autoregressive one.

Sometimes, there is such a situation in which there is no hint as to what character the model being searched should have or what kind of function should be applied. In such a case, the first step is usually the construction of a plot in the rectangular coordinate system *X, Y*.

---

[2] Econometricians are usually of the opinion that a model of the tendency of development is connected with time. In the engineering world the model parameter is frequently different, e.g. the number of tonnes of rock extracted or the number of cycles executed by a machine. These measures are more adequate because the reasons that generate the course of the process are physical phenomena, not time. Recently, in cosmology, a more and more popular opinion is that time does not exist. Time is a convenient measure but it was created by people. When measuring time, we compare only two courses of physical processes.

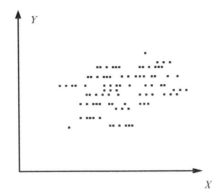

Figure 6.1. Empirical distribution of points that is probably a picture of a two-dimensional random variable.

In such a way, a diagram illustrating the distribution of the empirical points on the plane can be obtained.

If the scattering of points creates a 'cloud' such as is visible in Figure 6.1, then the further investigation can be conducted presuming that this distribution is a picture of a certain two-dimensional random variable.

Often the arrangement of empirical points on a plane 'suggests' which hypothetical function should be taken into consideration in the further part of the investigation (Figure 6.2a–6.2f).

Let us discuss what the procedure of estimation of the unknown structural parameters for hypothetical model actually looks like.

6.2 LINEAR REGRESSION

Let us start from the simplest case when the proposed regression[3] line is a simple straight line function:

$$y = \beta_1 x + \beta_0 + \xi \tag{6.2}$$

This function also has a very wide application in mining engineering, e.g. the speed of the corrosion of a shaft furnishing versus time can be described by a linear function (Carbogno et al. 2001); the resistance to the compression of soil samples depending on the depth of the sample location can also be modelled by this function (Bejamin and Cornell 1977), the utilisation rate for the means of transport in relation to its productivity can be designed using pattern (7.2) (Lin Zaikang et al. 1997), the unit energy of an excavation as a function of the rock compressive strength can be modelled using the linear function (Ceylanoğlu and Görgülü 1997) and the relationship between the peak cutting force and the cutting depth is also linear (Brown and Frimpong 2012).

The point of our interest will be an estimation of structural parameters β_0 and β_1 in order to obtain the analytical recipe of the right side of equation (7.2) that will give the *best* fit to the empirical data. Notice a certain subtleness. Due to the fact that the information in hand

---

[3] The term 'regression' is used in several different sciences such as: biology, economics, geography, psychology, geology and so on. In statistics, it is usually understood as the analysis or measure of the association between random variables.

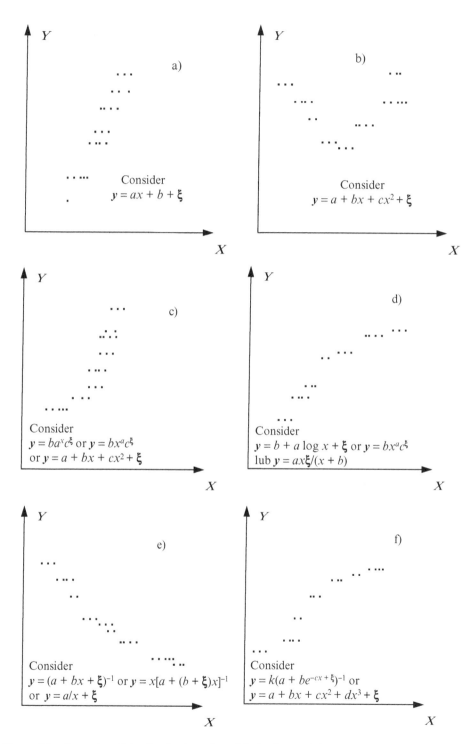

Figure 6.2.   Examples of the distribution of empirical points and the proposed analytical regression functions.

is in the form of a sample, we have no possibility of finding the real values of the model parameters; we can only estimate these parameters.

Denote these estimates by b_0 and b_1 appropriately and estimate the random component; the so-called residuals, by u. These residuals are defined as the differences:

$$u_i = y_i - (b_1 x_i + b_0) \tag{6.3}$$

and they take both negative and positive values.

If an assumption is made that it does not matter whether the difference value is negative or positive, then the best estimates of the unknown parameters of equation (6.3) will be such assessments for which the following sum

$$S = \sum_{i=1}^{n} u_i^2 \tag{6.4}$$

attains the minimum. Therefore, the following set of equations should be solved:

$$\begin{cases} \dfrac{\partial S}{\partial b_0} = 0 \\[2mm] \dfrac{\partial S}{\partial b_1} = 0 \end{cases} \tag{6.5}$$

which gives the following set of equations:

$$\begin{cases} b_1 \displaystyle\sum_{i=1}^{n} x_i^2 + b_0 \sum_{i=1}^{n} x_i = \sum_{i=1}^{n} y_i x_i \\[4mm] b_1 \displaystyle\sum_{i=1}^{n} x_i + b_0 n = \sum_{i=1}^{n} y_i \end{cases} \tag{6.6}$$

The above set is called the **set of normal equations** of the least squares method.

Move this consideration onto a theoretical ground.

There is a statistical population in which the values of the categories create a certain two-dimensional distribution. Let the relationship between these variables be given by formula (6.2). Thus, the following relationships hold:

$$\begin{cases} \beta_1 = \rho \dfrac{\sigma_Y}{\sigma_X} \\[4mm] \beta_0 = E(Y) - \rho \dfrac{\sigma_Y}{\sigma_X} E(X), \end{cases} \tag{6.7}$$

where: ρ is the linear correlation coefficient between variables X and Y.

$E(X)$, $E(Y)$ are the expected values of the random variables X and Y, respectively.

σ_X, σ_Y are the standard deviations of the random variables X and Y, respectively.

Because the only information is the sample taken, the estimators of the unknown structural parameters are obtained by solving the set of equations (6.6), namely:

$$\begin{cases} b_1 = R \dfrac{S_Y}{S_X} \\[4mm] b_0 = \bar{y} - b_1 \bar{x}, \end{cases} \tag{6.8}$$

where the symbols here are already well known.

Having such estimators the question arises as to what kind of properties do these relationships have?

Look at the stochastic 'mechanism' that generates the observations.

Assumptions of the classical least squares method are as follows (Goldberger 1966, Draper and Smith 1998):

a. Equation (6.2) means that each observation of y_i; $i = 1, 2, ..., n$ is the linear function of the observation x_i and the random component
b. The random component ξ is a random variable with a zero expected value and an unknown constant variance σ_ξ^2, i.e. $E(\xi) = 0$ and $\sigma_\xi^2 = const$ as well as $\sigma_\xi^2 > 0$
c. The random variables ξ_i and ξ_j are uncorrelated for $i \neq j$, thus cov $(\xi_i, \xi_j) = 0$
d. The variable x_i is non-random, thus x_i and ξ_i are independent for every i.

If the assumptions above are fulfilled, then the estimators just obtained are the best unbiased estimators with the minimum variance. If these assumptions are not satisfied, then the estimators have worse statistical properties and their application can mean that the estimates obtained will be of a low likelihood. If, for instance, these estimators are applied to safety problems, it may happen that a safety risk will be underestimated.

Often, an additional assumption (e) is made, which states that the random component distribution is normal, and then we decide on the classical model of normal linear regression (Goldberger 1966).

After the estimation of the structural parameters, an assessment of the random component should be made. The applied method of the least squares ensures that the mean value equals zero or the mean value that is assessed using the sample will be negligibly different than zero. The important information will be a measure of its dispersion.

The unbiased estimator of the unknown variance σ_ξ^2 of the random component in the model of the linear regression is determined by the function:

$$s_u^2 = \frac{1}{n-2}\sum_{i=1}^{n} u_i^2 \qquad (6.9)$$

Having the estimates (b_0, b_1) of the structural parameters (β_0, β_1) of the regression function that describes how these variables are mutually dependent and knowing the estimate of the unknown variance of the random component, the question can be formulated as to whether these estimates are significant.

If assumption (e) is a rational one, the significance can be easily verified.

Formulate a hypothesis that states that there is no linear relationship, H_0: $\beta_1 = 0$ between the variables that are being investigated; an alternative hypothesis rejects it.

It can be proved (Goldberger 1966) that the statistic:

$$t = \frac{b_1\sqrt{\sum_{i=1}^{n}(x_i - \bar{x})^2}}{s_u} \qquad (6.10)$$

has the Student's distribution with $n - 2$ degrees of freedom.

Therefore, if $|t|$ (6.10) is above the critical value taken from Table 9.3 for a presumed level of significance α, then the null hypothesis should be rejected. Otherwise, there is no basis to discard the verified conjecture.

The reasoning that is performed here can easily be generalised.

Maintaining assumption (e), the confidence interval for the parameter β_1 can be determined by applying the following formula:

$$b_1 + \frac{t_{1-\%,n-2}}{\sqrt{\sum_{i=1}^{n}(x_i - \overline{x})^2}} s_u < \beta_1 < b_1 - \frac{t_{1-\%,n-2}}{\sqrt{\sum_{i=1}^{n}(x_i - \overline{x})^2}} s_u \qquad (6.11)$$

presuming the level of probability $100(1 - \alpha)$. By calculating expression (6.11) and presuming the plus sign, one obtains the right-side boundary; presuming the minus sign the left-side boundary is obtained.

Testing the significance of the regression can be conducted by making use of the statistic *F* Snedecor's if the relationship between the random variables *t* and *F* is known (see the end of Chapter 1).

If positive information is obtained, i.e. there is a statistically significant linear relationship between the variables tested, then the standard deviations of the random variables of estimated parameters will be very useful information. They give information about how *good* the estimates are.

The standard deviation of parameter b_1 is given by formula:

$$S_{b_1} = \frac{s_u}{\sqrt{\sum_{i=1}^{n}(x_i - \overline{x})^2}} \qquad (6.12)$$

where as the standard deviation of parameter b_0 is:

$$S_{b_0} = s_u \sqrt{\frac{\sum_{i=1}^{n} x_i^2}{n\sum_{i=1}^{n}(x_i - \overline{x})^2}} \qquad (6.13)$$

The greater the standard deviation, the smaller the accuracy of the estimates. The standard deviation of the estimator is called its mean error.

Very important information for the researcher carrying out the investigation is in the sequence of the residuals: $u_i = y_i - y_i^{(t)} = y_i - (b_1 x_i + b_0)$, which is in fact the sequence of the differences between the empirical values of the variable being explained and its theoretical values. This sequence is a representation of the random component that is not directly observable.

The sum of these residuals should be zero or insignificantly different from zero. The sequence should be stationary, should have constant dispersion and a lack of autocorrelation in accordance with the assumptions that were made. It is recommended that it be checked whether these conditions are fulfilled in all of the cases that are being considered.

■ Example 6.1

A tribology investigation was carried out to analyse the wear process of the linings used in the disc brakes of a winder. The linings were made from different materials and some changes in the production process had been introduced.

One of the investigation results was the course of the wear process of the lining for the disc that was fluorescently nitrided versus the number of brakes that were executed by the tester.

The results of the investigation are presented in Figure 6.3.

The results of the investigation clearly indicated that the relationship between the number of brakes that were executed *x* and the linear loss of linings (measured in mm) is linear, which was expected based on the literature on the subject.

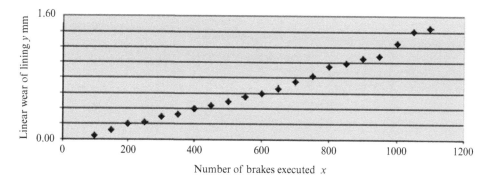

Number of brakes executed x

Figure 6.3.   The linear wear of the lining for the disc that was fluorescently nitrided versus the number of brakes that were executed by the tester.

Firstly, the linear correlation coefficient was calculated that gives:

$$R_{XY} = 0.990$$

This value is high, which 'suggests' that there is really a significant linear relationship between the variables that were investigated. Formally, a hypothesis was formulated that stated that there is no linear correlation between the variables versus an alternative hypothesis that rejects the statement of a null hypothesis.

For a presumed level of significance $\alpha = 0.05$ and a sample size $n = 21$, the critical value, which is 0.433, was taken from Table 9.13. Thus, the null hypothesis should be rejected in a favour of the alternative supposition. It can be stated that there is a strong linear relationship between the number of brakes that were executed and the linear loss of the lining.

The next step was the estimation of the structural parameters of the linear function.

Using the set of questions (6.6), the following estimates were obtained:

$$b_1 = 1.342 \times 10^{-3} \quad b_0 = -0.132$$

Thus, the relationship between the variables that were of interest can be expressed as:

$$y = 1.342 \times 10^{-3} x - 0.132 + u$$

The accuracy of the estimation was determined by two standard deviations:

$$S_{b_0} = 0.029 \qquad S_{b_1} = 4 \times 10^{-5}$$

The residuals are presented in Figure 6.4.
The mean loss and the corresponding standard deviation were as follows:

$$\bar{u} = -4.7 \times 10^{-4} \text{ mm} \qquad s_u = 0.06 \text{ mm}$$

The mean loss was not precisely zero because of the rounding up of some values. The small value of the standard deviation indicates that the theoretical function was properly selected and fit the empirical values well.

A sequence of the residuals was calculated and is presented in the Table below. This series was the object of further investigations.

Firstly, the stationarity of the sequence was tested using the Spearman's correlation coefficient.

The coefficient was calculated and the result was:

$$r_S = 0.056$$

This is very low value and it was suspected that this sequence was uncorrelated with the number of brakes executed. A null hypothesis was formulated stating that there was no linear correlation between the number of brakes and the goodness of fit of the theoretical function to the empirical values, $H_0: \rho = 0$ versus an alternative hypothesis rejecting it.

A level of significance was maintained as previously. For the known sample size and $\alpha = 0.05$, the critical value was 0.368 (Table 9.14).

The empirical value was significantly lower than the critical one; there was no ground to reject the verified hypothesis.

Let us conduct this investigation further by orientating it on dispersion testing, first of all.

i	u_i
0	0.048
1	0.051
2	0.063
3	0.026
4	0.029
5	$-8.05 \cdot 10^{-3}$
6	$-5.2 \cdot 10^{-3}$
7	-0.032
8	-0.039
9	-0.047
10	-0.074
11	-0.081
12	-0.058
13	-0.055
14	$7.6 \cdot 10^{-3}$
15	-0.02
16	-0.027
17	-0.064
18	0.039
19	0.132
20	0.105

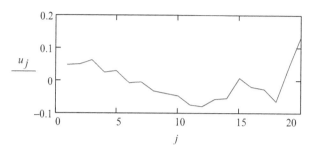

Figure 6.4.   The residuals of the function of the linear mass loss of disc brake lining.

By dividing the sequence of residuals in half and calculating the standard deviation for each sub-sequence (subsample), we have:

$$S_1 = 0.028 \text{ mm} \qquad\qquad S_2 = 0.052 \text{ mm}$$

These figures differ significantly at first glance. It is necessary to verify whether this difference is statistically important. The test that can be applied in this case is that one based on a comparison of the variances of random variables.

By calculating the variances, a null hypothesis was formulated that stated that these variances differ non-significantly, $H_0: \sigma_1^2 = \sigma_2^2$. Looking at Figure 6.4, we can suspect that the dispersion increases. Thus, an alternative hypothesis can be formulated as: $H_0: \sigma_1^2 < \sigma_2^2$.

The test is based on the **F** Snedecor's statistic because:

$$\frac{S_2^2}{S_1^2} = F(n_1 - 1, n_2 - 1)$$

where n_1, n_2 are the size of the first and the second subsample, respectively[4].

Calculating, we have $S_2^2/S_1^2 = 3.45$. Compare this value with the corresponding critical one for a level of significance $\alpha = 0.05$ and the subsamples sizes (9, 9) which is:

$$F_{0.05}(9, 9) = 3.18 \quad \text{(Table 9.6)}$$

The empirical value is distinctly above the critical one[5]. The null hypothesis should be rejected on the presumed level of significance. The dispersion in the second half of the observation is significantly greater than in the first half. It looks as though we are right to suggest that the dispersion increases with an increase in the number of brakes that were executed.

The last step in the investigations of the residuals can be autocorrelation testing.

Calculate the correlation coefficient between the values that are distant from each other by one, two and three steps. The results of calculations are as follows:

$$r_1^{(a)} = 0.751 \qquad r_2^{(a)} = 0.374 \qquad r_3^{(a)} = 0.042$$

---

[4] To be more precise, it is assumed that the subsamples are taken from a Gaussian distribution. It looks as though this assumption holds in the case of the residuals in this case.

[5] Remember that if the alternative hypothesis rejects only what the null hypothesis says, the critical value is the quantile $F_{0.025}(9, 9)$.

Let us check the null hypothesis that states a lack of autocorrelation of the order c of the investigated random variable, H_0: $\rho_c = 0$ versus an alternative hypothesis H_1: $\rho_c > 0$.

A measure that allows the null hypothesis to be tested—as we already know—is the statistic:

$$\chi^2(c) = (n-c)\left(r_c^{(a)}\right)^2$$

By making all of the necessary calculations and taking the critical values from the Chi-squared distribution for a presumed level of significance $\alpha = 0.05$, we have the following results:

$$11.283 \ (3.841) \qquad 2.658 \ (5.991) \qquad 0.761 \ (7.815)$$

where the numbers in brackets are the critical values.

Only an autocorrelation of the first order is statistically significant. This is important information because in the majority of cases the existence of the autocorrelation is connected with physical reasons. Rarely is the autocorrelation connected with a purely random arrangement of numbers. In the case being considered, it would be worthwhile to undertake an investigation to identify these reasons, i.e. the physical process that is generating the autocorrelation and very likely causing that the increases in the dispersion.

We can only state that when the lining is successively worn, important information for short-term prediction will be information on the wear at the current moment of time. The existence of the autocorrelation of residuals makes the statistical properties of the applied estimators to deteriorate but improves the process of inferences about the future. The problem of forecasting the degree of the wear of a lining can be significant for the functional reliability of the brake. However, what is more important is the vital information from the point of view of safety.   ◀

6.3 LINEAR TRANSFORMATIONS AND MULTIDIMENSIONAL MODELS

Our previous considerations can be generalised. The generalisation itself can be of a different nature. Let us list some of these. The generalisation can rely on:

a. A consideration of a function that is different than linear but that describes the course of the random variable being explained *well*
b. A consideration of a linear function with component variables that have errors in their values
c. A consideration of a linear function of one or more variables for which there is some additional information about these variables.

In this chapter our consideration will be orientated on the problem of the estimation of the function parameter based on a sample that is non-linear; however, it is of such a property that it can be transformed into a linear one. In such a case, we call it the **linearisation of regression function (linear transformation)**.

There is a certain class of functions for which the structural parameters can be easily estimated because of the possibility of transforming them into a linear function. Let us consider some examples.

Let the model being analysed be expressed by an exponential function of two variables for instance:

$$y = \alpha a^x b^z c^\xi \tag{6.14}$$

where α, a and b are the structural model parameters, x and z are the explanatory variables and ξ is the random component. In this case, the symbol c denotes a constant that is dependent on the assumed logarithm during linearisation.

The logarithms of both sides give:

$$\log y = \log \alpha + x \log a + z \log b + \xi \tag{6.15}$$

which allows it to be stated that the function is a linear one of the form:

$$v = a_0 + a_1 x + a_2 z + \xi \tag{6.16}$$

where:

$$v = \log y, \quad a_0 = \log \alpha \quad a_1 = \log a \quad a_2 = \log b \tag{6.16a}$$

Similarly, the power model:

$$y = \beta x^b z^c d^\xi \tag{6.17}$$

after taking the logarithms gives a linear model.

Such functions were applied in modelling the processes of the wear of hoist head ropes and for modelling the effective intensity of the wear of brake lining materials (Broś et al. 1976).

Another function, which after some transformations, gives a linear model is the function:

$$y\xi = \frac{ax}{x+b} \tag{6.18}$$

where a and b are the structural function parameters, x is the explanatory variable and ξ is the random component of the model.

By making the substitution:

$$z = (y\xi)^{-1} \tag{6.19}$$

we have

$$z = \frac{x+b}{ax} = \frac{1}{a} + \frac{b}{ax} \tag{6.20}$$

and further

$$v = x^{-1} \quad \lambda = a^{-1} \quad \delta = \frac{b}{a} \quad z_1 = z\xi \tag{6.20a}$$

which gives the linear function:

$$z_1 = (\lambda + \delta v)\xi \tag{6.21}$$

In some cases the following function can also be taken into consideration:

$$y\xi = \frac{x}{a_0 + a_1 x + a_2 w} \tag{6.22}$$

that linearisation gives:

$$(y\xi)^{-1} = \frac{a_0 + a_1 x + a_2 w}{x} = \frac{a_0}{x} + a_1 + a_2 \frac{w}{x} \qquad (6.23)$$

which yields

$$m_1 = (a_0 v + a_1 + a_2 h)\xi \qquad (6.24)$$

where:

$$\frac{1}{y} = \frac{m}{x} = v \qquad \frac{w}{x} = h \qquad m_1 = m\xi \qquad (6.24a)$$

It is assumed that after the linearisation of a given model, the appropriate conversion of the values of the observed variables occurs, and later the estimation of the structural parameters of a new (linearised) model is done. After these estimates are found, a return conversion is made in order to get the estimates of the structural parameters of the primary function.

There are two noteworthy problems associated with the above procedure.

Firstly, the method of least squares minimises the sum of squares of the deviations of the empirical values from the theoretical ones but only for the linearised function, not the primary one. Nonetheless, the solutions found are quite near to the real unknown solutions and for this reason, this procedure is widely applied; especially due to the simplicity of its calculation.

Secondly, there is more subtle and difficult problem that is connected with the random component. Notice that it is variously integrated into its model. Because this component is a random variable any transformation of it means that one achieves a different random variable and that its probability density function can be obtained by an appropriate transformation. However, this density function is often complicated. The problem of finding a probability distribution that can be commonly applied in statistics that the density function describes *well* arises—the function obtained on the way to linearisation. In some cases this problem causes serious difficulties.

Linearisation has found applications in the graphical methods of statistical inference that are connected with distribution functions by using probability papers. By using such a paper, we can:

a. Verify a statistical hypothesis in the form of a probability distribution function; however, without introducing the concept of level of significance
b. Estimate the unknown value of one or two of the structural parameters of the distribution function that is of interest.

Therefore, these methods concern one- or two-parameter distribution functions exclusively.

A **probability paper** of a given distribution of the random variable X that is characterised by the probability distribution function $F(x)$ is an appropriately selected orthogonal coordinate system with the abscissa $\varpi = \psi_1(x)$ and the ordinate $\eta = \psi_2[F(x)]$ in which the distribution function (or its complement—important in reliability analyses) is linearly determined by the formula:

$$\eta = a + b\varpi$$

The most frequently applied probability papers are for the:

a. Gaussian function
b. Lognormal function
c. Exponential function
d. Weibull function
e. Gumbel function[6]
f. Logistic function[7]
g. Rosin-Rammler distribution[8]
h. Power distribution[9].

If the distribution of the variable X that is being investigated has the assumed form, then the empirical points that are projected onto the probability paper should concentrate approximately on the straight line of the pattern:

$$\hat{\eta} = a^* + b^* \varpi$$

where a^* and b^* are estimates of the unknown values of the structural parameters a and b. These estimates can be obtained using, for instance, the method of least squares.

If, for example, the investigation concerns the Gaussian distribution, the following equations are applied:

$$\varpi = \psi_1(x) = x; \quad \hat{\eta} = \psi_2[F(x)] = \frac{x - m}{\sigma}; \quad \hat{\eta} = \frac{\varpi}{\sigma} - \frac{m}{\sigma}$$

It is also possible to construct a confidence region for the cumulative function $F(x)$ by using the quantiles of the Snedecor's statistic (see for instance Firkowicz 1970).

The use of probability papers was popular in the sixties and seventies of the 20th century, especially in reliability analyses. However, their popularity declined with the growing application of computers and advanced statistical programs.

Sometimes, a multinomial model is applied to get a statistically good description of data. It is necessary to remember that the higher the degree of the multinomial, it is necessary to have more structural parameters to assess and a larger sample size.

If for example the model is:

$$y = \alpha_2 x^2 + \alpha_1 x + \alpha_0 + \xi \tag{6.25}$$

the set of normal equations obtained from the application of the least squares method is:

---

[6] The cumulative distribution function of the **Gumbel distribution** is expressed by the formula: $\exp(-e^{-(x-\mu)/\beta})$. This distribution is used to model the distribution of the maximum or the minimum of a number of samples of various distributions. The Gumbel distribution (1954) is a particular case of the generalised extreme value distribution (also known as the Fisher-Tippett distribution). It is also known as the log-Weibull distribution and the *double exponential* distribution (which is sometimes used to refer to the Laplace distribution).

[7] The cumulative distribution function of the **logistic distribution** is expressed by the formula: $(1 + e^{\frac{x-\mu}{\tau}})^{-1}$. In some cases this distribution is applied in ore dressing to approximate the particle size distribution curve (Saramak and Tumidajski 2006).

[8] This distribution is sometimes applied in particle size analysis of comminution processes.

[9] This distribution is connected with the generalised gamma distribution proposed by Stacy (1962); however, a similar distribution had already been considered by Amoroso (1925). Amoroso-Stacy's distribution was intensively studied by two Polish engineers: Firkowicz (1969) and Ciechanowicz (1972) in connection with quality control and reliability.

$$a_2 \sum_{i=1}^{n} x_i^4 + a_1 \sum_{i=1}^{n} x_i^3 + a_0 \sum_{i=1}^{n} x_i^2 = \sum_{i=1}^{n} x_i^2 y_i$$

$$a_2 \sum_{i=1}^{n} x_i^3 + a_1 \sum_{i=1}^{n} x_i^2 + a_0 \sum_{i=1}^{n} x_i = \sum_{i=1}^{n} x_i y_i \qquad (6.26)$$

$$a_2 \sum_{i=1}^{n} x_i^2 + a_1 \sum_{i=1}^{n} x_i + a_0 n = \sum_{i=1}^{n} y_i$$

for the estimation of the unknown structural parameters.

The generalisation of the method for a case in which variable y is described by a multinomial of order k of the variable x does not cause any problems.

■ **Example 6.2**

The wear process of a hoist head rope was observed and the total number, n_t, of cracks in the wires was noted depending on time from the 86th day of its usage until the 468th day when the decision was made to withdraw the rope from further operation. Empirical points were inserted into the diagram and these points were connected by continuous lines—see Figure 6.5.

A power model was applied the first approximation function for the empirical data

$$N_t = \delta t^\gamma c^\xi$$

Note, that this model is a model that only describes the trend and has no physical sense. Its exclusive task is to describe the course of empirical data *well*.

A linearisation was made by applying a common logarithm and for this reason $c = 10$. The least squares method was applied and the following estimates were obtained:

- The estimate g of the exponent γ; $g = 2.46$
- $\log d = -4.079$, which gives $d = 8.33 \times 10^{-5}$ and is the estimate of the proportional coefficient δ.

A plot of the function:

$$N_t = dt^g$$

was inserted into Figure 6.5.

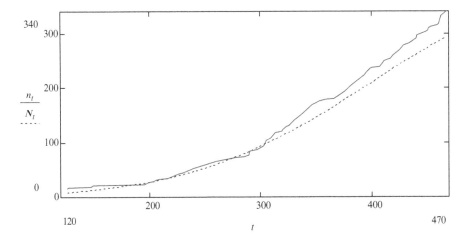

Figure 6.5. The total number of cracks in the wires of the rope versus the number t of days the rope was in operation; n_t—the empirical data, N_t—the theoretical data (the power function).

Looking at this Figure, it is easy to state that the goodness of fit is rather poor, especially where larger numbers of data are concerned.

Study the diagram of the residuals $u_t = N_t - n_t$ for this theoretical function. It is given in Figure 6.6.

Looking more carefully at this Figure, it is easy to perceive that there are long intervals in which the residuals have the same sign. This proves that the empirical model was not selected properly; the signs of the residuals should be mixed. Moreover, the investigation concerning the dispersion stability with time gives information that it is not fulfilled.

For these reasons, a different theoretical model should be applied.

Let us check the exponential function of the pattern:

$$N_t = \alpha \exp(\beta t^2 + \gamma t + \zeta) \tag{6.27}$$

where: α, β and γ are the structural function parameters and ζ is the random component.

Making the linearization, we have:

$$\ln N_t = \ln \alpha + \beta t^2 + \gamma t + \zeta$$

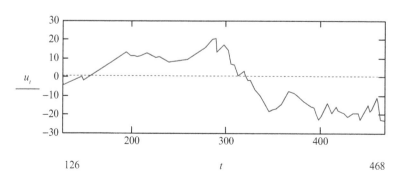

Figure 6.6.   The residuals $u_t = N_t - n_t$ chart for the power function as the theoretical model.

The second degree of polynomial is obtained in such a way. The set of equations (6.26) can be used to estimate the unknown structural parameters.

By making all of the necessary calculations, the following theoretical function is obtained:

$$N_t = 0.7478 \exp(1.976 \times 10^{-5} t^2 + 0.022t + \zeta) \qquad (6.27a)$$

Let us check whether this model describes the data *well*. This function and the empirical figures are visible in Figure 6.7.

Looking at this figure, we are inclined to say that the description does not satisfy our needs. Let us make sure by analysing the residuals $\tau_t = N_t - n_t$ for exponential function (6.27a). They are shown in Figure 6.8.

Both of the proposed approximation functions should be rejected due to their poor description of the data. Moreover, when the analysis of the residuals for both theoretical functions is conducted a little further, it turns out that these sequences have an autocorrelation up to the third order, which means that memories exist in both sequences.

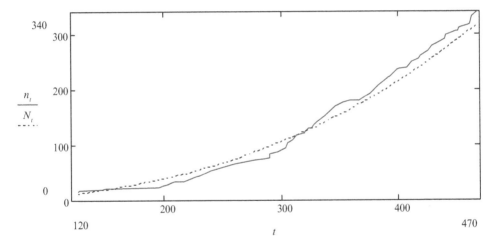

Figure 6.7.   The total number of cracks in the wires of the rope versus the number t of days of the rope's operation; n_t—empirical data, N_t—theoretical data (the power function).

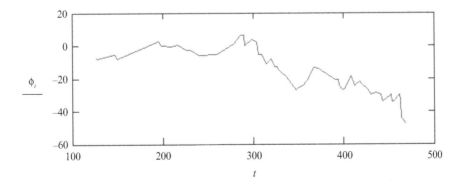

Figure 6.8.   The residuals $\tau_t = N_t - n_t$; chart for the exponential function as the theoretical model.

The results of the autocorrelation testing provide important information, namely:

- There is a memory in the wear process of the rope—the state of the rope at a given moment of time strongly depends on the state of the rope a while before
- The autoregression function should be considered as an approximation function.   ◄

6.4   AUTOCORRELATION AND AUTOREGRESSION MODELS

In theory, autoregressive models occur when the reversibility of stochastic stationary processes is analysed (e.g. Box and Jenkins 1976). In this chapter these models will be considered from an engineering point of view.

Presume that there is given laboratory data or data gathered during an investigation or information collected during operation research. The best situation is when data are in the form of a large sample. The point of interest is to find such a function that will describe the collected data *well*. The usual procedure is as follows. We select a function that has a certain number of structural parameters. These parameters are usually estimated using the least squares method. The next step is the calculation of the residuals—very often as a sequence of the differences between the empirical values and the corresponding theoretical ones. An examination of the residuals allows an assessment of how *good* the selected model is.

Consider the set of assumptions for the least squares method given at the beginning of this chapter. In engineering practice assumption (a) is usually fulfilled. Assumption (b) is fulfilled when the investigation concerns stationary processes. If the investigated process is non-stationary, assumption (b) is fulfiled in some cases, while in other cases it is not. There is also an area of engineering consideration in which the postulate of a lack of autocorrelation (assumption (c)) does not hold. In such a situation the problem of interdependence between variables arises.

Consider a model in which $E(\xi) = 0$ and the variance $\sigma_\xi^2 = const$ but there is an autocorrelation in the realisation of the random component.

Presume that the structural parameters were assessed using method that gave biased estimators and the sequence of the residuals is evaluated. Suppose further that the autocorrelation of the random component was traced and that the following relationship holds:

$$\xi_t = \sum_{i=1}^{L} \delta_i \xi_{t-i} + \chi_t \tag{6.28}$$

where $\{\chi_t\}$ is the pure random process and L is the natural number.

In practical applications the problem of how large the number L should be arises. Or to put it differently, how far back to go? Here it is worth referring to theoretical publications in this regard as well as to look at the practice of the application of such models that have a confirmed autocorrelation of a random variable in econometrics and technometrics.

In practice the deciding factor in how many elements of the right side of the equation (7.28) should be taken into account is information that the autocorrelation coefficients are significant.

However, a situation can occur in practice in which many autocorrelation coefficients are significant, sometimes up to fifth or even eighth order, inclusively (Czaplicki 2000). In such a situation a decision should be taken as to how many components of the sum the model should have (6.28).

Let us look at this problem from a theoretical point of view.

In 1925 Fisher suggested that in the autoregressive model, the coefficients for subsequent variables that are delayed in time decrease systematically with movement to more distant periods. This idea was revisited and developed by Koyck (1954) and Nerlove (1958). Koyck presumed that the coefficients decrease geometrically for elements connected with receding elements. Thus, model (6.28) should have two or three components and should neglect the random component.

It is worth noticing that each the process element, i.e. the value of the process at a given moment of time, depends on the values from previous moments and it does not matter how far back our inference is done. Thus, in the coefficients that are one or more steps back, there is information about the history. Therefore, a model that takes into account many steps back will have a weak rationale. Usually, in practice, applications have models with $L = 1$, $L = 2$ or $L = 3$, rarely more. A practical approach to the selection of the number of components of the autoregression model was presented in example 3.8 and will also be presented in example 6.4.

The unknown parameters δ_i are estimated by applying the method of least squares taking into consideration the residuals.

Presume now that model (6.28) is replaced by the model with the estimates of the unknown structural parameters and that this model will be used for forecasting. If the prediction is unbiased, we neglect the pure random component χ_t, which is the integral constituent of the model. This is the reason that errors are generated. Taking into account that these errors can accumulate, any inference too far into the future is inadvisable, especially if the variance of the component is high. However, if the prognosis will concern only one or two steps ahead, model (6.28) should be useful.

If the autocorrelation of the random component in the model being analysed was stated, and especially if the variance of this component is not constant with time, then the autoregressive model as a regression function:

$$Y_t = \sum_{k=1}^{U} v_k Y_{t-k} + \zeta_t \tag{6.29}$$

where v_k is the structural parameter of the model and ζ_t is the pure random component of the model should be considered.

Having the sample taken, we can estimate the unknown structural parameters, which is a minor task because the model considered is a linear one. The model has a physical sense and compatible dimensions. It is easy to construct a set of normal equations.

If, for example, the model concerns two steps back:

$$\boldsymbol{Y}_t = v_1 \boldsymbol{Y}_{t-1} + v_2 \boldsymbol{Y}_{t-2} + \boldsymbol{\zeta}_t \tag{6.30}$$

then we have the following set of equations:

$$\begin{cases} \hat{v}_1 \sum_t y_{t-1}^2 + \hat{v}_2 \sum_t y_{t-1} y_{t-2} = \sum_t y_t y_{t-1} \\ \hat{v}_1 \sum_t y_{t-1} y_{t-2} + \hat{v}_2 \sum_t y_{t-2}^2 = \sum_t y_t y_{t-2} \end{cases} \tag{6.31}$$

It is not a problem to find the estimators for the above structural parameters.

■ Example 6.3

The wear process of a hoist head rope was observed and the total number n_i of cracks in the wires was noted depending on the number of winds v that were executed. The data that were gathered are presented in a graphical form in Figure 6.9.

Two functions were taken into consideration in order to describe the data:

- The power function

$$\boldsymbol{N}_i = d v_i^g c^{\boldsymbol{\varepsilon}}$$

where: d and g are the structural function parameters and ε is the random component

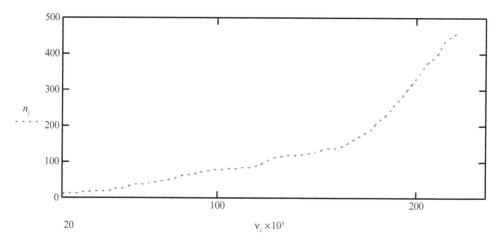

Figure 6.9.   The number n_i of total cracks in the wires for a hoist head rope versus the number v_i of winds that were executed.

- The exponential function

$$M_i = \alpha \exp(\beta v_i^2 + \gamma v_i + \pmb{\chi})$$

where: α, β and γ are the structural function parameters and $\pmb{\chi}$ is the random component.

The method for estimating the structural parameters is well known so let us look at the results of estimation, which are presented in a graphical form—Figure 6.10.

It is easy to see that both of the proposed theoretical functions describe the empirical data poorly.

Knowing that very often the autoregression function describes the empirical data *well*, the following function was considered

$$A_i = \pmb{n}_i = v_1 n_{i-1} + v_2 n_{i-2} + \zeta_i$$

An estimation of structural parameters v_1 and v_2 was done and the result in a shape of the plot A_i was inserted into Figure 6.10.

There is no doubt that the autoregression function describes the data *well*.

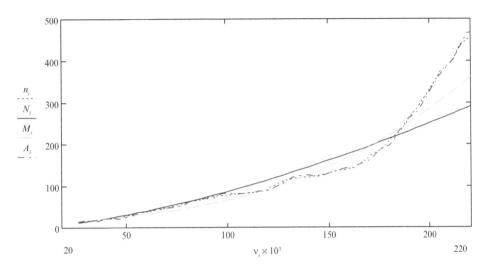

Figure 6.10. Diagram of the total number of cracks in the wires of a hoist head rope versus the number v_i of winds that were executed: n_i—the empirical data, N_i—the power function, M_i—the exponential function, A_i—the autoregression function.

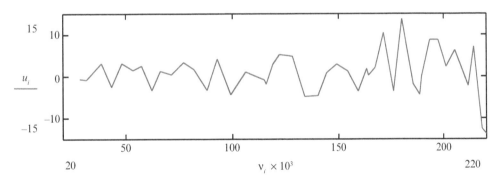

Figure 6.11. Residuals u_i generated using the autoregressive function.

Perform an investigation of the residuals $u_i = n_i - (\iota_1 n_{i-1} + \iota_2 n_{i-2})$, where ι_1 and ι_2 are the estimates of the parameters v_1 and v_2 that were obtained from the application of the least squares method. A sequence of these residuals is presented in Figure 6.11.

Looking at this figure it can be seen that:

1. The positive values are well mixed with the negative values
2. The values of the residuals probably increase when the total number of winds that are executed increases.

Calculate the two basic parameters for the sequence $\{u_i\}$, namely: the expected value and the corresponding standard deviation. Here we have:

$$\bar{u} = 0.62 \text{ cracks} \quad s_u = 5.16 \text{ cracks}$$

Remember the theory that the average value should be zero. It is different than that in this case. We can presume that this is due to rounding up and imperfect calculations. Therefore, formulate a hypothesis that states that the average value $\bar{u}$ is insignificantly different from zero, H_0: $m = 0$ (m is the average value in the population) versus an alternative hypothesis that rejects it.

The following statistic can be applied to verify the null hypothesis:

$$t = \frac{\bar{u}}{s_u}\sqrt{n}$$

which—if the null hypothesis is the true one—has the Student's distribution with $n - 1$ degrees of freedom.

After the calculation we have $t = 0.82$.

Presume a level of significance $\alpha = 0.05$. From Table 9.3 the critical value for a sample size of $n = 47$ is 2.01. Because the empirical value is below the critical value, there is no ground to reject the verified hypothesis. Our supposition was right.

Check supposition (2). Divide the sample in half and calculate the corresponding standard deviations for both subsamples. We have:

$$s_1 = 3.05 \text{ cracks} \quad s_2 = 6.76 \text{ cracks}$$

The numbers differ distinctly. A hypothesis H_0 is formulated that states that there is no significant difference between these parameters. If this is true, then the ratio of the variances has the F Snedecor's statistic. By calculating one obtains: $s_2^2/s_1^2 = 4.93$.

Formulate an alternative hypothesis H_1 that states that the dispersion in values for the second half of the observation is greater, H_1: $\sigma_2^2 > \sigma_1^2$. If so, the critical value taken from Table 9.6 for a presumed level of significance $\alpha = 0.05$ is $F_{0.05}(22, 23) = 2.03$.

The empirical value is definitely greater than the critical one. The alternative hypothesis is the true one. The dispersion in the second half of the data is significantly greater than in the first half. We are right to suspect that the dispersion of the random variable increases with an increase in the number of winds that are executed.

A more precise analysis of the diagram presented in Figure 6.9 allows one to notice that the nature of the course of the increasing number of cracks in wires up to $(160 \div 170) \times 10^3$ work cycles is gentle; the increment in the number of cracks is slow. However, the number of cycles grows as the speed of increment increases. Thus, a supposition can be formulated that something physically happened in the rope wear process that generated this increment. Ignoring the presumed essence of the event here[10], we can formulate a hypothesis that the nature of the wear process up to this figure is different than the one above this figure. If so, two different regression functions should be applied to describe the data. Following this path of reasoning, we should get two sequences of residuals and the test of the significance of the dispersion should be repeated separately for each case.

As the investigations show (Czaplicki 2010, p. 111 and more), this supposition was justified and resulted in a different evaluation of the reliability of the rope.   ◀

Consider the problem of the number of elements in the equation (6.28) in greater depth. This problem is important in time series analysis. It was found (see for instance Box and Jenkins 1976) that a function that can be useful to identify the trend of the lag in an autoregressive model is the **Partial Autocorrelation Function** (PACF). Once can determine the appropriate number of lags by plotting the partial autocorrelative function. This allows the stochastic 'mechanism' of the autoregression in the data analysed and the order of the autoregression

---

[10] A model of the fatigue wear of a wire rope used in hoisting installations in mines was presented in Czaplicki's book 2010.

model to be identified. This order, say p, is identified based on changes in the values of the partial autocorrelation φ_{ll} that is calculated by solving the Yule-Walker matrix equation:

$$
\begin{bmatrix}
1 & \rho_1 & \rho_2 & \cdots & \rho_{\lambda-1} \\
\rho_1 & 1 & \rho_1 & \cdots & \rho_{\lambda-2} \\
\cdots & \cdots & 1 & \cdots & \cdots \\
\rho_{\lambda-1} & \rho_{\lambda-2} & \rho_{\lambda-3} & \cdots & 1
\end{bmatrix}
\begin{bmatrix}
\varphi_{l1} \\
\varphi_{l2} \\
\cdots \\
\varphi_{ll}
\end{bmatrix}
=
\begin{bmatrix}
\rho_1 \\
\rho_2 \\
\cdots \\
\rho_l
\end{bmatrix}
\tag{6.32}
$$

where ρ_i is the **autocorrelation function** determined by the pattern:

$$
\rho_l = \frac{E[X(t+l)-m][X(t)-m]}{E[X(t)-m]^2}
\tag{6.33}
$$

If $\varphi_{ll} \neq 0$ for $l \leq p$ and $\varphi_{ll} = 0$ for $l > p$, then the order of the autoregression function should be p.

■ **Example 6.4** (based on Sokoła-Szewioła's dissertation 2011)

The process of the vertical displacement of a selected point located on the surface in the area that was under the direct influence of mining operations was noted in one of the underground coal mines in the Silesian Coal Basin. The point of interest was the increment of the displacement at a point in successive two-hour-long time intervals. The increment, which was denoted as Δw, was calculated from the formula:

$$
\Delta w_i = Z_{i+1} - Z_i \quad i = 1, 2, \ldots, n
$$

where Z_i was the observed height of the point in i-th measurement.

There are many factors that have an influence on rock displacement around drives and longwalls that are associated with the extraction of rock (the physical proximity of these openings, the speed of mining operation, the properties of the rocks surrounding the openings etc.). This rock movement finally reaches the surface and as a rule causes the subsidence of this surface.

The significance of these factors can vary over time. It is very hard to find any analytical model that takes into account all of these issues and that weighs their significance appropriately. Therefore, it was presumed that a good model to describe the vertical displacement of rock masses on the surface would be an autoregressive one described as:

$$
\Delta w_i = \sum_{j=1}^{p} d_j \Delta w_{i-j} + \varepsilon_i
$$

where: $d_j; j = 1, 2, \ldots, p$ are the structural function parameters
ε—the random component; $N_\varepsilon(0, \sigma_\varepsilon)$
p—the order of the autocorrelation.

Data were gathered, and an example of the daily distribution of the vertical displacement increments of the rock masses at the observed point in the period involved in the research is presented in Figure 6.12.

Simultaneously, any seismic activity of tremors with an energy $E \geq 7 \times 10^3$ J was recorded[11]. The results of the observation are presented in Figure 6.13.

---

[11] This figure was obtained from a separate study. This limited value is different for different mines.

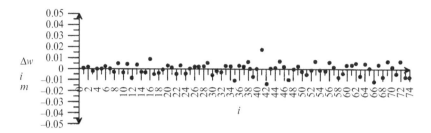

Figure 6.12.   Daily distribution of vertical displacement increments of the rock masses at the observed point in the period involved in the research (Sokoła-Szewioła 2011).

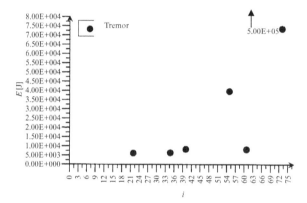

Figure 6.13.   Daily distribution of analysed tremors expressed in energy units within the period involved in the research (Sokoła-Szewioła 2011).

The author of the research presumed that the sequence that was observed was a realisation of a certain random process. Based on a previous investigation, she suspected that this process was a stationary one but it was necessary to verify this.

Recall here that the stochastic process $X(t)$ is stationary if:

1. $E\{X(t)\} = m = const$
2. $E\{[X(t) - m][X(s) - m]\}$ depends on the difference $(t - s)$ only.

Thus, as the first step, the stationarity of the sequence was analysed bearing the expected value in mind. Instead of the classical approach that applies a test based on the rank correlation coefficient, the author decided to solve the problem in a different way. She calculated the expected value of the random variable for subsamples of size l, $l < n$. Three sizes were presumed: $l = 5, 25, 50$. The following function was used as the estimator for the expected value:

$$m_k = \frac{1}{l} \sum_{i=k+1}^{k+l} \Delta w_i \quad l + k \leq n \quad i = 1, 2, ..., n = 132$$

The values of the parameter k varied: $k = 0, 1, 2, ..., 75$.

An example result of the estimation on the average value for $l = 50$ is shown in Figure 6.14.

Further analysis of the estimations showed that the average value did not depend on k for the presumed value of l. This indicated that the sequence could be treated as a stationary one as far as the expected value was concerned.

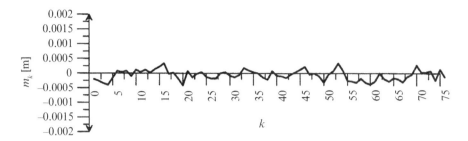

Figure 6.14.   Estimation of the average value for $k = 0, 1, 2, \ldots, 75$ and $l = 50$ (Sokoła-Szewioła 2011).

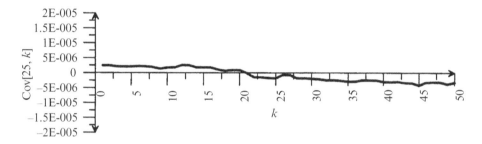

Figure 6.15.   The covariance diagram for $l = 25$ and $k = 0, 1, 2, \ldots, 50$ (Sokoła-Szewioła 2011).

The next point of the investigation was the calculation of the covariance according to the pattern:

$$\text{Cov}(l, k) = \frac{1}{n+1-l-k} \sum_{i=k}^{n-l} (\Delta w_{i+l} - \overline{\Delta w})(\Delta w_i - \overline{\Delta w})$$

where $\overline{\Delta w}$ is the average value. The estimation was done for the time interval $[k, n - l]$ and $[k + l, n]$.

An example plot of the covariance for $l = 25$ and $k = 0, 1, 2, \ldots, 50$ is presented in Figure 6.15.

Afterwards, the following values were found:

$$\max_{1 \leq x \leq 50} \text{Cov}(5, k) - \min_{1 \leq x \leq 50} \text{Cov}(5, k) = 6.3553 \times 10^{-6}$$
$$\max_{1 \leq x \leq 50} \text{Cov}(25, k) - \min_{1 \leq x \leq 50} \text{Cov}(25, k) = 6.7897 \times 10^{-6}$$
$$\max_{1 \leq x \leq 50} \text{Cov}(50, k) - \min_{1 \leq x \leq 50} \text{Cov}(50, k) = 5.7082 \times 10^{-6}$$

When looking at these numbers, a simple conclusion can be drawn that the covariance is constant over time.

The next characteristic that was taken into consideration was the process autocorrelation function. The following function was applied as its estimator:

$$\hat{\rho}_l = \frac{\sum_{i=1}^{n-l} (\Delta w_{i+1} - \overline{\Delta w})(\Delta w_i - \overline{\Delta w})}{\sum_{i=1}^{n-k} (\Delta w_i - \overline{\Delta w})^2}$$

The results of the calculation for $l = 1, 2, \ldots, 12$ are shown in Figure 6.16.

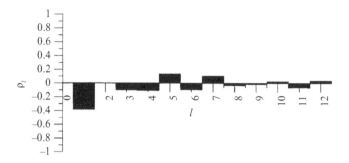

Figure 6.16. The autocorrelation diagram for l = 1, 2, …, 12 (Sokoła-Szewioła 2011).

Figure 6.17. Values of the partial autocorrelation φ_{ll} for l = 1, 2, …, 12 (Sokoła-Szewioła 2011).

The author of the research perceived that because the correlogram[12] became extinct when l increased; it indicated that the process was stationary one[13].

The subsequent step at this stage of the investigation was the identification of the autoregression order, i.e. it was necessary to recognise how many elements the autoregression formula should have.

Based on the changes in the values of the partial autoregression function φ_{ll} that were calculated by solving equation (6.32) for l = 1, 2, …, 12, it was found that the order of the autocorrelation should be: p = 4.

A juxtaposition of the values of the partial autoregression function is presented in the table below and the data is presented graphically in Figure 6.17.

l	1	2	3	4	5	6
φ_{ll}	−0.3812	−0.1658	−0.1958	−0.2869	−0.0839	−0.1710
l	7	8	9	10	11	12
φ_{ll}	−0.0627	−0.0589	−0.0654	−0.0328	−0.0697	−0.0574

After the investigations of the statistical properties of the data, an estimation of the structural function parameters was performed, which gave the model:

$$\Delta w_i = -0.532990 w_{i-1} - 0.3252678 w_{i-2} - 0.3325874 w_{i-3} - 0.2868859 w_{i-4} + \varepsilon_i$$

---

[12] A correlogram is a plot of the sample autocorrelation function. Its counterpart, which is important in geology, is the variogram, which is connected with spatial statistics.
[13] The value of the first autocorrelation coefficient is high and very likely is statistically significant.

The sequence of residuals ε_i was calculated and tested to determine whether they could be described by the Gaussian distribution $N(0, \sigma_\varepsilon)$. The Shapiro-Wilk test gave no ground to reject this hypothesis[14].

Generally, the autoregression function has information about the stochastic interdependence of the random variable. What is more, it carries important information. It is very likely that there is a physical background that generates such interdependence. The process that was analysed has a memory. Additionally, by analysing the magnitude of the particular parameters, we are able to assess how great the influence of the value of the process one or two or more steps back on the current state of the process is.

Where hoist head ropes are concerned, investigations have shown (Dłubała 2009) that the magnitude of rope wear at a given moment of time depends on the magnitude of the rope wear in the previous rope survey in almost 80% of the cases and in almost 20% of the cases on the rope wear two steps back. Moreover, an investigation can be made to test the strength of this memory. A question can be formulated as to whether the dependence that is found is steady in time. As investigations have proven (Czaplicki 2010), this memory strength is often constant in the processes of hoist rope wear. However, there are sometimes processes in which the importance of the most recent notations increases. Here once again, this can be an indication that a certain physical factor commences its action.

It seems that a recommendation can be formulated stating that the constant tracking and investigation of some physical processes is highly suggested because much important information is encoded in the realisations that are observed. And a task for the researcher is to catch, to identify and to explain their physical sense. Such information can be significant for the further inspection of some technical objects. Such information can also be beneficial to object's producer as it can indicate in which ways it can be improved.

6.5   CLASSICAL LINEAR REGRESSION FOR MANY VARIABLES

Consider a certain generalisation of the classical model of regression connecting two random variables. Instead of two variables, a certain number of random variables will be considered. Here we take into account one variable Y, which will be explained, and K variables x_i, $i = 0, 1, \ldots, K$ will be explanatory. The model has the form:

$$Y = \sum_{i=0}^{K} \beta_i x_i + \zeta \quad x_0 = 1 \tag{6.34}$$

Because this model is not so broadly applied in mining practice as the one considered in Chapter 6.2, the analysis presented here will be confined to some basic information. An example from an ore dressing area will be used to illustrate the study that is presented.

In a case with many variables and with a random sample of size n (thus we have $n \times K$ elements in the data), it is advisable to apply the matrix notation due to its simplicity.

Thus, relationship (6.34) can be expressed as:

$$
\begin{bmatrix} y_1 \\ \cdot \\ \cdot \\ \cdot \\ y_n \end{bmatrix}
=
\begin{bmatrix} x_{10} & \cdot & \cdot & \cdot & x_{1K} \\ \cdot & & & & \cdot \\ \cdot & & & & \cdot \\ \cdot & & & & \cdot \\ x_{n0} & \cdot & \cdot & \cdot & x_{nK} \end{bmatrix}
\begin{bmatrix} \beta_0 \\ \cdot \\ \cdot \\ \cdot \\ \beta_K \end{bmatrix}
+
\begin{bmatrix} \zeta_1 \\ \cdot \\ \cdot \\ \cdot \\ \zeta_n \end{bmatrix}
\tag{6.35}
$$

---

[14] Incidentally, due to the fact that the process is stationary in a weak-sense and is normal, it means that it is of a strict stationarity. This is because the distribution of a normal process is specifically determined by means of the average value and the autocorrelation.

or in short

$$Y = XB + Z \tag{6.36}$$

where: Y—the vector of the observations of the variable being explained
$\quad\quad X$—the matrix of the observations of the explanatory variables
$\quad\quad B$—the vector of the structural function parameters
$\quad\quad Z$—the random component vector.

Traditionally, the basis for selecting the best fit hyperplane to the empirical data is the method of least squares. A vector of the estimates of the structural parameters can be obtained from the equation[15]:

$$b = (X'X)^{-1}X'Y \tag{6.37}$$

Therefore, one can obtain the vector of the theoretical values of the variable being explained $Y^{(t)}$ from the formula:

$$Y^{(t)} = Xb \tag{6.38}$$

whereas the residual vector is determined by the difference:

$$u = Y - Y^{(t)} \tag{6.39}$$

The assumptions of the model are as follows:

a. $Y = XB + Z$
b. $E Z = 0$ $\hfill (6.40)$
c. $E Z Z' = \sigma^2 I$ $\hfill (6.41)$
d. X is the matrix of fixed elements in repeated attempts
e. $r(X) = 1 + K \le n,$ $\hfill (6.42)$

where: I is the unit matrix and $r(X)$ means the rank of the matrix X.

According to assumption (a), each observation of the variable y is the linear function of the observations of variables x and the random component.

Assumption (b) says that each random component has an expected value of zero.

The next assumption (c) gives two pieces of information. Firstly, that the variance σ^2 is constant. Secondly, that the random components are not correlated.

Assumption (d) says that the explanatory variables are not random, which means that X and Z are independent. Moreover, this statement informs us that there is no linear relationship between variables x, which is a necessary condition for estimation.

In the linear regression model, the best unbiased linear estimator of the structural vector parameters is the vector obtained by applying the least squares method. It has the minimum variance and the variance-covariance matrix is given by:

$$V = s^2(X'X)^{-1} \tag{6.43}$$

where s^2 is the estimate of the unknown variance of the random component.

If an in-depth analysis is not conducted, at least three measures of the stochastic structure are calculated, namely:

---

[15] Notice that during the calculation, a column of ones appears in matrix X because of the free term.

- the residual variance:

$$S_u^2 = \frac{1}{n-K-1} u' u \tag{6.44}$$

- the goodness of fit factor:

$$\varphi^2 = \frac{\sum_{j=1}^{n}\left(y_j - y_j^{(t)}\right)^2}{\sum_{j=1}^{n}\left(y_j - \bar{y}\right)^2} \qquad \bar{y}\text{—the arithmetic mean} \tag{6.45}$$

- the multiple correlation coefficient:

$R_m = \sqrt{1-\varphi^2}$ (compare (5.12) and (5.13))

The interpretation of these measures and their properties has already been discussed[16].

■ **Example 6.5** (based on Gawenda 2004)

In Gawenda's Ph.D. dissertation (2004), the problem of the analysis and proper description of the process of the comminuition of porphyry in a jaw crusher was studied. Based on the paper of Cardou (Cardou et al. 1993), it was stated that the grain composition during crushing in jaw crushers can be described *well* by the modified Weibull distribution given by the formula:

$$G(z) = 1 - \exp\left[-c\left(\frac{z}{d_{max} - z}\right)^{\varpi}\right] \qquad z \geq 0; c > 0, d_{max} > 0, \varpi > 0$$

where c, ϖ, d_{max} are the structural parameters estimated based on the empirical data.

Many issues connected with such a description were considered. Among other things was the problem of the selection of a model to describe the relationship between parameter ϖ and two basic technical parameters of a jaw crusher, namely: x_1 the input slot width to the crusher and x_1 the jaw pitch. The experiment was carried out and its results are shown in the table below.

---

[16] For more on this topic see for instance Draper and Smith (1998) or Goldberger (1966).

	x_1	x_2	ϖ
1	25	13.4	0.810
2	25	13.4	0.860
3	25	13.4	0.771
4	25	8	0.899
5	25	8	0.864
6	25	8	0.855
7	20	8	0.916
8	20	8	0.869
9	20	10.7	0.899
10	15	10.7	1.028
11	15	10.7	0.957
12	15	10.7	0.842
13	25	10.7	0.850
14	25	10.7	0.904
15	25	10.7	0.840
16	20	10.7	0.835
17	20	10.7	0.915
18	20	10.7	0.893
19	20	5.3	0.981
20	20	5.3	0.919
21	20	5.3	0.898
22	15	5.3	1.025
23	15	5.3	1.068
24	15	5.3	0.973
25	15	2.7	0.965
26	15	2.7	1.158
27	15	2.7	0.965

Gawenda (2004).

Ignoring the method of reasoning made by the author of the research here, let us find a model to describe the relationship between these variables. It was known from previous investigations that the relationship should be linear one.

Thus, the proposed model was as follows:

$$\varpi = \beta_1 x_1 + \beta_2 x_2 + \beta_0 + \zeta$$

Using equation (6.37), the following estimates are obtained:

$$b_0 = 1.17873; \quad b_1 = -0.00817; \quad b_2 = -0.001226$$

which means that the model is:

$$\varpi = -0.00817 x_1 - 0.01226 x_2 + 1.17873 + \zeta$$

Now assess the residuals. Their average value is 0 and the standard deviation is 0.0553. Looking at these figures we are inclined to say that the model describes the data *well*.

Calculate the goodness of fit factor. By applying formula (6.45), we get $\varphi^2 = 0.374$. Having this parameter, the multiple correlation coefficient is:

$$R = \sqrt{1 - \varphi^2} = 0.791$$

This value looks high; however, it is necessary to check whether it is significant.

Formulate a null hypothesis that states that there is no significant correlation between the investigated variables versus an alternative hypothesis rejecting it. Presume a level of significance $\alpha = 0.05$

The sample size is 27 and the total number of variables 3. Thus, the critical value taken from Table 9.15 is 0.470.

The empirical value is clearly above the critical one. Thus, there is a reason to reject the verified hypothesis. The model gets its substantiation.

The variance-covariance matrix is as follows:

$$\mathbb{V} = \begin{bmatrix} 3.007 \times 10^{-3} & 6.423 \times 10^{-5} & -1.704 \times 10^{-4} \\ 6.423 \times 10^{-5} & 2.359 \times 10^{-5} & -1.266 \times 10^{-5} \\ -1.704 \times 10^{-4} & -1.266 \times 10^{-5} & 1.359 \times 10^{-5} \end{bmatrix}$$

Readers may comment on these parameters themselves.   ◄

6.6   REGRESSION WITH ERRORS IN VALUES OF RANDOM VARIABLES

A few generalisations were made in the consideration that was conducted starting from the traditional model of regression. In this chapter a new important generalisation will be made—one that is important from an engineering point of view.

Return for a moment to the simplest linear model that can be determined by formula (6.2). Assumptions (a)–(d), which were listed on p. 174, were associated with this model. Let us make a break in these assumptions that comes closer to engineering reality. Presume that the observations of the explanatory variable are not accurate, but were made with a certain error. For an engineer, it is obvious that when looking at the digits given by a measuring device or when looking at a gauge pointer, the information is read with some preciseness. In many econometric problems the issue of the accuracy of some variables is very difficult to formalise.

In mathematical statistics there is an extensive chapter that deals with **total least squares** in which data modelling takes into consideration observational errors for both types of variables. The method deals with both linear and non-linear models[17].

Because the level of considerations here is a basic one, we confine our study to a simple case in which only two variables are taken into account and the errors are exclusively connected with the explanatory variable. Readers who are interested in this topic should become familiar with Gillard's elaboration (2006) and Fuller's monograph (2006); for non-linear models, the book by Carroll et al. (2006) is recommended.

Presume now that instead of the observation of variable x a different variable is perceived that is the sum of two variables:

$$X = x + \boldsymbol{u} \tag{6.46}$$

where $\boldsymbol{u}$ is a certain random variable. It can be presumed that no systematic errors are being made and therefore

$$\boldsymbol{u} : N(0, \sigma_u) \tag{6.47}$$

---

[17] In statistics there is a special chapter that is dedicated to the errors-in-variables model, which tries to find the line of the best fit for a two-dimensional set of data only. This is the so-called Deming regression, although the model was considered much earlier by Adcock (1878) and Kummell (1879). These ideas were almost entirely unknown until the appearance of Koopmans' work (1937). Later, only Deming (1943) propagated this idea.

The observed variable X is sometimes called the **manifest variable** or the indicator variable. The unobservable variable x is called a **latent variable**. Models with a fixed x are called functional models[18]. Models with a random x are called structural models.

Now, it is necessary to determine the mutual relationship between the variables that are the components of the model.

Assume that the error in measurement of variable X is independent from:

a. The real value of this variable and
b. Independent from the random variable that is a component of the model.

Let us analyse what kind of repercussions are connected with such a research situation.

Construct the variance–covariance matrix, similar to the one made in formula (6.43). We have:

$$\begin{bmatrix} \sigma^2_{yy} & \sigma^2_{Xy} \\ \sigma^2_{Xy} & \sigma^2_{XX} \end{bmatrix} = \begin{bmatrix} \beta_1^2\sigma^2_x + \sigma^2_\xi & \beta_1\sigma^2_x \\ \beta_1\sigma^2_x & \sigma^2_x + \sigma^2_u \end{bmatrix} \tag{6.48}$$

Let us define the regression coefficient based on the information contained in the sample taken:

$$\hat{R} = \frac{\sum_{j=1}^n (X_j - \bar{X})(y_j - \bar{y})}{\sum_{j=1}^n (X_j - \bar{X})^2} \tag{6.49}$$

where all of the variables with a straight accent above the mean are the average values.

Calculate its expected value. Having in mind that the random variables are normal, we have:

$$E(\hat{R}) = \beta_1 \frac{\sigma^2_x}{\sigma^2_x + \sigma^2_u} \tag{6.50}$$

Looking more carefully at the formula above, it is easy to see that the regression coefficient has been attenuated (its value is reduced) by the presence of a measurement error. It is assumed (see for instance Fuller 2006) that the basic measure of this attenuation is the quotient:

$$\kappa_{xX} = \frac{\sigma^2_x}{\sigma^2_X} \tag{6.51}$$

i.e. the ratio of the variances of the variable x and X. This measure is called the **reliability ratio**. However, it is not a reliability measure of a technical item that could be expected by an engineer. It is a determined statistical measure that is associated with the theory and practice of measurement[19].

---

[18] There is again some inconsistency in understanding terms. The term 'functional' as presented here comes from econometrics; engineers will have a different association with it.

[19] In genetics this ratio is a measure of heritability. An observed characteristic of a plant or an animal, the X value, is called the phenotype and the unobserved true genetic makeup of the individual, the x value, is called the genotype. The phenotype is the sum of the genotype and the environmental effect, where the environmental effect is the measurement error.

If this measure can be estimated, then the estimator of the structural parameter standing with the explanatory variable is given by the formula:

$$\hat{\beta}_1 = \frac{\hat{R}}{\kappa_{xX}} \tag{6.52}$$

The second parameter can be estimated using the pattern:

$$\hat{\beta}_0 = \bar{y} - \hat{\beta}_1 \bar{X} \tag{6.53}$$

Now define the variance–covariance matrix connected with the defined estimators. It is determined as the following:

$$\begin{bmatrix} \frac{1}{n}S_v^2 + \bar{X}^2\hat{V}(\hat{\beta}_1 | X) & -\bar{X}\hat{V}(\hat{\beta}_1 | X) \\ -\bar{X}\hat{V}(\hat{\beta}_1 | X) & \hat{V}(\hat{\beta}_1 | X) \end{bmatrix} \tag{6.54}$$

where:

$$S_v^2 = \frac{1}{n-2} \sum_{i=1}^{n} (y_i - \hat{\beta}_0 - \hat{\beta}_1 \bar{X})^2 \tag{6.55}$$

and it is the residual variance of the model

$$\hat{V}(\hat{\beta}_1 | X) = \frac{S_m^2}{\kappa_{xX}^2 \sum_{i=1}^{n} (X_i - \bar{X})^2} \tag{6.56}$$

and it is the unbiased estimator of the conditional variance of $\hat{\beta}_1$

$$S_m^2 = \frac{1}{n-2} \sum_{i=1}^{n} \left[y_i - \bar{y} - (X_i - \bar{X})\hat{R} \right]^2 \tag{6.57}$$

Unfortunately, it is seldom the case when the reliability ratio is known in engineering practice. At most, you can meet with cases in which we are able to determine its approximate value, but we are not able to assess the error that results from this approximation.

Consider now a different case.

Examination often allows it to be realised that the values of the variable that are read off or calculated are flawed. This means that the data are imprecise. Concern arises that the model that is created will be less accurate. In engineering practice in some areas of consideration, e.g. in issues that are associated with safety problems, we cannot afford to commit significant errors. Thus, by all means, it is advisable to assess the scale of the error of a given model.

Often an engineer who is conducting research is able to assess the accuracy that is connected with reading off the values of physical magnitude. But, he/she has no idea about repercussions that will be generated if the regression model that is selected is applied. What is more, he/she does not know how the accuracy of the estimation of the variable being explained will change because of the model that is applied.

Therefore, presume that the standard deviation of the measurement error is known. Note that even if this parameter is unknown, the measurement can usually be repeated many times and then the estimation of this parameter can be achieved.

Thus, we can assume that the value of σ_u is known.

Let us make an estimation of the empirical product moments that apply the formulas:

$$m_{Xy} = \frac{1}{n-1}\sum_{i=1}^{n}\left(X_i - \bar{X}\right)\left(y_i - \bar{y}\right) \tag{6.58a}$$

$$m_{yy} = \frac{1}{n-1}\sum_{i=1}^{n}\left(y_i - \bar{y}\right)\left(y_i - \bar{y}\right) \tag{6.58b}$$

$$m_{XX} = \frac{1}{n-1}\sum_{i=1}^{n}\left(X_i - \bar{X}\right)\left(X_i - \bar{X}\right) \tag{6.58c}$$

Now, we are able to construct the estimators of the most important regression parameters. They are as follows (compare Fuller 2006):

$$\hat{\sigma}_x^2 = m_{XX} - \sigma_u^2 \tag{6.59}$$

$$\hat{\beta}_1 = \frac{m_{Xy}}{m_{xx} - \sigma_u^2} \tag{6.60}$$

$$\hat{\sigma}_\xi^2 = m_{yy} - \hat{\beta}_1^2\hat{\sigma}_x^2 \tag{6.61}$$

$$\hat{\beta}_0 = \bar{y} - \hat{\beta}_1\bar{X} \tag{6.62}$$

Look at the two relationships (6.59) and (6.61) that determine the variances. It is the same element in both of these formulas—the variance of the measurement error. It is known that the variance is almost always positive. This condition allows the upper limit of this error to be determined. This problem is important in engineering practice.

By analysing formula (6.59), we come to the conclusion that

$$\sigma_u^2 \le m_{XX}$$

whereas by appropriately transforming pattern (6.61). it can be stated that:

$$\sigma_u^2 \le m_{XX} - \frac{m_{Xy}^2}{m_{yy}}$$

which is a stronger restriction than the previous one.

And further

$$\sigma_u \le \sqrt{m_{XX} - \frac{m_{Xy}^2}{m_{yy}}} \tag{6.63}$$

The right side of inequality (6.63) determines the maximum of the allowable average measurement error that can be taken into consideration when examining this regression analysis. Express this limit as the percentage of the average value of the variable x that yields:

$$\frac{\sigma_u}{\bar{X}}100 = \frac{100}{\bar{X}}\sqrt{\left(m_{XX} - \frac{m_{Xy}^2}{m_{yy}}\right)} \tag{6.64}$$

Pattern (6.64) determines the upper limit of the measurement error which cannot be exceeded in this regression analysis as a percentage.

The matrix of the variance–covariance of the structural parameter estimators (6.60) and (6.62) is given by the formula:

$$\begin{bmatrix} \dfrac{1}{n}S_v^2 + \bar{X}^2\hat{V}(\hat{\beta}_1) & -\bar{X}\hat{V}(\hat{\beta}_1) \\[2mm] -\bar{X}\hat{V}(\hat{\beta}_1) & \hat{V}(\hat{\beta}_1) \end{bmatrix} \tag{6.65}$$

where now:

$$S_v^2 = \frac{1}{n-2}\sum_{i=1}^{n}\left(y_i - \bar{y} - \hat{\beta}_1(X_i - \bar{X})\right)^2 \tag{6.66}$$

$$\hat{V}(\hat{\beta}_1) = \frac{1}{n-1}\frac{m_{XX}S_v^2 + \hat{\beta}_1^2\sigma_u^4}{\hat{\sigma}_x^4} \tag{6.67}$$

and the random variable

$$v = \xi + \beta_1 u \tag{6.68}$$

■ Example 6.6

In the article by Pyra et al. (2009), the discussion was related to selected problems of blasting and the impact of shock waves that are generated by setting off explosives. This problem was studied in connection with a surface mining operation. Some statistical data were analysed, the relationship between the speed of propagation read off from a measuring device and the air pressure shock wave, among other things. This pressure was also taken from the gauge.

Let us use the data which are in this article. This information is presented in Figure 6.18.

We made only one correction here assuming that the point (0, 0) belongs to the data.

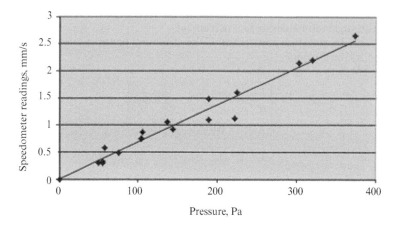

Figure 6.18.   Relationship between the speed of propagation read off from a measurement device and the air pressure shock wave (Pyra et al. 2009).

For the information presented in Figure 6.18, we have:

- The pressure characterised by two parameters:

$$\bar{X} = 151.1 \, \text{Pa} \qquad S_x = 104.9 \, \text{Pa}$$

- The speed also characterised by two parameters:

$$\bar{y} = 1.03 \, \text{mm/s} \quad S_y = 0.73 \, \text{mm/s}$$

Presume now, for the time being, that both variables are given precisely, i.e. without errors.

If so, the regression function of these variables can be found by applying the classical regression model.

This path of reasoning gives the following estimates of the unknown values of the structural parameters:

$$b_1 = 6.84 \times 10^{-3} \quad b_0 = 1.7 \times 10^{-3}$$

A graph of the function:

$$y(x) = b_1 x + b_0$$

is shown by a continuous line in Figure 6.18.

Let us make an estimation of the standard deviation of the random component ξ of the model:

$$S_\xi = \sqrt{\frac{1}{n-2} \sum_{i=1}^{n} \left(y_i - y(x_i) \right)^2} = 0.14 \, \text{mm/s}$$

Presume now that the variable x—the pressure—is estimated imprecisely and that the standard error of estimation is known, is stable and equals σ_u.

Make an estimation of the limit in the accuracy of the measurement.

The estimation of the empirical product moments is as follows:

$$m_{Xy} = 75.28 \qquad m_{yy} = 0.53 \qquad m_{XX} = 1.101 \times 10^4$$

and we are able to estimate the matrix of the product moments:

$$\begin{bmatrix} m_{yy} & m_{yX} \\ m_{Xy} & m_{XX} \end{bmatrix}$$

Taking into account relationship (6.63), we have

$$\sigma_u \leq 20.18 \text{ Pa}$$

which represents 13.4% of the average value of the variable X.

Assume for the purposes of our analysis that the standard deviation that determines the accuracy of the measurement of the variable x equals: 4; 8 and 12% of the average value of X.

The results of estimations of the structural parameters of the regression function and the standard deviations are presented in the table below where:

b_{11}—The estimation of the structural parameter standing by the variable X
b_{00}—The estimation of the free term
S_x—The estimation of the standard deviation of the variable x
S_ξ—The estimation of the standard deviation of the random component ξ
S_v—The estimation of the standard deviation of the random variable v.

σ_u	b_{11}	b_{00}	S_x	S_ξ	S_v
formula	(6.60)	(6.62)	(6.59)	(6.61)	(6.66)
Pa	10^{-3} mm/s Pa	mm/s	Pa	mm/s	mm/s
$0.04\,\bar{X}$	6.86	−0.002	104.76	0.13	0.14
$0.08\,\bar{X}$	6.93	−0.01	104.23	0.11	0.15
$0.12\,\bar{X}$	7.05	−0.03	103.35	0.06	0.15

The plots of the function for three analysed cases are shown Figure 6.19.

Let us analyse the most important information contained in this figure.

Firstly, the value of parameter b_{11} changes; its value increases with an increment in the measurement error, which causes the second parameter to decrease.

Secondly, when the accuracy of the measurement worsens, the regression function performs a rotation around a certain point. This point is determined by the average values of both variables.

It follows that any analysis or prediction of the variable values that are being explained that are not much different than the mean value will be correct even when the measurement error is larger. For values that are far from the average, the likelihood of statistical inference gets worse.

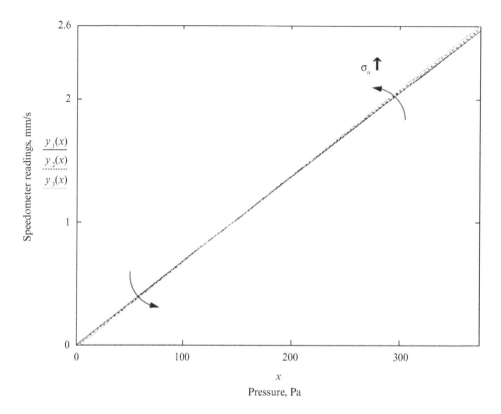

Figure 6.19. Regression function: the speed y of propagation vs. the pressure x with measurement errors for three levels of accuracy; $y_1(x)$ for $\sigma_u = 0.04\bar{X}$, $y_2(x)$ for $\sigma_u = 0.08\bar{X}$ and $y_3(x)$ for $\sigma_u = 0.12\bar{X}$.

A slight decrease of the standard deviation S_x that increases the measurement error is obvious taking into account equation (6.59). Along with an increase in the size of the error there is a decrease in the importance of the explanatory variable.

Similarly, as in the classical regression analysis, an investigation can be made to check the significance of the regression that has been done, i.e. verification of the hypothesis that states that parameter β_1 is zero, H_0: $\beta_1 = 0$. However, the verification procedure is more subtle than the previous one.

It can be proved (Fuller 2006) that the statistic

$$t = \frac{\hat{\beta}_1 - \beta_1}{\sqrt{\hat{V}(\hat{\beta}_1)}} \qquad (6.69)$$

has approximately the standardised normal distribution $N(0,1)$. However, it is recommended that this variable be treated as a variable of the Student's distribution with $n - 2$ degrees of freedom. This approximation is fully justified (ibidem) when the following condition is fulfilled:

$$\frac{\sigma_u^4}{(n-1)\left(m_{XX} - \sigma_u^2\right)} < 0.001 \qquad (6.70)$$

This means that it is preferred when the sample size is large and the variance of the random variable *u* small. If this does not hold, any interference based on statistic (6.69) should be treated as one with a reduced likelihood.

Notice that so far the consideration does not deal with the issue of the latent variable *x*. A case often occurs in which we are interested in finding this variable. And here two cases should be considered.

Looking at equation (6.48), it can be assumed that:

a. The latent variable is deterministic
b. The latent variable is random.

Analysis of both cases will lead to the construction of two different procedures.
Consider case (a).
Let us change our notation a little. The model can be described as:

$$\begin{bmatrix} y_i - \beta_0 \\ X_i \end{bmatrix} = \begin{bmatrix} \beta_1 \\ 1 \end{bmatrix} x_i + \begin{bmatrix} \xi_i \\ u_i \end{bmatrix} \tag{6.71}$$

Note, that the values β_0, β_1, σ_ξ and σ_u are known.

Equation (6.71) is the classical linear regression model in which it can be assumed that x_i is the unknown parameter that is being estimated. Using the formulas that were developed on the basis of regression theory, we can state that the best unbiased estimator for the latent variable in a general case is given by the expression:

$$\ddot{x}_i = [(\hat{\beta}_1, 1) \Sigma_\varepsilon^{-1} (\hat{\beta}_1, 1)']^{-1} (\hat{\beta}_1, 1) \Sigma_\varepsilon^{-1} (y_i - \hat{\beta}_0, X_i)' \tag{6.72}$$

where Σ_ε is the variance-covariance matrix of the two-dimensional random variable $\varepsilon_i = (\xi_i, u_i)$.

If the matrix Σ_ε is a diagonal[20] one $\Sigma_\varepsilon = diag(\sigma_\xi'^2, \sigma_u^2)$, then equation (6.72) is reduced:

$$\ddot{x}_i = \frac{\hat{\beta}_1(y_i - \hat{\beta}_0)\sigma_u^2 + X_i \sigma_\xi^2}{\hat{\beta}_1^2 \sigma_u^2 + \sigma_\xi^2} \tag{6.72a}$$

The variance of estimator (6.72) is determined by the formula:

$$V(\ddot{x}_i - x_i) = [(\hat{\beta}_1, 1) \Sigma_\varepsilon^{-1} (\hat{\beta}_1, 1)']^{-1} \tag{6.73}$$

whereas estimator (6.72a) has a variance that is given by:

$$V(\ddot{x}_i - x_i) = \left[\frac{\hat{\beta}_1^2}{\sigma_\xi^2} + \frac{1}{\sigma_u^2} \right]^{-1} \tag{67.3a}$$

Consider case (b)—the latent variable has a stochastic character.

The number of random variables of the model increased. The variance-covariance matrix now has the form:

---

[20] Recall, a diagonal matrix is a matrix in which the entries outside the main diagonal are all zero. The diagonal entries themselves may or may not be zero.

$$\begin{bmatrix} \beta_1^2\sigma_x^2 + \sigma_\xi^2 & \beta_1\sigma_x^2 + \sigma_{\xi u}^2 & \beta_1\sigma_x^2 \\ \beta_1\sigma_x^2 + \sigma_{\xi u}^2 & \sigma_x^2 + \sigma_u^2 & \sigma_x^2 \\ \beta_1\sigma_x^2 & \sigma_x^2 & \sigma_x^2 \end{bmatrix} \tag{6.74}$$

The unbiased estimator of the latent random variable is the statistic:

$$\tilde{X}_i = \tilde{\gamma}_0 + \tilde{\gamma}_1 y_i + \tilde{\gamma}_2 X_i \tag{6.75}$$

where:

$$\tilde{\gamma}_0 = (1 - \tilde{\gamma}_2)\bar{X} - \tilde{\gamma}_1\bar{y} \tag{6.76}$$

$$(\tilde{\gamma}_1, \tilde{\gamma}_2)' = \mathbf{m}_{zz}^{-1}[\mathbf{m}_{zx} - (\sigma_{\xi u}, \sigma_{uu})'] \tag{6.77}$$

$$\mathbf{m}_{zz} = \frac{1}{n-1}\sum_{i=1}^{n}(\mathbf{Z}_i - \mathbf{Z}_{\acute{s}r})(\mathbf{Z}_i - \mathbf{Z}_{\acute{s}r}) \tag{6.78}$$

$$\mathbf{z}_i = (y_i, X_i) \quad \mathbf{z}_{\acute{s}r} = (\bar{y}, \bar{X}) \tag{6.79}$$

Let us write the equation so that the estimations of the parameters $\tilde{\gamma}_1$ and $\tilde{\gamma}_2$ are achieved in the following form:

$$\begin{bmatrix} \tilde{\gamma}_1 \\ \tilde{\gamma}_2 \end{bmatrix} = \begin{bmatrix} m_{yy} & m_{yX} \\ m_{Xy} & m_{XX} \end{bmatrix}^{-1} \begin{bmatrix} m_{yX} \\ \sigma_x^2 \end{bmatrix} \tag{6.80}$$

Construction of estimator (6.75) in the explicit form can be achieved based on equations (6.76) and (6.80).

The variance of the estimator (6.75) can be assessed by applying formula:

$$\hat{V}(\tilde{x}_i - x_i \mid \mathbf{Z}_i) \cong \sigma_u^2 - (\sigma_{\xi u}^2, \sigma_u^2)\mathbf{m}_{zz}^{-1}(\sigma_{\xi u}^2, \sigma_u^2)' \tag{6.81}$$

□ **Example 6.4 (cont.)**

Making use of the information obtained earlier during the analysis in example 6.4 let us find the estimator for the latent variable assuming $\sigma_u = 0.12\bar{X}$.

Presume that the latent variable is deterministic.
Using formula (6.72a), we have:

$$\ddot{x}_i = 114.27 y_i + 0.195 X_i + 3.428$$

Presume now that the latent variable is random.
Making use of formula (6.80) and (6.76) we have:

$$\tilde{x}_i = 147.13 y_i - 0.036 X_i + 4.296$$

As you can see, the formulas are slightly different.

Now look at how the values of the variable x look depending on whether the mine records are treated as:

1. Correct and without errors
2. Deterministic but with random errors with a known variance
3. Randomly inaccurate.

In Figure 6.20 in which, in order to be more communicative, the values of each set are connected by straight lines so that all three sets of data are visible.

The information contained in this figure comprises:

1. The results of measurements; these outcomes are treated as correct and without errors
2. The calculated values of the pressure that correspond with these outcomes assuming that the measurements have random errors that are normally distributed with a zero expected value and a finite constant variance
3. The calculated values of the pressure assuming that the measurements are randomly imperfect.

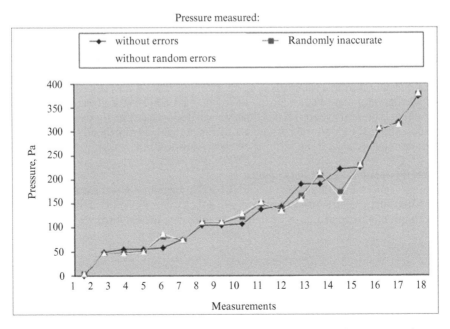

Figure 6.20. Pressure in consecutive measurements depending on how the data are treated.

Some comments on the information contained in Figure 6.20:

- All three courses of the pressure are similar
- The data gathered at the mines represent a kind of point of reference although the statement that the data are correct and without errors raises objections
- Both courses with random components are very similar
- By analysing both sets of data with random components more comprehensively, it can be stated that greater dispersion is connected with the information when three random variables are considered (the greatest number of stochastic components amongst the models that were considered)
- It should be taken into account that the limitation in measurement imperfection is rather sharp, thus the differences in the courses of the latent variable are small
- By analysing which assumption on randomness is closer to the reality in the mine, it looks as though the postulation on the randomly imperfect measurement of the explanatory variable is closer to reality in the mine.   ◄

6.7   LINEAR REGRESSION WITH ADDITIONAL INFORMATION

The idea of this regression relies on taking an advantage of any supplementary information that the researcher sometimes has—additional in the sense that it is information outside the sample. This information can be of a different nature and it can come from different sources. It can originate from:

- Theoretical considerations
- Information from earlier investigations and analyses
- Previously taken samples.

In engineering practice we often make use of information from research that had been conducted earlier. Therefore, we are able to formulate some conditions that have to be fulfilled by the structural parameters of the regression function. In some cases these conditions are determined by strict relationships; in other cases the conditions can have a different form, e.g. some of the parameters are defined only over certain intervals.

The basic purpose of this type of regression is to find the increment in the efficiency of the statistical inference by using this *a priori* information. In some cases the Bayesian approach is applied; information from a sample is coupled with information that comes from outside the sample. We are not going to consider the Bayesian approach here; it goes rather beyond the basic level that is presented here. At this time, we will discuss a few methods that allow *a priori* information about the structural parameters of the model being analysed to be used during their estimation within the framework of classical statistical inference.

Additional information—exact linear limits
In this model one assumption is that the additional information has the form of linear relationships that link the structural parameters together—all of them or only some of them. The information in hand can be expressed in the following matrix form:

$$g = G\,B \tag{6.82}$$

where: g—a known vector of size $(J \times 1)$, $J \le K$
   G—a known matrix of size $(J \times K)$
   B—the vector of the structural function parameters.

The information that the number of additional conditions is J is contained in equation (6.82) and obviously this number does not exceed the number of structural parameters.

These conditions can have different forms, e.g.

a. Some values of the parameters are known, for instance

$$\beta_2 = \beta_2^* \qquad \text{and then} \qquad (6.83)$$

$$g = [\beta_2^*] \qquad G = [0\ 1\ 0\ \dots\ 0], \qquad (6.83a)$$

b. Some ratios of the parameters are known, for instance

$$\frac{\beta_1}{\beta_2} = f_1 \quad \frac{\beta_3}{\beta_2} = f_2 \qquad \text{and then} \qquad (6.84)$$

$$g = \begin{bmatrix} 0 \\ 0 \end{bmatrix} \qquad G = \begin{bmatrix} 1 & -f_1 & 0 & 0 & \dots & 0 \\ 0 & -f_2 & 1 & 0 & \dots & 0 \end{bmatrix} \qquad (6.84a)$$

c. Some linear combinations of the parameters are known, for instance

$$\sum_{i=1}^{K} \beta_i = 1 \qquad \text{and then} \qquad (6.85)$$

$$g = [1] \qquad G = [1\ 1\ 1\ \dots\ 1] \qquad (6.85a)$$

Let us pay more attention to the condition above. In some cases not all outcomes contained in the sample are treated in the same way during the estimation; a case may occur in which some observations are more important or have a greater likelihood than the others. In such a case weights are usually assigned to each outcome according to some presumed principle. The sum of these weights is close to unity. Such an attitude is regularly applied when an estimation of the abundance of deposits, for example gold, is carried out. This method is called kriging and relies on assigning appropriate weights to particular measurement points located in the area being estimated in order to minimise the mean squares error of estimation. In this technique greater weights are allocated to the points that lie near the testing point. This method gives the best unbiased linear estimations of the variable being analysed[21].

It can be proved (Goldberger 1966) that estimator b_1 of the unknown values of parameters B can be determined by the formula:

$$b_1 = b + (X'X)^{-1}G' [G\ (X'X)^{-1}G']^{-1}(g - Gb) \qquad (6.86)$$

The estimator is unbiased. It is easy to see that this statistic differs from estimator b that is obtained from the classical regression by the expression that is the linear function $(g - Gb)$, which means that the estimator without additional information does not fulfil the defined conditions.

---

[21] For more on this topic see Hustrulid and Kuchta (2006, Chapter 3.10).

The matrix of variance–covariance is given by the pattern:

$$V_1 = V - VG'(GVG')^{-1} GV \tag{6.87}$$

where $V = \sigma^2(X'X)^{-1}$ and it is the variance–covariance matrix (6.43).

It is worth noting that we have achieved a profit in effectiveness. Matrix V is the positively defined matrix. The second matrix on the right side of equation (6.87) is the non-negative defined matrix. This means that every diagonal element of matrix V_1 is smaller or equal to the corresponding element of matrix V. Thus, the variance of each coordinate of vector b_1 is not greater than the variance of the corresponding variance of the coordinate of b.

Now define two vectors of the residuals:

$$u_1 = Y - X b_1 \tag{6.88}$$

$$u_2 = X(b - b_1) \tag{6.89}$$

The first vector is the vector of the differences between the observed values of the variable being explained and the corresponding theoretical values that are obtained from the estimation of the structural model parameters using the conditional method.

The second vector is the vector of the differences between the theoretical values of the variable being explained that are obtained from the estimation of the structural model parameters using the unconditional method and the theoretical values obtained by applying the conditional method.

It was proved (Dziembała 1972) that the residual variance of variable u_1 is given by the formula:

$$S_{u_1}^2 = \frac{1}{n-(K-J)} u_1' u_1 = \frac{1}{n-(K-J)} (u'u + u_2' u_2) \tag{6.90}$$

The first component of the sum is known as the residual variance which is obtained from the estimation of the parameters using the unconditional least squares method, whereas the second component is the so-called differentiating variance. The root square of the variance gives information about how the theoretical values differ from each other on average; the theoretical values obtained from the unconditional method and the corresponding theoretical values obtained from the conditional method.

The following sum should be considered as the second basic measure of the stochastic structure:

$$\varphi^2 + \varphi_1^2 = \frac{\sum_{t=1}^{n}\left(y_t - y_t^{(t)}\right)^2}{\sum_{t=1}^{n}\left(y_t - \bar{y}\right)^2} + \frac{\sum_{t=1}^{n}\left(y_t^{(t)} - y_t^{(tw)}\right)^2}{\sum_{t=1}^{n}\left(y_t^{(tw)} - \bar{y}^{(tw)}\right)^2} \tag{6.91}$$

where $\bar{y}^{(tw)}$ is the arithmetic mean of the theoretical values that are obtained from the conditional method.

The first component of the sum is known as the goodness-of-fit factor (6.45), while the second component is the so-called inconsistency factor. This is supported on the [0, 1] interval and it informs which part of the variance of the variable is the differentiating variance. The closer the value φ_1^2 to zero, the smaller the difference between estimators b and b_1.

Dziembała proposed (ibidem) that the following factor be analysed additionally:

$$r_{d1} = \frac{\sum_{t=1}^{n}(y_t - \bar{y})\left(y_t^{(t)} - y_t^{(tw)}\right)}{\sqrt{\sum_{t=1}^{n}(y_t - \bar{y})^2 \sum_{t=1}^{n}\left(y_t^{(t)} - y_t^{(tw)}\right)}} \tag{6.92}$$

It is a kind of factor to measure the goodness-of-fit of the proposed model. It is supported on the same interval $[0, 1]$ and for $r_{d1} = 1$, we have the ideal compatibility between the empirical and theoretical values. When r_{d1} decreases, the goodness-of-fit also decreases.

Sometimes, in some practical cases, a slightly different approach can be more advantageous. This relies on first using the conditions at hand to eliminate some structural parameters and second to applying the least squares method to the reduced model that is obtained. At the final stage of analysis, we should return to the conditions in order to estimate the rest of the parameters. For example, presume that the model of interest is:

$$y = \beta_1 x_1 + \beta_2 x_2 + \varepsilon$$

provided that

$$\beta_1 + 2\beta_2 = 1$$

Substituting we have:

$$y = (1 - 2\beta_2)x_1 + \beta_2 x_2 + \varepsilon$$

and further

$$y - x_1 = \beta_2(x_2 - 2x_1) + \varepsilon$$

This equation can be written as

$$y^* = \beta_2 x^* + \varepsilon$$

By applying the least squares method, one obtains an estimate b_2^* and next we make use of the condition to obtain

$$b_1^* = 1 - 2b_2^*$$

Additional information—estimates of structural parameters

Assume now that the additional information is in the form of the unbiased estimators of some structural parameters of the linear model. Such a situation occurs, for instance, when there is a continuation of a previous investigation.

Consider the relationship as divided into blocks:

$$Y = X_1 B_1 + X_2 B_2 + E \tag{6.93}$$

Presume that vector b_1^*, which is the unbiased estimator of vector B_1, is known, i.e.

$$b_1^* = B_1 + h \tag{6.94}$$

where h is the vector of size $(p \times 1)$; $1 \le p \le K$ and it is the vector of the random errors of the estimates components of vector b_1^*.

Assume also that

$$E(h) = o \quad E(h\ E') = o \tag{6.95}$$

Information from the sample taken and the additional information can be connected and therefore the following equation can be written:

$$\begin{bmatrix} y \\ b_1^* \end{bmatrix} = \begin{bmatrix} X \\ G \end{bmatrix} B + \begin{bmatrix} E \\ h \end{bmatrix} \tag{6.96}$$

where $G = [I\ O]$.

The variance-covariance matrix of the 'enlarged' random component is:

$$E \begin{bmatrix} E \\ h \end{bmatrix} [E'\ h'] = \begin{bmatrix} \sigma^2 I & O \\ O & W \end{bmatrix} \tag{6.97}$$

where

$$W = E(h\ h') \tag{6.97a}$$

Knowing vector $W = E(h\ h')$, it can be proved (Goldberger 1966) that the best unbiased estimator of vector B is the statistic:

$$b^{**} = (\sigma^{-2} X'X + G'W^{-1}G)^{-1} (\sigma^{-2} X'y + G'W^{-1} b_1^*) \tag{6.98}$$

Consider the following example. Assume that vector B_1 contains only one element, which means that $G = [1\ 0\ ...\ 0]$ (compare formula (6.83a)). In such a case the matrix W is a scalar. Denote it as $W = \sigma_w^2$.

Calculating we have:

$$G'W^{-1}G = \frac{1}{\sigma_w^2} \begin{bmatrix} 1 & 0 & ... & 0 \\ 0 & 0 & ... & 0 \\ . & . & . & . \\ . & . & . & . \\ 0 & 0 & ... & 0 \end{bmatrix} \quad \text{and} \quad G'W^{-1}b_1^* = \frac{1}{\sigma_w^2} \begin{bmatrix} b_1^* \\ 0 \\ . \\ . \\ 0 \end{bmatrix} \tag{6.99}$$

The normal equations corresponding with this case can be shown as:

$$\begin{bmatrix} \Sigma x_1^2 + \varphi & \Sigma x_1 x_2 & ... & \Sigma x_1 x_K \\ \Sigma x_2 x_1 & \Sigma x_2^2 & ... & \Sigma x_2 x_K \\ . & . & . & . \\ . & . & . & . \\ \Sigma x_K x_1 & \Sigma x_K x_2 & ... & \Sigma x_K^2 \end{bmatrix} \begin{bmatrix} b_1^{**} \\ b_2^{**} \\ . \\ . \\ b_K^{**} \end{bmatrix} = \begin{bmatrix} \Sigma x_1 y + \varphi b_1^* \\ \Sigma x_2 y \\ . \\ . \\ \Sigma x_K y \end{bmatrix} \tag{6.100}$$

whereas

$$\varphi = \frac{\sigma^2}{\sigma_w^2} \tag{6.101}$$

The only difference between the above set of equations and the set of the normal equations that are obtained from the least squares method is:

– The first entry upper left in the matrix $x'x$
– The first upper entry in the vector $x'y$.

The method just considered requires accurate knowledge about variance σ^2 and the matrix w to the proportional coefficient. If these parameters are unknown, their estimation can be used, however the solutions obtained will only be approximate. For example, by applying the unbiased estimator of variance, σ^2 can be obtained using the least squares method (without any conditions) and the unbiased estimator of w can be achieved from the regression in which b_1^* was determined[22].

Intuitively, it can be stated that using conditional estimators makes sense only if their variances are small. If, however, their dispersions are high, then this *a priori* information is poor and it does not mean much in the interference. Therefore, it can be omitted.

▪ Example 6.6

Data for this example are taken from Ścieszka's investigation (1971) and his dissertation (1972).

In order to identify factors that have a significant influence on the wear process running in the brake linings of winders, the following function was tested as the theoretical model:

$$\boldsymbol{I}_g = kp^a v^b e^{\xi}$$

where: I_g—the intensity of mass wear
  p—the unit pressure on brake lining
  v—the sliding speed
  k, a, b—the structural parameters of the model
  ξ—the random component.

---

[22] For more on this topic see, for instance, Goldberger 1966.

This function was linearised:

$$\ln I_g = \ln k + a \ln p + b \ln v + \xi$$

and becomes a recipe for a linear function of the form:

$$Y = a_0 + a_1 X_1 + a_2 X_2 + u$$

The investigation of the wear process was carried out in the stand at the Mining Faculty. The data were gathered and natural logarithms were calculated from which the following sets of outcomes were obtained:

$$
y = \begin{bmatrix}
2.0669 \\
2.5096 \\
2.6927 \\
3.0421 \\
2.1747 \\
2.7973 \\
3.1046 \\
3.3656 \\
2.4723 \\
3.0445 \\
3.6689 \\
4.0490 \\
3.2308 \\
3.7171 \\
4.3624 \\
4.7741
\end{bmatrix}
\qquad
x = \begin{bmatrix}
5.9130 & 1.6094 \\
6.6063 & 1.6094 \\
7.0116 & 1.6094 \\
7.2608 & 1.6094 \\
5.9130 & 2.3026 \\
6.6063 & 2.3026 \\
7.0116 & 2.3026 \\
7.2608 & 2.3026 \\
5.9130 & 2.9957 \\
6.6063 & 2.9957 \\
7.0116 & 2.9957 \\
7.2608 & 2.9957 \\
5.9130 & 3.6889 \\
6.6063 & 3.6889 \\
7.0116 & 3.6889 \\
7.2608 & 3.6889
\end{bmatrix}
$$

By making an appraisal of the structural model parameters using the least squares method, the following estimates were achieved (pattern (6.37)):

$$
a = \begin{bmatrix} a_1 \\ a_2 \\ a_0 \end{bmatrix} = \begin{bmatrix} 0.9616 \\ 0.6893 \\ -5.0746 \end{bmatrix}
$$

The residual variance (pattern (6.44)) is:

$$S_u^2 = 0.0363$$

The variance-covariance matrix (pattern (6.43)) is:

$$
v = \begin{bmatrix}
8.728 \times 10^{-3} & 0 & -0.058 \\
0 & 3.778 \times 10^{-3} & -0.01 \\
-0.058 & -0.01 & 0.42
\end{bmatrix}
$$

The goodness of fit factor (pattern (6.45)):

$$\varphi^2 = 0.0532$$

and the multiple correlation coefficient

$$R = \sqrt{1 - \varphi^2} = 0.973$$

This value is high and looks substantial. Let us check it statistically. A hypothesis is formulated that states that there is no significant correlation between the random variable being explained and the explanatory variables, H_0: $\rho = 0$ against and alternative hypothesis is that rejects it.

Presume a level of significance in our reasoning $\alpha = 0.05$. We have the sample size $n = 16$ and the number of variable $K = 3$. Thus, the critical value for the multiple correlation coefficient is (Table 9.15) 0.608. The empirical value is high and is above the critical value. There is a ground to reject the basic hypothesis. The model can be accepted.

Notice that the analysis presented here is carried out in the area of linearised function, not the original function.

A year before this investigation a similar examination had been carried out (Ścieszka 1971) in which the main point of interest was sliding speed and the form of the function was also a power one. At the end of the investigation, the estimate of the power exponent obtained whose value was 0.6508 and the variance equalled 0.0615.

Looking at both estimates of this power exponent, it looks as though they are similar. Let us check how our assessment will change if we include the information from the previous investigation into our procedure.

Consider formula (6.99). Construction of both patterns takes into account information about the first structural parameter; in our case, it is the second parameter. So, rearranging both patterns appropriately and using formula (6.98), we have:

$$\mathbf{a}^* = \begin{bmatrix} a_0^* \\ a_1^* \\ a_2^* \end{bmatrix} = \begin{bmatrix} 0.9616 \\ 0.6871 \\ -5.0687 \end{bmatrix}$$

By comparing both sets of estimates of the structural parameters, we observe that the differences are very small, as was expected.

Let us make an assessment of the residual variance. By applying formula (6.90), we get:

$$\overline{\overline{S}}_u^2 = 0.0337$$

The accuracy of our inference increases somewhat. However, this increment does not change the estimate in the multiple correlation coefficient (a change is observed for further decimal places). ◄

CHAPTER 7

Special topic: Prediction

7.1 INTRODUCTION AND BASIC TERMS

It is a good practice that the basic terms, together with their definitions, which are associated with this field are presented before any study of problems from a given area. Usually such an approach makes the considerations that are conducted unambiguous and communicative. But in some cases, although rare, this principle fails.

Before we clarify such terms as prediction, prognosis, predictor and a few other expressions that are connected with forecasting, let us commence our consideration from the term 'anticipation'.

Anticipation (prescience) is an inference about unknown events that is based on events that have happened and that are known. These unknown events can be located in the future; they can also be located in past, but the main feature is that they are unknown (unidentified), e.g. evaluating the abundance of a mineral deposit through test drillings.

An inference on events that will happen in the future that is based on information from the past is called **predicting** (forecasting) **the future**. This act of predicting can be rational or irrational.

A rational forecast is when the inference is based on a logical process that runs from premises, i.e. from a set of facts that belong to the past along with their proper interpretation towards their conclusions.

We call this a scientific prediction on future events when the process of the inference is based on the rules of science.

Among the problems that are connected with a general prediction[1] of the future is a subarea that is associated with the use of logical tools for forecasting, and this scope is called **scientific forecasting** or a **scientific prediction**. It is characterised by an approach that that is based on a research process that comprises learning about the past, i.e. gathering the data and diagnosing them by applying an appropriate method—appropriate in the sense that the model possesses the property of using the data concerning the past to infer future events. This property is usually expressed by a suitable mathematical component in the model being used that indicates that the inference concerns the future. Moreover, this component should change with time because usually the further into the future that we are discussing, the less precise is our inference[2]. However, in some engineering areas a degenerated approach to prediction is presented that relies on the assumption that what has happened in the past will be repeated unchanged in the future. And then it is not necessary to have any component that is associated with forecasting, and the term 'prognosis' is also not required. In such a case, it is hard to call it a prediction, and only modelling remains[3]. Such an approach can often be found, for instance in reliability books (see for instance Dovitch (1990), Smith (2007), O'Connor (2005)); however the term 'prediction' is there used.

---

[1] Cramér said 'prediction is the practical aim of any form of science' (1999, p. 339).
[2] We ignore here the precision of displacement of—for example—planets; the prediction horizon can be extended very far and the precision of the inference will still be high.
[3] O'Connor writes openly: 'reliability prediction, i.e. modelling' (2005, p. 75).

An essential term that is associated with prediction is the word '**prognosis**'.
Let us cite some of the scientific definitions of this term[4].

'Any judgement in which truthfulness is a random event of known and high enough probability' (Hellwig 1963, Cieślak 2001, p. 20).
'Future value of a stochastic process' (Gichman and Skorochod 1965).
'A function' (Goldberger 1966).
'Prognosis is a number' (Benjamin and Cornell 1970).
'It is a particular (numerical) result of an inference process of the future' (Pawłowski 1973).
'A random variable' (Czaplicki 1976).
'A judgment of unknown states of physical objects' (Rybicki 1976).
'A specific result of prediction' (Greń 1978).
'Prognosis is a state of the predicted variable belonging to the future' (Cieślak 2001, p. 37).
'Prognoses are made by means of models, judgements, and models and judgements, and they are final product of the forecasting process' (Kasiewicz 2005).
'Estimated future value of the sequence' (Bielińska 2007).

There is no doubt that a mining engineer may feel lost in this area; however, this diversity of definitions can be partly justified by the multiplicity of prognostic situations, the different goals of prediction and the variety of methods applied.

A typical engineering situation in which problems of prediction have been considered in many specialised articles and conference papers since the early 1970s is reliability although here the situation is no better when a general approach is concerned. Melchers (1999) in his extensive book that had the word 'prediction' in the title used this term only twice in the whole book. In books concerning reliability in mining engineering (Dhillon 2008 and Czaplicki 2010a), the problem of prediction does not occur; in many reliability books, prediction in fact only means modelling.

There is also a problem of how to understand a formula that allows a prognosis to be obtained. This 'recipe' is termed a '**predictor**'.[5]

'A quantity giving a prognosis is a stochastic process' (Gichman and Skorochod 1965).
'A quantity giving a prognosis is a random variable' (ibid em).
'It is a function' (Benjamin and Cornell 1970).
'Operator, function' (Greń 1978).
'Functional' (Pawłowski 1973).
'Stochastic functional' (Czaplicki 1976).
'Random variable, functional, element or transformation' (Rybicki 1976).

What a wealth of approaches, concepts and ideas. However, an engineer likes to move on the solid ground of exact sciences.

Thus, let us now create some order in the material above.

1. Not wanting to formulate another definition of prognosis and prediction, let us presume the following statements for our further study:

    • Prediction is the process of inference on an unknown current state or future state of the external world
    • The result of this process of inference is prognosis
    • Prognosis can be a number, a function (in a particular random variable or stochastic process); it can also be a matrix (deterministic or stochastic)

---

[4] On the Web the word 'prognosis' is frequently associated with medicine.
[5] Sometimes predictor means a formula for determining additional values or derivatives of a function from the relationship of its given values or is a term used for an independent variable.

2. A recipe that is used in prediction is every functional that ascribes a prognosis to each model that is applied to define the course of the variable being predicted
3. Each procedure in prognostic inference is based on some assumptions; two main ones are[6]:

   a. There is information on the course of the magnitude that will be predicted; this information concerns its past, or it can be current information and it is in fact a sample in a statistical sense
   b. The properties of this course are stable overtime

4. Each prognosis should be calculated together with the appropriate measure of the accuracy of the prediction (this is a so-called basic postulate).

An analytical description of a **predictor** $\mathfrak{I}$ can be expressed as:

$$\mathfrak{I} = \Phi_T[G(Y)] \tag{7.1}$$

where: Φ—operation that should be performed to get a prognosis
   G—model describing the course of the variable that is being predicted
   Y—the variable that is being predicted
   T—the period for which the prediction is being done.

Looking more carefully at formula (7.1), two additional terms should be determined.

An approach to the construction of a prognosis that relies on selection of a form of operator Φ will be called the **principle of prediction**.

The application of a particular principle of prediction to one class of the models that describe the course of the variable being forecasted will be called the **method of prediction**.

In econometrics one method of prediction, which relies on determination of the expected value of the variable being predicted, is very often applied. This principle is based on the reasoning that errors in plus and errors in minus create a balance and no systematic error[7] will be made. Consider, for example, a model that describes the course of variable Y that is given by the formula:

$$Y(t) = \sum_{i=1}^{k} \alpha_i X_i(t) + \xi_t$$

and we are interested in obtaining the prognosis for variable $Y(t = T)$ presuming an unbiased principle of prediction. If so, expression (7.1) takes the form:

$$\Phi_T\{G[Y(T)]\} = E\left\{\sum_{i=1}^{k} \alpha_i X_i(T) + \xi_{t=T}\right\} = \sum_{i=1}^{k} \alpha_i E[X_i(T)] + E(\xi_{t=T})$$

If the least squares method was applied to estimate the structural parameters, the random variable ξ will have the normal distribution of zero expected value. If, additionally, there is no autocorrelation in the random component, we have:

$$\sum_{i=1}^{k} \alpha_i E[X_i(T)] + E(\xi_{t=T}) = \sum_{i=1}^{k} \alpha_i E[X_i(T)]$$

Notice a certain subtleness. Usually, we don't know the exact values of the explanatory variables that will be taken by them in the future. Thus, we have to treat these variables as

---

[6] In some particular cases, this list is longer, e.g. there is knowledge on values of explanatory variables for the period being predicted (see for instance Pawłowski 1973).

[7] In prediction theory, we say an error of the predictor, error of the prediction procedure (sometimes called error of prediction for short) and error of prognosis. The first one has a theoretical meaning only, the second one is connected with the prediction method that is applied, and the last one is the difference between the prognosis and the real value of the variable being investigated.

random and the above pattern is correct for this reason. Nevertheless, if these future values are known (from a theoretical consideration, for instance, or if we knowingly and intentionally presume these values), then the model used for prediction is simplified and the problem of forecasting becomes minor.

However, in engineering considerations an approach that is based on expected values is correct in some cases but in some other cases it is not correct. An error of prognosis, for instance, that relies on underestimating the wear magnitude of a hoist head rope operating in a mine shaft can have serious repercussions compared to overestimating this rope wear. A long list of similar examples can be enumerated in mining engineering[8].

In prediction theory there is a well-known principle that if a prognosis can only be done once and will not be repeated, it is better to select a mode of the variable being forecasted because it has the greatest chances of occurring.

In some cases weights are given to any result for which overestimating or underestimating the variable forecasted will probably occur and these weights are important components of the prediction procedure.

In some other cases, it is better not to consider the point estimation (prediction) of the variable of interest but a prediction interval paying special attention to any area that is connected with a possible crossing over of the limited value by the forecasted variable.

In a case in which the postulates associated with the prediction are concerned, often an additional postulate is formulated that concerns the desirability of the efforts necessary to achieve a high degree of effectiveness of the prediction (see for instance Pawłowski 1973); however, it is not entirely correct. It was proved that when an optimal decision is being taken, a prognosis that is less accurate can be more useful; one with a lower likelihood (Winters 1960). It was also proved that sometimes a situation can happen in which the use of the prognosis that was obtained is useless or can even be adverse (Sadowski 1977).

7.2 SUBJECT OF PREDICTION

Taking into account a *meritum* of the consideration of the problems of prediction in the engineering world, problems of prediction can be divided into two groups:

- Forecasting of future realisations of random variables, stochastic processes and (rarely) random fields[9,10]
- Forecasting of future realisations of the parameters and characteristics of random variables, stochastic processes and random fields.

Where mining engineering is concerned, the first group of problems are mainly connected with the diagnostic and operational issues of a technical object as well as with econometric issues.

In mine practice, an engineer must make a decision of whether to allow the further use of a hoist head rope in the main mine shaft by predicting the state (degree of wear) of this rope in the near future (at least until the date of the next inspections). The process of rope wear has a stochastic character. When forecasting the future surface subsidence, an engineer must make a decision about whether to continue the current underground extraction process or to change it significantly. The process of surface deformation has a random character. When estimating the course of the commodity price of a mineral on the world market, a mining

---

[8] Errors in prediction are generally divided into *ex-ante* and *ex-post* if the criterion for the division is time.

[9] However, sometimes, we are interested in a single number only.

[10] A method of prediction of non-stationary–stationary stochastic field was presented for example in Czaplicki's article (1977a).

engineer must make a decision to continue the operation of a mine or to change its course significantly (to speedup or to slow down). These changes have a stochastic nature. When observing the development of a pit, an engineer may come to the conclusion that the removal of the overburden is too slow and that it threatens the level of production and that he must make a decision to accelerate removal of overburden in order to avoid economic losses; all based on a prediction. Thus, all of these problems can be considered on the ground of the extrapolation of stochastic processes; in some special cases—stochastic fields. Observe that in mining engineering prediction is usually directly associated with decision making, which as a rule has significant after-effects either in economics or in safety or both.

Sometimes tracking changes in economic magnitudes that are connected with a mining time series are obtained and their analysis, decomposition and the identification of particular components are difficult, time consuming and require good knowledge and proficiency in the statistical area. An example of such a procedure will be given in Example 7.1, which is a continuation of Example 4.8.

The second group of problems are mainly connected with reliability prediction. When observing the realisation of a given diagnostic parameter of a technical object, it is easy to come to the conclusion that its changes have a random character. Some parameters are so important that if they go above or below defined limits, they can have serious repercussions. Mining operations are often connected with safety problems and there is no way to ignore them.

Depending on the character of the observed variable whose value will be forecasted, we say it is a prediction of:

- Stochastic chains
- Time series
- Stochastic streams (fluxes)
- Processes of changes of states
- Continuous random processes
- Stochastic fields.

The prediction of stochastic chains is mainly used in forecasting any realisations of wear processes as well as in the renewal processes of technical objects and in predicting the occurrence of special events. A basic model here is the Markov chain.

By predicting the processes of the wear and renewal of technical objects and having some *a priori* information, it is possible to construct a prognosis (Koźniewska and Włodarczyk 1978) of:

- the predicted distribution of the number of technical objects of a given age interval at a given moment in time
- the predicted number of a given subpopulation at a given moment in time
- the predicted number of renewals at a given moment in time.

This type of prediction is a **passive** one. The term 'passive prognosis' usually means such a prognosis for which the future realisation of a variable is a direct consequence of the actions of a certain stochastic 'mechanism' and this action is stable over a certain period of time. In some cases, the point of interest is a so-called **active** prognosis that is based on a random process that is the source of information and that is accompanied by decisions to change the course of the process in the proper direction. An active prognosis helps a suitable decision to be made. Such a forecast is not a prognosis in the common sense (Koźniewska and Włodarczyk 1978).

When data are collected through discrete notations, then the information has the character of a time series. In such a case, the prediction is a procedure composed of several stages. The researcher who has the data selects the appropriate method of the extrapolation of the time series and then follows the steps that are associated with this method. Frequently, the core of consideration is either such a mathematical depiction of the course of the investigated variable that faithfully describes its path and that enables forecasting, or a decomposition of the time series is done before the forecast that gives the appropriate tool for prediction. An example of such decomposition was discussed in Example 3.8.

Actually, there are many methods for the prediction of a time series, which are mainly applied in econometrics (see for instance: Holt (1957)[11], Brown (1959), Hellwig (1963), Box and Jenkins (1976), Clements and Hendry (1998), Armstrong (2001), Chatfield (2001), Bowerman et al. (2005)). In some cases these methods are useful for mining engineering when the study concerns the future courses of the prices of mineral commodities on the world market, the demand for a given mineral commodity etc. The models that are applied in these cases are usually in the form of a classical mathematical description of trends that are accompanied by a conventional analysis of the statistical properties of the data. An example of such a way of prediction will be presented in the next subchapter.

In some particular cases when there is no possibility to apply such models and for which there is also no way to apply a cause-and-effect pattern because of many stochastically dependent variables, a model of autoregression is used, often with success. Obviously, memory in sequence should be observed. Such an approach to inference in the future will be shown in Example 7.2.

Problems with forecasting stochastic streams usually comprise issues such as:

– the prediction of the parameters and characteristics of impulse processes (impulse frequently means failure)
– the prediction of the parameters and characteristics of the times between impulses.

The most commonly investigated topics in this class of prediction problems are: the prediction of the number of failures in a given time period and the prediction of the time to the nearest failure (e.g. Cunningham et al. (1973), Czaplicki (1976)).

The basic information that is needed to study stochastic streams is the identification of:

– the essential properties of the stream being investigated (singularity or not, stationarity or not, the existence of memory or not)
– the distribution of the time between successive impulses
– the stream parameter (in some cases this can be a function)
– the distribution of the number of impulses that can appear at a given moment in time
– the stochastic relationships that exist between the above-mentioned stream characteristics.

In the theory of exploitation the term 'operation process of a technical object' is usually understood as the process of changes of states (Czaplicki 2010). Its past, current and future properties are of interest from the operational and reliability points of view. An interval prediction of parameters for the work-repair process of the Markov type will be given in Example 7.3 as an example of such an issue.

The area of prediction also includes issues that are connected with forecasting in durability investigations. The most interesting problems are: the prediction of the length of the investigation, the problem of how many failures will be recorded during the investigation and the application of the prediction to shorten the length of the investigation[12].

7.3 EXAMPLES

We will discuss three examples of prediction—three different models to describe the courses of the forecasted variables from three different areas of mining engineering. The first two

---

[11] See Gelper et al. 2008.
[12] See for instance Czaplicki (1980) and (1981a).

cases are taken from practice, the third, which is theoretical, concerns the reliability problems of single items and their systems whose process of changes of states is of the Markov type.

The first two studies were especially selected in order to show not a quick simple reasoning but a procedure in which many steps are needed and for which sophisticated statistical tools are sometimes required in order to achieve the final result. Such a scheme is closer to the reality of mining engineering and has a didactic nature.

■ **Example 7.1** (Continuation of **Example 3.8**; based on Manowska's dissertation 2010)

Our consideration is connected with the mass of hard coal that was sold in Poland around the end of the 20th century for which the final formula that describes the time series that was observed after its decomposition will be presented. This pattern was as follows:

$$y(t) = \hat{y}_S(t) + y_{C1}(t) + y_{C2}(t) + \xi_t$$

Continuing the previous analysis, a study of the pure random component ξ_t should now be made. The properties of this component should be identified by analysing the time series of the residuals (here: differences) that are determined by general pattern:

$$u(t) = y(t) - [\hat{y}_S(t) + y_{C1}(t) + y_{C2}(t)]$$

An analysis of the time series of residuals should comprise several points, namely:

a. Verification hypothesis to determine whether the series is a stationary one by applying, for instance, a test based on the Spearman's rank correlation coefficient
b. If the series is stationary, calculating the average value of the residuals
c. Verification of a hypothesis that states that this calculated mean is insignificantly different from zero[13]; e.g. by applying the test using the Student's t statistic
d. Calculation of the standard deviations of the residuals for the first and for the second half of the sample; verification of a hypothesis that states that the calculated values differ insignificantly from each other
e. If both standard deviations differ insignificantly from each other, calculation of the standard deviation of the residuals for the whole sample
f. Verification of the hypothesis that states that there is no autocorrelation between the sequences of the residual values.

Let us briefly discuss the above-listed points.

If the procedure of the decomposition of the time series was made properly, we can expect that the series of residuals is stationary. This gives meaning to point (b) and we can be certain that verification of the hypothesis that states that this calculated mean is not significantly different than zero gives a positive result, i.e. there is no systematic error in the deviations.

Point (d) is a more subtle point of the analysis being conducted. However, taking into account that the cyclic component that was traced consisted of two pieces, we may suspect that the dispersion of the residuals should be stable over time; however, this is only a supposition. This gives meaning to the next point (e).

The last but one point (f) concerning autocorrelation in the sequence of the residuals is free of any suggestion in this regard. The result of the investigation is unknown and the only solution is to apply the appropriate statistical test. The outcome of this study is important if a prediction of the time series is to be done.

---

[13] As a rule the least squares method is applied to estimate the systematic component. It should ensure that the average value of the residuals equals zero. If this value is different, it is usually only slightly different, which is a result of rounding during calculation.

Let us analyse these points one by one.

The sequence of the residuals in relation to time is presented in Figure 7.1.

Let us first check whether this sequence is independent of time. Formulate a hypothesis H_0 that states that there is no dependence of time versus the alternative hypothesis that rejects it.

By applying the test for stationarity based on the Spearman's rank correlation coefficient, we have (pattern (3.18) and (3.19)):

$$r_S = 1 - \frac{6 \times 594052}{160(160^2 - 1)} = 0.1298$$

Let us compare this empirical value with the critical one. By applying formula (3.23) we have:

$$r_S(\alpha = 0.05; n = 160) = \frac{u_{1-\alpha}}{\sqrt{n-1}} = \frac{1.645}{\sqrt{159}} = 0.1305$$

By comparing these two values, we see that the empirical one does not exceed the critical one, which means that there is no ground to reject the hypothesis H_0. Thus, we can conclude that the sequence of residuals is a stationary one. However, because both values are quite close to each other, the result obtained should be treated cautiously.

By calculating the average value of the residuals, we have:

$$\bar{u} = -41.5 \text{ tonnes}$$

Formulate a hypothesis H_0 that states that this value differs insignificantly from zero versus the alternative hypothesis that rejects it. Presume the level of significance $\alpha = 0.05$.

If it is presumed that the residuals can be satisfactorily described by the normal distribution, then the statistic that can be applied is from the formula:

$$t = \frac{\bar{u} - 0}{s_u} \sqrt{n}$$

where s_u is the estimate of the standard deviation of the residuals that has the Student's distribution with $n-1$ degrees of freedom provided that the basic hypothesis is true.

By calculating, we obtain:

$$t = \frac{\bar{u} - 0}{s_u} \sqrt{n} = \frac{-41.5}{731} \sqrt{160} = -0.718$$

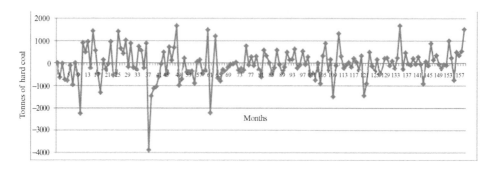

Figure 7.1.   The sequence of residuals versus time.

In Table 9.3 we look for such a value for the presumed level of significance α that corresponds with the sample size $n - 1 = 159$. We have: $t(\alpha = 0.05; n - 1 = 159) = 1.98$. The empirical value in modulo is distinctly lower than the critical one, thus there is no ground to reject the verified hypothesis, which means that the sequence of residuals can be treated as having a zero expected value.

Analyse the dispersion of the residuals. Let us check whether it is constant over time.

Divide the sequence in half and calculate the standard deviations for each half separately. We have:

$$s_u^{(I)} = 872.1 \text{ tonnes} \qquad s_u^{(II)} = 527.7 \text{ tonnes}$$

The difference in the values between these parameters is substantial. If a hypothesis is formulated that states that this difference is insignificant and the ratio of the variances is used (as in Chapter 3.4) as the verifying statistic, we can immediately come to the conclusion that this ratio will have a large value. However, this ratio is the F Snedecor's statistic and for a large sample size its critical value is quite close to unity; just a little above 1. Thus, a clear and immediate conclusion can be formulated that the dispersion of residuals decreases over time.

Compare this outcome with the original data. The information that was gathered concerned the period of restructuring in the Polish hard coal mining industry and the total mass of coal sold decreased over time. The situation was due to the transition from the old system to a new one, which now has a clear economic background. Therefore, it can be expected that the trend should slowly go towards a certain stabilised level and the dispersion should decrease over time. For this reason the dispersion of the residuals should also decline, again, to a certain level.

Continue the analysis of the dispersion. Divide the period of observation into four disjoint periods, each period forty elements in length. Calculate the standard deviations for each period. We have:

$$1028 \quad 1027.6 \quad 513.9 \quad 588.7 \text{ tonnes}$$

It looks as though the standard deviation is stable for the first half of the sequence observed (2×40) as well as in the second half. The first two standard deviation values are almost identical. Check whether the second pair can be treated as the same values statistically.

Calculate the ratio of the variances:

$$\frac{588.7^2}{513.9^2} = 1.31$$

Compare this empirical value with the critical one presuming a level of significance $\alpha = 0.05$ as usual. We have the critical value: $F_{\alpha = 0.05}(r_1 = 40 - 1, r_2 = 40 - 1) = 1.68$ for the same sample sizes in both cases.

The empirical value does not exceed the critical one, and therefore there is no foundation to reject the verified hypothesis that states that these two standard deviations differ in significantly from each other. Thus, we can actually state that the dispersion of the residuals was stabilised after approximately 80 months of observation.

This result has serious repercussions. It seems reasonable to state that a further consideration should be conducted based on the last 80 notations only and that these data should be used to:

1. Conduct a preliminary analysis consisting of:

   a. the identification of the trend
   b. the identification of the cyclic components

   c. the determination of the residuals together with an analysis of their properties

   d. the construction of a model of the time series that was investigated that can be used for prediction

2. Estimation of the distribution of the *ex-post* error of the prognosis based on data from the past

   a. the selection of the size of a subsample from which forecasting will be commenced

   b. the identification of the model for prediction

   c. the selection of the range (reach) of prediction

   d. the calculation of the prognoses

   e. the calculation of the *ex-post* errors

   f. the repetition of calculations adding successive elements of the time series (up to the last but one notation)

   g. the identification of the statistic properties of the *ex-post* error versus the range of prediction

3. Prediction of the variable that is being investigated

The way of reasoning is now extended and it looks as though there will be many phases to pass.

Thus, a new sample size to be analysed should consist of approximately 80 elements—the most recent notations. However, taking into account that a possible cyclic component consists of a 12-month period (and probably 2 times 6), we should presume that the sample size should be a multiple of 12 elements. Therefore, our new sample size has $6 \times 12 = 72$ months and covers the months in the years 2002 to 2007. If the sample size is different in its length because it is not a multiple of the period, the sum of residuals very likely will not be zero.

The outcomes taken into further analysis are presented in Figure 7.2.

The linear trend that can be associated with the time series above can be expressed by the formula:

$$\hat{y}_S(t) = -14.4\,t + 8462 + u_t$$

for which the structural parameters were estimated by applying the least squares method.

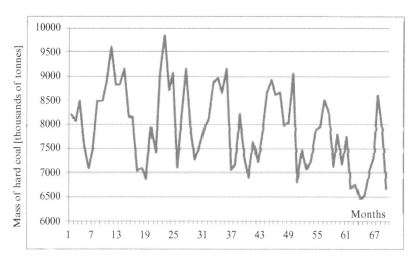

Figure 7.2.   Mass of hard coal sold versus time.

We can now verify whether the pattern found is significant from a statistical point of view. By calculating the linear correlation coefficient between random variables x and y, we have:

$$R_{x,y} = -0.370$$

Compare this figure with the critical one assuming no stochastic dependence between the random variables being tested. The critical number taken from Table 9.13 is $\cong 0.217$. Because the empirical value in modulo is clearly above the critical one, we are can assume that the trend can be depicted by the function $\hat{y}_S(t)$. However, the value of the coefficient is low.

Relate this formula with the one that takes into account all of the data (160 elements; Example 3.8). The slope angle of the trend is now flatter than before, which means that the situation on the market is becoming stabilised.

This result also gave the ground to conduct further reasoning and the residuals were calculated removing the trend from the data. The series of these residuals (first ones) are presented in Figure 7.3.

The cyclic component, which was estimated applying the same method as previously, is also visible in this picture.

The pattern that determined this component was:

$$\hat{y}_{C1}(t) = \sin\left(2\frac{\pi}{T_0}t + \varphi\right) A$$

where: $T_0 = 12$ $\varphi = 2.2$ $A = 721$

The information contained in the formula above is important. The period of the component is twelve months, which means that the periodicity is connected with a calendar year. This information strengthens the previous conclusion that the situation is approaching stabilisation[14].

A further investigation was focused on tracing any second periodic component. The sequence that was tested was obtained from a calculation of the differences:

$$y(t) = \hat{y}_S(t) - \hat{y}_{C1}(t)$$

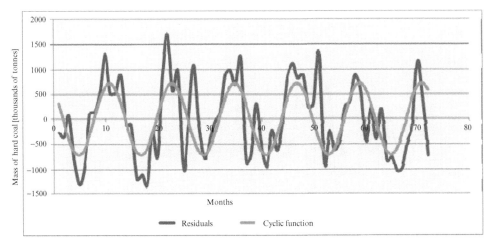

Figure 7.3.   Time series of the first differences and the first cyclic function.

---

[14] In many countries the amount of coal sold versus time has a cyclic component of a 12-month period.

Again, a cyclic component was found by applying the same method of analysis as previously. A model of this component was:

$$\hat{y}_{C2}(t) = \sin\left(2\frac{\pi}{T_0}t + \varphi\right)A$$

where: $T_0 = 6$ $\varphi = 4.28$ $A = 338$

And here again it is easy to read the outcome obtained—with the exception of one year, the period exists as a half year cycle.

The second residuals and the second cyclic function are presented in Figure 7.4.

The model of random process observed is now:

$$y(t) = \hat{y}_S(t) + \hat{y}_{C1}(t) + \hat{y}_{C2}(t) + \xi_t$$

The sequence of the third residuals was calculated removing both the trend and the cyclical constituents from the data. This sequence is shown in Figure 7.5. It is the realisation of the pure random component of the model. Let us examine its properties.

Let us first check whether this sequence is independent of time. Formulate a hypothesis H_0 that states that there is no dependence of time versus the alternative hypothesis that rejects it.

By applying the test for stationarity based on the Spearman's rank correlation coefficient, we have (patterns (3.18) and (3.19)):

$$r_S = 1 - \frac{407748}{72(72-1)} = -0.093$$

This figure is small and we can suspect that is not significant. Verify it formally. A statistical hypothesis that states that there is no dependence between the values of the variable occurring in time and time was formulated. The critical value for the test can be obtained using the formula:

$$r_S(\alpha = 0.05; n = 80) = \frac{u_{1-\alpha}}{\sqrt{n-1}} = \frac{1.645}{\sqrt{79}} = 0.185$$

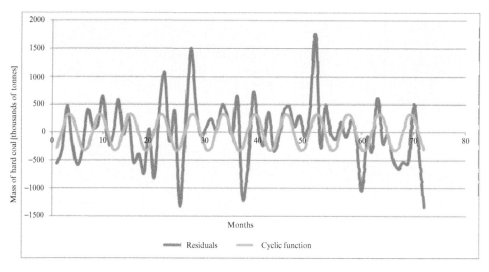

Figure 7.4.   Time series of the second differences and the second cyclic function.

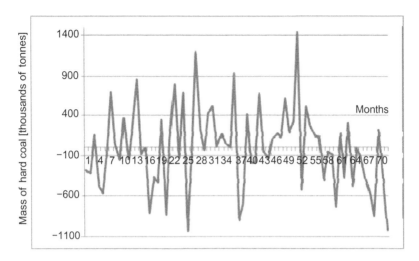

Figure 7.5. Time series of the third differences.

The empirical value in modulo is below the critical one, therefore there is no basis to reject the verified hypothesis. We can assume that the sequence of the third residuals is independent of time.

Calculate the average value of the residuals. We have:

$$\bar{u} = 0 \text{ tonnes}$$

Calculate the standard deviation of the residuals. Here we have:

$$s_u = 502.4 \text{ tonnes}$$

Analyse the dispersion of the residuals. Let us check whether it is constant over time.

Divide the sequence in half and calculate the standard deviations for each half separately. We have:

$$s_u^{(I)} = 503.5 \text{ tonnes} \quad s_u^{(II)} = 501.1 \text{ tonnes}$$

We can assume that the dispersion is constant over time.

The last but one property of the residuals to check is their autocorrelation.

By applying formula (3.47), one gets an estimation of autocorrelation of the first order:

$$r_1^{(a)} = \frac{\sum_{i=1}^{N-1}(u_i - \bar{u}_1)(u_{i+1} - \bar{u}_2)}{\sqrt{\sum_{i=1}^{N-1}(u_i - \bar{u}_1)^2 \sum_{i=1}^{N-1}(u_{i+1} - \bar{u}_2)^2}} = -0.031$$

Calculate the value of the Breusch-Godfrey statistic:

$$\chi^2(c) = (N - c)(r_c^{(a)})^2 = 71(-0.031)^2 = 0.068$$

The verified hypothesis is $H_0 : \rho_c = 0$, i.e. there is no autocorrelation of the order $c = 1$ in the tested random variable.

Read the critical value. For the given order $c = 1$ and presuming the level of significance $\alpha = 0.05$ (Table 9.4).

$$\chi_\alpha^2 (c = 1) = 3,84$$

The critical value is above the empirical one. We can assume that the null hypothesis is true—there is no autocorrelation in the sequence being analysed.

The final property to be checked is whether the sequence of residuals can be satisfactorily described by the normal distribution.

The sample size is large and for this reason the Chi-squared test of goodness of fit can be applied. An estimation of the basic statistical parameters—the average value and the standard deviation—was just done. Thus, the theoretical probability distribution is specified. Following all of the further steps described in Chapter 4.2, the Chi-squared statistic (4.32) is finally:

$$\chi^2 = 10.21$$

The verified hypothesis states that the theoretical Gaussian probability distribution of the specified parameters describes the empirical distribution *well*. Presume a level of significance $\alpha = 0.05$. For the number of degrees of freedom $r - 2 - 1 = 5$, the critical value taken from Table 9.4 is: 11.07.

The empirical value is below the critical one—conclusion: there is no ground to reject the verified hypothesis. We may assume that the distribution $N(0; 502.4)$ describes the distribution of residuals *well*.

Now, we have the theoretical model fully defined. It is given by the formula:

$$y(t) = \hat{y}_S(t) + \hat{y}_{C1}(t) + \hat{y}_{C2}(t) + \xi_t$$

where:

$$\hat{y}_S(t) = -14.4\,t + 8462$$

$$\hat{y}_{C1}(t) = 721\sin\left(2\frac{\pi}{12}t + 2.2\right)$$

$$\hat{y}_{C2}(t) = 338\sin\left(2\frac{\pi}{6}t + 4.28\right)$$

$$\xi : N(0; s_u) = N(0; 502.4)$$

and there is no autocorrelation in the realisation of the residuals.

The preliminary analysis (I) is now finished. The model specified above can be used for prediction.

Now, there is no problem to forecast the further realisation of the random variable that is being discussed, say for one, two or three months in the future because the forecasted random variable is the normal one and therefore the unbiased prediction and the most likely prediction are identical. As an *ex-ante* accuracy measure of the prediction, the standard deviation s_u can be used.

There is also a possibility to improve the prediction inference.

The sample size, which is the main source of information, is large and for this reason we can use a certain part of it to learn about the accuracy of the prognoses.

The idea in this regard is as follows.

Presume that we have information that the realisation of the random variable of interest is less, say, without the 12 last notations, which means that the sample size comprises $72 - 12 = 60$ elements, which are associated with the first 60 months. Using the previous method, we are able to:

– find the trend
– find both cyclic components
– identify the pure random component

for this shorter sequence.

The theoretical model for the random variable investigated is found in this way and by making use of it, we can predict the future realisation of the random variable. However, we are in possession of information about the real value that the variable has taken. Thus, the error of the prognosis can be calculated.

This procedure can be repeated presuming that the sample size consists of 61, 62, ... up to 72 elements. Twelve patterns for the trend will be found in this way. We can presume that the cyclic components are the same. Some important information can be obtained using this way of reasoning. Two of the most important pieces of information are:

– the evolution of the trend
– the error of the prognosis.

Let us devote our attention to predicting only one month ahead. Having the twelve prognoses and twelve real values that were taken by the random variable, the twelve errors of the prognosis (prognosis, because it concerns still the same random variable) can be calculated. Each error is a number but we can treat this information as information about the realisation of a certain random variable. We make a randomisation of this parameter of prediction in this way. If so, the average value of the error can be estimated as well as the corresponding standard deviation. These parameters are *ex-post* measures of the accuracy of the prediction for the method of forecasting that was applied. What is important here is that the calculation of the error concerns the most recent notations.

Let us do the calculation.

Firstly, calculate twelve formulas for the trend. The proportional coefficient (slope of a straight line) increases in modulo, which means that the trend becomes more horizontal. The two trend functions that differ most in their formulas are presented in Figure 7.6. Such a change is obvious because the situation on the market developed to stability.

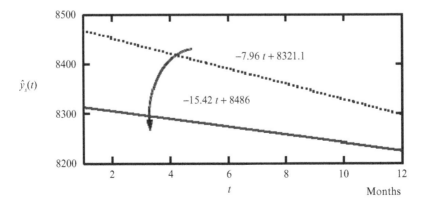

Figure 7.6. Trend functions.

By presuming the same cyclic functions, one can calculate the prognoses. These are shown in the table below in the first column. In the next column are the real values that were taken by the random variable. The errors of prognosis are given in the last column.

Prognosis t	Real values t	Error of prognosis t
7871.6	7796.5	−75.1
7744.3	7169.7	−574.6
7646.4	7752.1	105.7
7364.7	6699.3	−665.4
6933.1	6775.2	−157.9
6738.3	6464	−274.3
6983.0	6539.8	−443.2
7578.9	7018.4	−560.5
8152.4	7324.1	−828.3
8334.3	8606	271.7
8149.5	7857.4	−292.1
7760.0	6688.2	−1071.8

Look at Figure 7.7. Both realisations of the sequences—the prognosis and the real values are visible in it.

Check the sequence of the errors of prognosis. This sequence has the following properties:

a. There is no ground to reject the hypothesis that states its stationarity
b. The growing dispersion in it is statistically insignificant
c. There is no autocorrelation in it.

If so, calculate the mean and the corresponding standard deviation. They are as follows:

$$\overline{\Delta} = -380.5\,t \quad \textbf{and} \quad s_\Delta = 388.1\,t$$

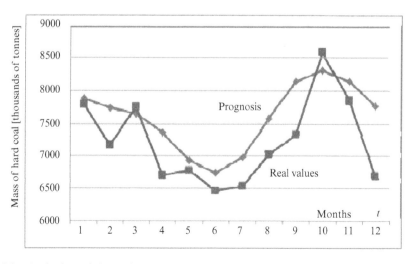

Figure 7.7. Real values of the total mass of hard coal sold in Poland and the corresponding prognosis.

Looking at these figures, we can state that:

– Because the prognosis is below the real value taken by the random variable on average, we can suspect that our next prognosis will also underestimate the forecasted value; the mean error in this regard is 380 *t*
– The *ex-post* error of the prognosis is 388 *t* and this is information about the predicted accuracy of our inference for the future[15].

Notice that this error is 23% smaller than the *ex-ante* one, which was 502 *t*. This means that this way of reasoning is more effective than the previous one statistically. Recall, our inference is based on the most recent (fresh) notations.

▪ **Example 7.2** (based on Sokoła-Szewioła's dissertation 2011)

In Example 6.4 the final model was presented that describes the vertical displacement of mining terrain situated in the area under the direct influence of a mining operation that was being conducted in one of the underground coal mines in the Silesian Coal Basin. This model was developed after the empirical and theoretical study and its ultimate form was:

$$\Delta w_i = -0.5329900 w_{i-1} - 0.3252678 w_{i-2} - 0.3325874 w_{i-3} - 0.2868859 w_{i-4} - \varepsilon_i$$

Remember that Δw_i is an increment in the height of the point that is being observed in *i*-th measurement[16].

This model was the basis for the prediction. It corresponds with notation *G* in formula (7.1). The point of interest was to forecast one step ahead. Because notations were done every two hours, the question was what the increment would be for two hours counting from the moment of last notation. It was presumed that measurements $i-1$, $i-2$, ... were the history and for this reason the *i*-th measurement already belongs to the future.

Such a formulated prediction problem has repercussions in relation to the model selected. All of the addends on the right side of the equation above are deterministic and known values except for the random component. So, the whole stochastic nature of this equation is connected with this component. Knowing that it has a zero expected value, we can be sure that there will be no systematic error connected with the prediction. Thus, the prediction will be unbiased. Moreover, because the Shapiro-Wilk test that was applied during the statistical analysis gave no ground to reject the hypothesis that the residuals can be described by the Gaussian distribution, then calculated prognoses will have the greatest chances of being realized; the mode equals the expected value.

Hence the predictor is:

$$\mathfrak{I} = \Phi_T[G(Y)] = \Delta w_i^{(P)} \Rightarrow$$
$$= E(-0.5329900 w_{i-1} - 0.3252678 w_{i-2} - 0.3325874 w_{i-3} - 0.2868859 w_{i-4} + \varepsilon)$$
$$= -0.5329900 w_{i-1} - 0.3252678 w_{i-2} - 0.3325874 w_{i-3} - 0.2868859 w_{i-4} = \Delta w_i^{(P)}$$

Notice the difference. The symbol $= \Delta w_i^{(P)}$ denotes the random variable which is predicted. The symbol $= \Delta w_i^{(P)}$ in this case means a prognosis concerning *i*-th measurement and this prognosis is a number. However, if the prediction is repeated, the model applied is identical and almost all of the data used in the prediction are the same, then the following prognoses that are calculated can be treated as the realisation of a certain random variable.

---

[15] Often, errors of prognosis are expressed in percentages that relate to real values.
[16] Notice that all of the components of the sum are of the same signs; they are negative. This mathematical regularity has a physical background—the successive subsidence of rock masses.

Moreover, an important conclusion can be formulated that **the model that generates a decrement in the height of a point that is located on the surface—which is under the direct influence of the mining operation that is being conducted in the underground mine—is an autoregressive one that describes a stationary Gaussian process with memory and constant dispersion.** This was proved in the cited dissertation by investigating different points in different mines.

Now pay attention to the standard deviation. In the case being considered, it has an important role because this parameter is the same for the right side of the equation and for the left side, i.e.

$$\sigma(\Delta w) = \sigma(\varepsilon)$$

and this is a measure *ex-ante* of the accuracy of the prediction.

Using the data gathered, the standard deviation was estimated obtaining:

$$s(\varepsilon) = 0.00542969 \cong \sigma(\varepsilon)$$

Taking into account that a large sample was taken, it can be assumed that the real value is close to the calculated one[17].

But the main purpose of the study was to forecast sudden great tremors of high energy that would cause the sudden relatively significant displacement of rock masses and subsidence of the surface[18]. The author of the dissertation assumed that this should be a presage of a coming tremor of this type and she tried to trace it in the sequences of these very small values of ground subsidence. She knew that the appearance of such a precursor should have a stochastic nature. She did not expect that it would be possible to predict this great tremor using any analytical model but her attention was focused on the accuracy of the prognosis—an error value, i.e. the error of the prognosis. When making a prediction differences between prognoses and real values were regularly smaller or greater but sometimes, rarely, a great divergence was noted. And this divergence should be the precursor being searched for—it was assumed.

Thus, having such rich data, successive prognoses were calculated and the difference between the real value and the projected one were also calculated. It was presumed that all of the errors of the prognoses lying in the area $\pm 3\sigma(\Delta w)$ could be neglected.

The successive errors of prognoses, the zone determined by boundaries $\pm 3s(\varepsilon)$ and additionally, the tremors of high energy ($E \geq 7 \times 10^3$ J) are presented in Figure 7.8.

Looking at this figure, we can come to the conclusion that Sokoła-Szewioła was right. During the period of observation, six tremors were noted and only one was unpredicted—the preceding error of prognosis remains in the zone.

Similar results were obtained when data from a few different mines were analysed; however, the assumed level of energy was different. The specific conditions that exist in rock masses connected with geology parameters as well as with mining operations being conducted determine the level of energy that should be considered in a given mine.

A separate set of problems is connected with the decisions about where to set a measurement point, how to detect rock displacements, what kind of devices should be applied, how often measurements should be done and how many etc.

At the end of considerations connected with the above example, let us pay attention to some generalisation possibilities.

---

[17] Such a large number of digits was needed because of the dimensions that were applied; this estimate was converted into the length dimension: $\sigma(\varepsilon) \approx 1.63$ mm.

[18] The term 'relatively significant' is specifically used here; 'significant' in relation to the prognosis that was made. In some cases no subsidence was noted, again, in contradiction to the forecast.

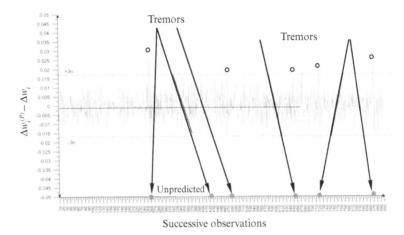

Figure 7.8.   Errors in the prognoses, the zone determined by boundaries ±3s(ε) and tremors of high energy (Sokoła-Szewioła 2011).

Presume that interest is focused on a prognosis two steps ahead. Immediately, the inference is less certain because the model for prediction is:

$$\Delta w_{i+1}^{(P)} = -0.5329900 w_i^{(P)} - 0.3252678 w_{i-1} - 0.3325874 w_{i-2} - 0.2868859 w_{i-3} + \varepsilon$$

This means that the right side of the equation has two random variables, which results in an increment in the variance of the prognosis expressed by the left side of the equation. The value of the *ex-ante* measure of the accuracy of the prediction increases although the calculation itself is simple. This way of reasoning in forecasting uses the so-called chain principle of prediction[19] at is connected with recursive equations.   ◄

Let us extend the reasoning presented in the example above.

Notice at first that the sequence that was the basis for the analysis and for further prediction was a sequence of differences. This means that the original components of these differences were in the primal information. However, the sequence of this primordial data Z_i; $i = 1, 2 ..., n$ is nonstationary; subsidence almost always increases over time (sometimes it becomes zero and rarely does the ground rise). Instead of reasoning in the area of nonstationary processes, the author reduced the problem to a stationary process by calculating the sequence of the first differences. It is obvious that the analysis and prediction of a stationary process is much easier to conduct compared to a nonstationary one.

The idea of converting data that comprise a description of a nonstationary process to a stationary one, especially when prediction is concerned, was popular in the late 1960s and 1970s. The usual way of reasoning was to simply calculate the differences in adjacent series of elements (those that were nonstationary) as was done in Example 6.4. A new sequence of first differences often occurred in the stationary one. However, sometimes the test that was applied for stationarity rejected the hypothesis that stated the stationarity of the sequence of differences. In such a case, a new sequence of differences was calculated; differences between differences. In most cases, this new sequence had a stationary character. Generally, it was recommended to make such a 'differentiation' until a stationary sequence was obtained. When this was achieved, prediction was

---

[19] H. Wold seems to be the father of this principle (1964) but some researchers assign paternity to Tinbergen (1951).

done and a prognosis was calculated in the area of stationary processes. The last but one step was to return to the original sequence with the prognosis attained, after any necessary mathematical transformations. A condition was formulated that the prognosis that was transferred to the original space should correspond with the properties of the nonstationary process. A final step was the calculation of the accuracy of the prognosis and an analysis of the result attained.

It was obvious that such a way of reasoning was not suitable for all cases. This way of analysis is not adequate if the stochastic process that is being analysed has an explosive character. Thus, the terms 'quasi-stationary stochastic process' or 'almost stationary stochastic process' were formulated.

A stochastic process $\mathbf{Y}_t$ that has the following property:

$$\underset{t}{\wedge}\|\mathbf{Y}_t\| < (1+\omega)^t \vartheta$$

where: $\|\mathbf{Y}_t\|$ means the length of the stochastic vector of $\mathbf{Y}_t$
$\qquad$ ϑ—constant
$\qquad$ ω—any positive number

can be called a **quasi-stationary stochastic process**[20].

A characteristic feature of such a process is that any changes have an evolutionary character without a rapid pace of growth of the length of the vector $\mathbf{Y}_t$ as time increases.

Presume now that m random variables $\mathbf{Y}_i$; $i = 1, 2, ..., m$ are observed that are stochastic copies of each other and that a sample of size N was taken. Presume additionally that all moments $t < t_0$ belong to the past and all moments $t \geq t_0$ belong to the future.

If so, the following matrix can be constructed with all of the known elements:

$$\Upsilon = \begin{pmatrix} y_1(t_0 - N) & y_2(t_0 - N) & ... & y_m(t_0 - N) \\ y_1(t_0 - N+1) & y_2(t_0 - N+1) & ... & y_m(t_0 - N+1) \\ ... & ... & ... & ... \\ y_1(t_0 - 1) & y_2(t_0 - 1) & ... & y_m(t_0 - 1) \end{pmatrix}$$

All of the row vectors of this stochastic matrix are stationary whereas the column vectors can be either stationary or nonstationary. Assume that the column vectors are realisations of quasi-stationary random processes.

Now, if $\Upsilon(t_0 - s)$ denotes any quasi-stationary stochastic vector and if $\mathbf{x}(t_0 - s)$ denotes the stochastic vector of differences, then the formula that allows the stationary vector to be obtained is given by the pattern:

$$\mathbf{x}(t_0 - s) = \sum_{k=0}^{\rho} (-1)^k \binom{\rho}{k} \Upsilon(t_0 - k - s) \qquad (7.2)$$

where: ρ is the multiplicity of calculation of the differences needed to obtain the stationary sequence: $s = 1, 2, ..., N - \rho$.

The number ρ also informs about the number of lost elements from the original sample of size N because of the calculation of the differences.

If we are interested in a prognosis at a moment just at t_0, then using formula (7.2), we can state that

$$\Upsilon(t_0)^{(P)} = \mathbf{x}(t_0)^{(P)} - \sum_{k=0}^{\rho} (-1)^k \binom{\rho}{k} \Upsilon(t_0 - k) \qquad (7.3)$$

where symbol (P) denotes the prognosis.

---

[20] Compare Zadora (1974).

By looking at formula (7.3), we can state that by having the prognosis of the stationary process for moment t_0, we are able to get the prognosis for the quasi-stationary process for the same moment.

This way of reasoning can be extended (Czaplicki, Ph.D. dissertation 1975) but it will not be presented here because it is significantly beyond the basic level of considerations which is presented here.

▪ Example 7.3

In Chapter 4.1 the problem of interval estimation was considered and the set of formulas that allows the confidence intervals for the parameters of technical object which the process of changes of states is of work-repair type, to be obtained. The process was presumed to be the Markov process. This means that the times of both states are independent and exponentially distributed. The reason that these patterns are presented is the fact that many of the technical objects in mining have operational processes that can be satisfactorily described by such a model.

By following this line of reasoning, it appears to be advantageous to present a similar set of formulas but which are orientated to the future.

The process whose parameters are the subject of our interest is stationary, which means that these parameters are independent of time. If so, it can be assumed that the interval estimations (based on the sample taken) that are obtained remain valid for the future provided that neither the method of operation nor the exploitation conditions do not change significantly. This is very simplified approach to prediction.

However, there is a possibility to improve this reasoning.

Assume that we are interested in the prediction intervals of these parameters for m future process cycles and that we have information on the process comprising n process cycles. We know that the random variable $2\lambda\sum_{i=1}^{n}t_{wi}$ has a χ^2 distribution with $2n$ degrees of freedom. Analogically, it can be stated that the random variable $2\lambda\sum_{i=1}^{m}t_{wi}^{(P)}$ has a χ^2 distribution with $2m$ degrees of freedom and the upper script (P) denotes the future (predicted) value.

Consider the equation:

$$\frac{m\sum_{i=1}^{n}t_{wi}}{n\sum_{i=1}^{m}t_{wi}^{(P)}} = F(2n, 2m)$$

Because the formula:

$$\frac{n}{\sum_{i=1}^{n}t_{wi}} = \hat{\lambda}_{(n)}$$

is the estimator of the intensity of the failures of a sample of size n, then the following equation can be constructed:

$$P\left\{F_1(2n, 2m) < \frac{\hat{\lambda}_{(m)}^{(P)}}{\hat{\lambda}_{(n)}} < F_2(2n, 2m)\right\} = 1-\alpha$$

where $\hat{\lambda}_{(m)}^{(P)}$ concerns the future as before.

Thus, the prediction interval for the intensity of failures for m future process cycles is determined by the pattern:

$$P\{\hat{\lambda}_{(n)}F_1(2n, 2m) < \hat{\lambda}_{(m)}^{(P)} < \hat{\lambda}_{(n)}F_2(2n, 2m)\} = 1-\alpha \qquad (7.4)$$

Analogically, the prediction interval for the intensity of repair for m future process cycles is determined by the pattern:

$$P\{\hat{\beta}_{(n)}F_1(2n, 2m) < \hat{\beta}_{(m)}^{(P)} < \hat{\beta}_{(n)}F_2(2n, 2m)\} = 1 - \alpha \qquad (7.5)$$

To construct the prediction intervals for the expected values of times of states is easy and—having equation (1.119) in mind—the following equations hold:

$$P\{\bar{T}_{w(n)}F_1(2m, 2n) < \bar{T}_{w(m)}^{(P)} < \bar{T}_{w(n)}F_2(2m, 2n)\} = 1 - \alpha \qquad (7.6)$$

$$P\{\bar{T}_{r(n)}F_1(2m, 2n) < \bar{T}_{r(m)}^{(P)} < \bar{T}_{r(n)}F_2(2m, 2n)\} = 1 - \alpha \qquad (7.7)$$

where obviously:

$$\bar{T}_{w(n)} = \frac{1}{n}\sum_{i=1}^{n} t_{wi} \qquad \bar{T}_{r(n)} = \frac{1}{n}\sum_{i=1}^{n} t_{ri}$$

Now consider the quotient of two F statistics:

$$\frac{\hat{\lambda}_{(m)}^{(P)}}{\hat{\lambda}_{(n)}} : \frac{\hat{\beta}_{(m)}^{(P)}}{\hat{\beta}_{(n)}} = \frac{F(2n, 2m)}{F(2n, 2m)} \qquad (7.8)$$

Denote this random variable by $\vartheta(2n, 2m)$. If so, equation (7.8) can be expressed as:

$$\hat{\kappa}_{(m)}^{(P)} = \hat{\kappa}_{(n)}\vartheta(2n, 2m) \qquad (7.9)$$

Now, we can obtain the prediction interval for the repair rate from the expression:

$$P\{\hat{\kappa}_{(n)}\vartheta_1(2n, 2m) < \hat{\kappa}_{(m)}^{(P)} < \hat{\kappa}_{(n)}\vartheta_2(2n, 2m)\} = 1 - \alpha \qquad (7.10)$$

Similarly, the prediction interval for the steady-state availability can be achieved using the equation:

$$P\left\{\frac{\hat{A}_{(n)}}{\hat{A}_{(n)} + (\hat{A}_{(n)} - 1)\vartheta_2(2n, 2m)} < \hat{A}_{(m)}^{(P)} < \frac{\hat{A}_{(n)}}{\hat{A}_{(n)} + (\hat{A}_{(n)} - 1)\vartheta_1(2n, 2m)}\right\} = 1 - \alpha \qquad (7.11)$$

Note, that

$$\vartheta_1(2n, 2m)\,\vartheta_2(2n, 2m) = 1 \qquad (7.12)$$

By constructing the prediction intervals for M future cycles of the process of the series system consisting of k identical elements, one obtains the following formulas, which allow these intervals to be found:

• The prediction interval for the intensity of failures from the equation:

$$P\{\hat{\lambda}_{(N)}F_1(2Nk, 2Mk) < \hat{\lambda}_{(M)}^{(P)} < \hat{\lambda}_{(N)}F_2(2Nk, 2Mk)\} = 1 - \alpha \qquad (7.13)$$

- The prediction interval for the expected value of work time from the equation:

$$P\{\bar{\boldsymbol{T}}_{w(N)}F_1(2Mk, 2Nk) < \bar{\boldsymbol{T}}_{w(M)}^{(P)} < \bar{\boldsymbol{T}}_{w(N)}F_2(2Mk, 2Nk)\} = 1 - \alpha \qquad (7.14)$$

- The prediction interval for the intensity of repair from the equation:

$$P\{\hat{\boldsymbol{\beta}}_{(N)}F_1(2N, 2M) < \hat{\boldsymbol{\beta}}_{(M)}^{(P)} < \hat{\boldsymbol{\beta}}_{(N)}F_2(2N, 2M)\} = 1 - \alpha \qquad (7.15)$$

- The prediction interval for the expected value of repair time from the equation:

$$P\{\bar{\boldsymbol{T}}_{r(N)}F_1(2M, 2N) < \bar{\boldsymbol{T}}_{r(M)}^{(P)} < \bar{\boldsymbol{T}}_{r(N)}F_2(2M, 2N)\} = 1 - \alpha \qquad (7.16)$$

The prediction interval for the repair rate of the system can be determined using the ratio of two statistics $\boldsymbol{F}$ (not identical ones this time), namely:

$$\frac{\hat{\boldsymbol{\lambda}}_{(M)}^{(P)}}{\hat{\boldsymbol{\lambda}}_{(N)}} : \frac{\hat{\boldsymbol{\beta}}_{(M)}^{(P)}}{\hat{\boldsymbol{\beta}}_{(N)}} = \frac{F(2Nk, 2Mk)}{F(2N, 2M)} \qquad (7.17)$$

Denote this random variable by $\boldsymbol{\Omega}(2Nk, 2Mk, 2N, 2M)$. If so, the following equation can be constructed:

$$P\{\hat{\boldsymbol{\kappa}}_{(N)}\boldsymbol{\Omega}_1(2Nk, 2Mk, 2N, 2M) < \hat{\boldsymbol{\kappa}}_{(M)}^{(P)} < \hat{\boldsymbol{\kappa}}_{(N)}\boldsymbol{\Omega}_2(2Nk, 2Mk, 2N, 2M)\} = 1 - \alpha \qquad (7.18)$$

which allows the prediction interval for the repair rate of the system to be obtained.

The prediction interval for the steady-state availability of the system can be obtainedfrom the pattern:

$$P\left\{\frac{\hat{A}_{(N)}}{\hat{A}_{(N)} + (\hat{A}_{(N)} - 1)\boldsymbol{\Omega}_2(2Nk, 2Mk, 2N, 2M)} < \hat{A}_{(M)}^{(P)} < \frac{\hat{A}_{(N)}}{\hat{A}_{(N)} + (\hat{A}_{(N)} - 1)\boldsymbol{\Omega}_1(2Nk, 2Mk, 2N, 2M)}\right\}$$
$$= 1 - \alpha$$

$$(7.19)$$

After certain modifications, the interval estimations can be obtained for a series system that is constructed from non-identical elements (Czaplicki 1977b).

Constructing the prediction interval for the repair rate of a series system consisting of k of the same elements and for the steady-state availability of this type of system, it is necessary to know the values of the random variable $\boldsymbol{\Omega}(\delta_1, \delta_2, \delta_3, \delta_4)$, which is the quotient of two $\boldsymbol{F}$ statistics $F(\delta_1, \delta_2)$ and $F(\delta_3, \delta_4)$ for the presumed level of probability[21] α. A particular case of this random variable is the random variable ϑ that appears in formulas (7.10) and (7.11), and this variable is the quotient of two $\boldsymbol{F}$ statistics with the same degrees of freedom. For the two most frequently applied probability levels $\alpha = 0.95$ and $\alpha = 0.99$, the values ϑ_α are given in Table 9.17.

---

[21] The probability density function of this random variable is given by the formula:

$$f(\boldsymbol{\Omega}) = \frac{\delta_1^{\frac{\delta_1}{2}}\delta_2^{\frac{\delta_2}{2}}\delta_3^{\frac{\delta_3}{2}}\delta_4^{\frac{\delta_4}{2}}}{\Gamma\left(\frac{\delta_1}{2}\right)\Gamma\left(\frac{\delta_2}{2}\right)\Gamma\left(\frac{\delta_3}{2}\right)\Gamma\left(\frac{\delta_4}{2}\right)}\Gamma\left(\frac{\delta_1+\delta_2}{2}\right)\Gamma\left(\frac{\delta_3+\delta_4}{2}\right)\boldsymbol{\Omega}^{\frac{\delta_1-\delta_2}{2}-1}\int_0^\infty \frac{F^{\frac{\delta_1-\delta_2}{2}-1}dF}{(\delta_1 F+\delta_2)^{\frac{\delta_1-\delta_2}{2}}(\delta_3 F+\delta_4)^{\frac{\delta_3-\delta_4}{2}}}$$

If the critical values of the $\boldsymbol{\Omega}_\alpha$ distribution are needed, the following approximation can be applied:

$$\Omega_\alpha \cong \exp\left[u_\alpha \sqrt{\frac{k+1}{Nk} + \frac{k+1}{Mk}} \right] \tag{7.20}$$

where u_α is the quantile of the order α of the standardised normal distribution.

The above approximation is obtained from the following reasoning. By calculating the logarithms of both sides of the equation (7.15), one obtains:

$$\ln\Omega = \ln(\hat{\lambda}_{(M)}^{(P)} / \hat{\lambda}_{(N)}) - \ln(\hat{\beta}_{(M)}^{(P)} / \hat{\beta}_{(N)}) \tag{7.21}$$

Both logarithms of the right side of the equation are the Fisher's z transformation of the statistic F. For a large M, N and k become normal distributions with a zero expected value; where the first logarithm has the variance $((Nk)^{-1} + (Mk)^{-1})$ while the second one has the variance[22] $(N^{-1} + M^{-1})$. For his reason, the random variable $\ln\Omega$ has an approximately normal distribution with a zero expected value and the variance $((Nk)^{-1} + (Mk)^{-1} + N^{-1} + M^{-1})$.

When the variable ϑ is being analysed, the approximation has the form:

$$\vartheta_\alpha \cong \exp\left[u_\alpha \sqrt{\frac{2}{n} + \frac{2}{m}} \right] \tag{7.22}$$

At the end of these considerations, it looks as though it is worth recalling the obvious stipulation that the information about the past should comprise a longer period than that concerning the future.

The presented relationships fulfil the mathematical requirements but now it is necessary to refer to engineering practice. There is information about 'when the number of process cycles becomes large ...'. In some cases, this constraint is difficult to fulfil in practice. There are some pieces of equipment for which the exploitation process can be satisfactorily modelled by the Markov process but these are of very high reliability, e.g. many main belt conveyors. Failures are recorded only a few times per year on average and therefore, in order to get a large sample where the process cycles are concerned, it is necessary to observe a dozen years or more. However, very often such conveyors do not work in one place for such a long period. A cardinal feature of mine transport routes is their changeability, especially where their length is concerned. From time to time, new pieces are added or sometimes some pieces are withdrawn. This problem is connected with both a large sample size and with how far statistical inference should be conducted in cases for which prediction is being analysed. All researchers should keep these things in mind when undertaking a study in the field of forecasting.   ◀

---

[22] See formula (1.125).

CHAPTER 8

Explanations of some important terms

Correlation—a certain kind of stochastic relation between two or more random variables; it relies on such a regularity that any changes in the values of one variable are accompaniedby systematic stochastic changes in the values of the second variable or other variables.

Correlation coefficient—a basic measure of the strength of the stochastic relationship between the random variables being investigated

Critical region of test—a region of the rejection of the verified statistical hypothesis; the region is determined by the test that is applied.

Degrees of freedom (the number)—the number of independent outcomes of the observation reduced by the number of relationships that connect these outcomes among themselves. When an estimation of the structural parameters of the regression function is done and their significance is tested, then the number of degrees of freedom equals the number of observations reduced by the number of parameters that are being estimated.

Error of type I (error of the first kind)—an error that may be made during the verification of a statistical hypothesis; it relies on the rejection of the true hypothesis.

Error of type II (error of the second kind)—an error that may be made during the verification of a statistical hypothesis; it relies on the approval of the false hypothesis.

Estimation—the process of finding an estimate; an approximate calculation; in mathematical statistics a chapter on such a statistical inference that deals with the assessment of the unknown values of parameters in the general population based on a sample taken from it.

    Interval estimation—such a method of the assessment of the unknown value of the general population parameter that relies on the construction of a numerical interval of a random length that covers the unknown value of the parameter with a presumed *a priori* level of the probability.

    Point estimation—such a method of the assessment of the unknown value of the general population parameter that this value is assumed to be the value of its estimator based on data taken from the sample.

Histogram—a graphical representation of a function that assigns a number of observations falling within it to each discrete interval (bin) of the investigated variable; a histogram may also be normalised in order to display the relative frequencies that show the proportion of cases that fall into each of several categories, with the total area equaling 1.

Hypothesis (statistical)—any supposition concerning an unknown distribution or parameters of the general population for which this supposition is verified based on a random sample taken from this population.

    Null hypothesis—the basic statistical hypothesis that is being verified.

    Alternative hypothesis—any statistical hypothesis that is a competitive one to the null hypothesis.

Intensity of failures

1. For **nonrenewable** technical objects (i.e. working to the first failure occurrence), it is the **conditional probability** that a failure will occur in time unit $t + \Delta t$ provided that until time t no failure has occurred.
2. For **renewable** technical objects, it is the **unconditional probability** that a failure will occur; it is the reciprocal of the expected work time between two successive failures; dimension: failure per unit of time.

Interval, confidence—a random interval determined by the estimator distribution; this interval has such a property that it **covers**, with a presumed *a priori* probability (level of confidence), the unknown value of the population parameter or parameters.

Least squares method—a method for the estimation of the parameters of the regression function based on empirical data shown in the form of a sequence of numbers. The method relies on such a selection of the values of the parameters of the approximate function so as to achieve the minimum of the sum of squares of the deviations of the empirical points from the corresponding theoretical points as determined by the approximation function. The method of least squares is a standard approach to the approximate solution of over-determined systems, i.e. sets of equations in which there are more equations than unknowns.

Level of confidence—the presumed, near to 1, probability that the confidence interval **covers** the unknown value of the general population. Most frequently, the presumed level of significance is 0.10; 0.05 or 0.01.

Level of significance—the probability, presumed near zero, of making an error of type I; most frequently the level is presumed to be 0.1; 0.05 or 0.01.

Parameters (of the general population)—parameters of the distribution of the random variable being tested from the general population; they characterise this distribution in a synthetic way. Usually parameters are divided into several groups: *measures of an average level* (e.g. mean, median), *measures of dispersion* (e.g. variance, standard deviation), *measures of asymmetry*, *measures of concentration* and other statistical measures (e.g. correlation); these measures are applied during the statistical study of a population.

Population, general population—a set, any finite or infinite aggregation of elements (homogeneous, but not identical) from which samples are drawn.

Regression—the analysis or measure of the association between random variables; the analysis that allows the mean value of the variable being explained by one or more explanatory variables to be described.

Sample, random—an observed part of the general population; the sample that was taken in a random way and that will be used to make an inference on a determined regularity in the whole population.

Sample, representative—a random sample for which the structure is not significantly different than the structure of the general population for the variable that is being investigated and which is taken into consideration.

Sequence—a function from a subset of the natural numbers to a different set (numbers, points, functions etc.).

Standard estimation error—the square root of the variance, i.e. the standard deviation in the estimator distribution; the estimator that is used to assess the unknown value of a parameter of the general population.

Statistic—any function of outcomes of the sample, a random variable; theoretically, any function of a number of random variables, which are usually identically distributed, that may be used to estimate a population parameter.

Statistics—
1. formerly, knowledge on the state
2. the discipline of science that deals with the quantitative methods for investigating mass phenomena
3. the numerical facts or data themselves.

Time series—a sequence of ordered observations of the variable of interest, which is connected with different or identical moments in time or time periods; a significant feature of a time series is its arrangement according to the occurrence of the observations in time.

Test (statistical)—a rule of conduct that allows the decision to be made to reject the verified hypothesis or not, based on a sample taken.

Test of significance—a test that allows, with a presumed high probability, for the rejection of the verified hypothesis if it is false; it does not allow it to be stated that the verified hypothesis is a true one.

Variable (explanatory)—a variable that is used to explain the formation of another variable or variables that are being investigated and that are logically associated with it.

Variable (being explained)—a dependent variable; a variable whose formation is explained by another variable or variables that are being investigated and that are logically associated with it.

CHAPTER 9

Statistical tables

Table 9.1. Distribution function $\Phi(z)$ of standardised normal distribution $N(0, 1)$.

z	0.00	0.01	0.02	0.03	0.04	0.05	0.06	0.07	0.08	0.09
0.0	0.5000	0.5040	0.5080	0.5120	0.5160	0.5199	0.5239	0.5279	0.5319	0.5359
0.1	0.5398	0.5438	0.5478	0.5517	0.5557	0.5596	0.5636	0.5675	0.5714	0.5753
0.2	0.5793	0.5832	0.5871	0.5910	0.5948	0.5987	0.6026	0.6064	0.6103	0.6141
0.3	0.6179	0.6217	0.6255	0.6293	0.6331	0.6368	0.6406	0.6443	0.6480	0.6517
0.4	0.6554	0.6591	0.6628	0.6664	0.6700	0.6736	0.6772	0.6808	0.6844	0.6879
0.5	0.6915	0.6950	0.6985	0.7019	0.7054	0.7088	0.7123	0.7157	0.7190	0.7224
0.6	0.7257	0.7291	0.7324	0.7357	0.7389	0.7422	0.7454	0.7486	0.7517	0.7549
0.7	0.7580	0.7611	0.7642	0.7673	0.7704	0.7734	0.7764	0.7794	0.7823	0.7852
0.8	0.7881	0.7910	0.7939	0.7967	0.7995	0.8023	0.8051	0.8078	0.8106	0.8133
0.9	0.8159	0.8186	0.8212	0.8238	0.8264	0.8289	0.8315	0.8340	0.8365	0.8389
1.0	0.8413	0.8438	0.8461	0.8485	0.8508	0.8531	0.8554	0.8577	0.8599	0.8621
1.1	0.8643	0.8665	0.8686	0.8708	0.8729	0.8749	0.8770	0.8790	0.8810	0.8830
1.2	0.8849	0.8869	0.8888	0.8907	0.8925	0.8944	0.8962	0.8980	0.8997	0.9015
1.3	0.9032	0.9049	0.9066	0.9082	0.9099	0.9115	0.9131	0.9147	0.9162	0.9177
1.4	0.9192	0.9207	0.9222	0.9236	0.9251	0.9265	0.9278	0.9292	0.9306	0.9319
1.5	0.9332	0.9345	0.9357	0.9370	0.9382	0.9394	0.9406	0.9418	0.9429	0.9441
1.6	0.9452	0.9463	0.9474	0.9484	0.9495	0.9505	0.9515	0.9525	0.9535	0.9545
1.7	0.9554	0.9564	0.9573	0.9582	0.9591	0.9599	0.9608	0.9616	0.9625	0.9633
1.8	0.9641	0.9649	0.9656	0.9664	0.9671	0.9678	0.9686	0.9693	0.9699	0.9706
1.9	0.9713	0.9719	0.9726	0.9732	0.9738	0.9744	0.9750	0.9756	0.9761	0.9767
2.0	0.9772	0.9778	0.9783	0.9788	0.9793	0.9798	0.9803	0.9808	0.9812	0.9817
2.1	0.9821	0.9826	0.9830	0.9834	0.9838	0.9842	0.9846	0.9850	0.9854	0.9857
2.2	0.9861	0.9864	0.9868	0.9871	0.9875	0.9878	0.9881	0.9884	0.9887	0.9890
2.3	0.9893	0.9896	0.9898	0.9901	0.9904	0.9906	0.9909	0.9911	0.9913	0.9916
2.4	0.9918	0.9920	0.9922	0.9925	0.9927	0.9929	0.9931	0.9932	0.9934	0.9936
2.5	0.9938	0.9940	0.9941	0.9943	0.9945	0.9946	0.9948	0.9949	0.9951	0.9952
2.6	0.9953	0.9955	0.9956	0.9957	0.9959	0.9960	0.9961	0.9962	0.9963	0.9964
2.7	0.9965	0.9966	0.9967	0.9968	0.9969	0.9970	0.9971	0.9972	0.9973	0.9974
2.8	0.9974	0.9975	0.9976	0.9977	0.9977	0.9978	0.9979	0.9979	0.9980	0.9981
2.9	0.9981	0.9982	0.9982	0.9983	0.9984	0.9984	0.9985	0.9985	0.9986	0.9986
3.0	0.9987	0.9987	0.9987	0.9988	0.9988	0.9989	0.9989	0.9989	0.9990	0.9990
3.1	0.9990	0.9991	0.9991	0.9991	0.9992	0.9992	0.9992	0.9992	0.9993	0.9993
3.2	0.9993	0.9993	0.9994	0.9994	0.9994	0.9994	0.9994	0.9995	0.9995	0.9995
3.3	0.9995	0.9995	0.9995	0.9996	0.9996	0.9996	0.9996	0.9996	0.9996	0.9997
3.4	0.9997	0.9997	0.9997	0.9997	0.9997	0.9997	0.9997	0.9997	0.9997	0.9998

For $z < 0$ values $\Phi(z)$ are calculated applying formula $\Phi(z) = 1 - \Phi(-z)$.

Table 9.2. Quantiles of the standardised normal distribution $N(0, 1)$.

p	0.000	0.001	0.002	0.003	0.004	0.005	0.006	0.007	0.008	0.009	p
0.50	0.000 000	0.002 507	0.005 013	0.007 520	0.010 027	0.012 533	0.015 040	0.017 547	0.020 054	0.022 562	0.50
0.51	0.025 069	0.027 576	0.030 084	0.032 592	0.035 100	0.037 608	0.040 117	0.042 626	0.045 135	0.047 644	0.51
0.52	0.050 154	0.052 664	0.055 174	0.057 684	0.060 195	0.062 707	0.065 219	0.067 731	0.070 243	0.072 756	0.52
0.53	0.075 270	0.077 784	0.080 298	0.082 813	0.085 329	0.087 845	0.090 361	0.092 879	0.095 396	0.097 915	0.53
0.54	0.100 434	0.102 953	0.105 474	0.107 995	0.110 516	0.113 039	0.115 562	0.118 085	0.120 610	0.123 135	0.54
0.55	0.125 661	0.128 188	0.130 716	0.133 245	0.135 774	0.138 304	0.140 835	0.143 367	0.145 900	0.148 434	0.55
0.56	0.150 969	0.153 505	0.156 042	0.158 580	0.161 119	0.163 658	0.166 199	0.168 741	0.171 285	0.173 829	0.56
0.57	0.176 374	0.178 921	0.181 468	0.184 017	0.186 567	0.189 118	0.191 671	0.194 225	0.196 780	0.199 336	0.57
0.58	0.201 893	0.204 452	0.207 013	0.209 574	0.212 137	0.214 702	0.217 267	0.219 835	0.222 403	0.224 973	0.58
0.59	0.227 545	0.230 118	0.232 693	0.235 269	0.237 847	0.240 426	0.243 007	0.245 590	0.248 174	0.250 760	0.59
0.60	0.253 347	0.255 936	0.258 527	0.261 120	0.263 714	0.266 311	0.268 909	0.271 508	0.274 110	0.276 714	0.60
0.61	0.279 319	0.281 926	0.284 536	0.287 147	0.289 760	0.292 375	0.294 992	0.297 611	0.300 232	0.302 855	0.61
0.62	0.305 481	0.308 108	0.310 738	0.313 369	0.316 003	0.318 639	0.321 278	0.323 918	0.326 561	0.329 206	0.62
0.63	0.331 853	0.334 503	0.337 155	0.339 809	0.342 466	0.345 126	0.347 787	0.350 451	0.353 118	0.355 787	0.63
0.64	0.358 459	0.361 133	0.363 810	0.366 489	0.369 171	0.371 856	0.374 544	0.377 234	0.379 926	0.382 622	0.64
0.65	0.385 320	0.388 022	0.390 726	0.393 433	0.396 142	0.398 855	0.401 571	0.404 289	0.407 011	0.409 735	0.65
0.66	0.412 463	0.415 194	0.417 928	0.420 665	0.423 405	0.426 148	0.428 894	0.431 644	0.434 397	0.437 154	0.66
0.67	0.439 913	0.442 676	0.445 443	0.448 212	0.450 986	0.453 762	0.456 542	0.459 326	0.462 113	0.464 904	0.67
0.68	0.467 699	0.470 497	0.473 299	0.476 104	0.478 914	0.481 727	0.484 544	0.487 365	0.490 189	0.493 018	0.68
0.69	0.495 850	0.498 687	0.501 527	0.504 372	0.507 221	0.510 073	0.512 930	0.515 792	0.518 657	0.521 527	0.69
0.70	0.524 401	0.527 279	0.530 161	0.533 049	0.535 940	0.538 836	0.541 737	0.544 642	0.547 551	0.550 466	0.70
0.71	0.553 385	0.556 308	0.559 237	0.562 170	0.565 108	0.568 052	0.570 999	0.573 952	0.576 910	0.579 873	0.71
0.72	0.582 842	0.585 815	0.588 793	0.591 777	0.594 766	0.597 760	0.600 760	0.603 765	0.606 775	0.609 791	0.72
0.73	0.612 813	0.615 840	0.618 873	0.621 912	0.624 956	0.628 006	0.631 062	0.634 124	0.637 192	0.640 266	0.73
0.74	0.643 345	0.646 431	0.649 524	0.652 622	0.655 727	0.658 838	0.661 955	0.665 079	0.668 209	0.671 346	0.74

0.75											0.75
0.75	0.674 490	0.677 640	0.680 797	0.683 961	0.687 131	0.690 309	0.693 493	0.696 685	0.699 884	0.703 089	0.75
0.76	0.706 303	0.709 523	0.712 751	0.715 986	0.719 229	0.722 479	0.725 737	0.729 003	0.732 276	0.735 558	0.76
0.77	0.738 847	0.742 144	0.745 450	0.748 763	0.752 085	0.755 415	0.758 754	0.762 101	0.765 456	0.768 820	0.77
0.78	0.772 193	0.775 575	0.778 966	0.782 365	0.785 774	0.789 192	0.792 619	0.796 055	0.799 501	0.802 956	0.78
0.79	0.806 421	0.809 896	0.813 380	0.816 875	0.820 379	0.823 894	0.827 418	0.830 953	0.834 499	0.838 055	0.79
0.80	0.841 621	0.845 199	0.848 787	0.852 386	0.855 996	0.859 617	0.863 250	0.866 894	0.870 550	0.874 217	0.80
0.81	0.877 896	0.881 587	0.885 290	0.889 006	0.892 733	0.896 473	0.900 226	0.903 991	0.907 770	0.911 561	0.81
0.82	0.915 365	0.919 183	0.923 014	0.926 859	0.930 717	0.934 589	0.938 476	0.942 376	0.946 291	0.950 221	0.82
0.83	0.954 165	0.958 124	0.962 099	0.966 088	0.970 093	0.974 114	0.978 150	0.982 203	0.986 271	0.990 356	0.83
0.84	0.994 458	0.998 576	1.002 712	1.006 964	1.011 034	1.015 222	1.019 428	1.023 651	1.027 893	1.032 154	0.84
0.85	1.036 433	1.040 732	1.045 050	1.049 387	1.053 744	1.058 122	1.062 519	1.066 938	1.071 377	1.075 837	0.85
0.86	0.080 319	0.084 823	0.089 349	0.093 897	0.098 468	0.103 063	0.107 680	0.112 321	0.116 987	0.121 677	0.86
0.87	0.126 391	0.131 131	0.135 896	0.140 687	0.145 505	0.150 349	0.155 221	0.160 120	0.165 047	0.170 002	0.87
0.88	0.174 987	0.180 001	0.185 044	0.190 118	0.195 223	0.200 359	0.205 527	0.210 727	0.215 960	0.221 227	0.88
0.89	0.226 528	0.231 864	0.237 235	0.242 641	0.248 085	0.253 565	0.259 084	0.264 641	0.270 238	0.275 874	0.89
0.90	1.281 552	1.287 271	1.293 032	1.298 837	1.304 685	1.310 579	1.316 519	1.322 505	1.328 539	1.334 622	0.90
0.91	0.340 755	0.346 939	0.353 174	0.359 463	0.365 806	0.372 204	0.378 659	0.385 172	0.391 744	0.398 377	0.91
0.92	0.405 072	0.411 830	0.418 654	0.425 544	0.432 503	0.439 531	0.446 632	0.453 806	0.461 056	0.468 384	0.92
0.93	0.475 791	0.483 280	0.490 853	0.498 513	0.506 262	0.514 102	0.522 036	0.530 068	0.538 199	0.546 433	0.93
0.94	0.554 774	0.563 224	0.571 787	0.580 467	0.589 268	0.598 193	0.607 248	0.616 436	0.625 763	0.635 234	0.94
0.95	1.644 854	1.654 628	1.664 563	1.674 665	1.684 941	1.695 398	1.706 043	1.716 886	1.727 934	1.739 198	0.95
0.96	0.750 686	0.762 410	0.774 382	0.786 613	0.799 118	0.811 911	0.825 007	0.838 424	0.852 180	0.866 296	0.96
0.97	0.880 794	0.895 698	0.911 036	0.926 837	0.943 134	0.959 964	0.977 368	0.995 393	2.014 091	2.033 520	0.97
0.98	2.053 749	2.074 855	2.096 927	2.120 072	2.144 411	2.170 090	2.197 286	2.226 212	0.257 129	0.290 368	0.98
0.99	0.326 348	0.365 618	0.408 916	0.457 263	0.512 144	0.575 829	0.652 070	0.747 781	0.878 162	3.090 232	0.99

Table 9.3. Critical values of the Student's t-distribution.

v \ α	0.2	0.1	0.05	0.02	0.01	0.002	0.001	α \ v
1	3.078	6.314	12.706	31.821	63.657	318.309	636.619	1
2	1.886	2.920	4.303	6.965	6.925	22.327	31.598	2
3	1.638	2.353	3.182	4.541	5.841	10.212	12.941	3
4	1.533	2.132	2.776	3.747	4.604	7.173	8.610	4
5	1.476	2.015	2.571	3.365	4.032	5.893	6.859	5
6	1.440	1.943	2.447	3.143	3.707	5.208	5.959	6
7	1.415	1.895	2.365	2.998	3.499	4.785	5.405	7
8	1.397	1.860	2.306	2.896	3.355	4.501	5.041	8
9	1.383	1.833	2.262	2.821	3.250	4.297	4.781	9
10	1.372	1.812	2.228	2.764	3.169	4.144	4.587	10
11	1.363	1.796	2.201	2.718	3.106	4.025	4.437	11
12	1.356	1.782	2.179	2.681	3.055	3.930	4.318	12
13	1.350	1.771	2.160	2.650	3.012	3.852	4.221	13
14	1.345	1.761	2.145	2.624	2.977	3.787	4.140	14
15	1.341	1.753	2.131	2.602	2.947	3.733	4.073	15
16	1.337	1.746	2.120	2.583	2.921	3.686	4.015	16
17	1.333	1.740	2.110	2.567	2.898	3.646	3.965	17
18	1.330	1.734	2.101	2.552	2.878	3.611	3.922	18
19	1.328	1.729	2.093	2.539	2.861	3.579	3.883	19
20	1.325	1.725	2.086	2.528	2.845	3.552	3.850	20
21	1.323	1.721	2.080	2.518	2.831	3.527	3.819	21
22	1.321	1.717	2.074	2.508	2.819	3.505	3.792	22
23	1.319	1.714	2.069	2.500	2.807	3.485	3.767	23
24	1.318	1.711	2.064	2.492	2.797	3.467	3.745	24
25	1.316	1.708	2.060	2.485	2.787	3.450	3.725	25
26	1.315	1.706	2.056	2.479	2.779	3.435	3.707	26
27	1.314	1.703	2.052	2.473	2.771	3.421	3.690	27
28	1.313	1.701	2.048	2.467	2.763	3.408	3.674	28
29	1.311	1.699	2.045	2.462	2.756	3.396	3.659	29
30	1.310	1.697	2.042	2.457	2.750	3.385	3.646	30
40	1.303	1.684	2.021	2.423	2.704	3.307	3.551	40
60	1.296	1.671	2.000	2.390	2.660	3.232	3.460	60
120	1.289	1.658	1.980	2.358	2.617	3.160	3.373	120
∞	1.282	1.645	1.960	2.326	2.576	3.090	3.291	∞

Table 9.4.   Critical values of the Chi-squared distribution.

ν \ α	0.9995	0.999	0.995	0.99	0.975	0.95	0.90	0.10	0.05	0.025	0.01	0.005	0.001	0.0005	α \ ν
1	0.0^6393	0.0^5157	0.0^4393	0.0^3157	0.0^3982	0.0^2393	0.0158	2.706	3.841	5.024	6.635	7.879	10.828	12.116	1
2	0.0^2100	0.0^2200	0.0100	0.0201	0.0506	0.103	0.211	4.605	5.991	7.378	9.210	10.597	13.816	15.202	2
3	0.0153	0.0243	0.0717	0.115	0.216	0.352	0.584	6.251	7.815	9.348	11.345	12.838	16.266	17.730	3
4	0.0639	0.0908	0.207	0.297	0.484	0.711	1.064	7.779	9.488	11.143	13.277	14.860	18.467	19.997	4
5	0.158	0.210	0.412	0.554	0.831	1.145	1.610	9.236	11.070	12.832	15.086	16.750	20.515	22.105	5
6	0.299	0.381	0.676	0.872	1.237	1.635	2.204	10.645	12.592	14.449	16.812	18.548	22.458	24.103	6
7	0.485	0.598	0.989	1.239	1.690	2.167	2.833	12.017	14.067	16.013	18.475	20.278	24.322	26.018	7
8	0.710	0.857	1.344	1.646	2.180	2.733	3.490	13.362	15.507	17.535	20.090	21.955	26.125	27.868	8
9	0.972	1.153	1.735	2.088	2.700	3.325	4.168	14.684	16.919	19.023	21.666	23.589	27.877	29.666	9
10	1.265	1.479	2.156	2.558	3.247	3.940	4.865	15.987	18.307	20.483	23.209	25.188	29.588	31.420	10
11	1.587	1.834	2.603	3.053	3.816	4.575	5.578	17.275	19.675	21.920	24.725	26.757	31.264	33.136	11
12	1.934	2.214	3.074	3.571	4.404	5.226	6.304	18.549	21.026	23.336	26.217	28.300	32.909	34.821	12
13	2.305	2.617	3.565	4.107	5.009	5.892	7.042	19.812	22.362	24.735	27.688	29.819	34.528	36.478	13
14	2.697	3.041	4.075	4.660	5.629	6.571	7.790	21.064	23.685	26.119	29.141	31.319	36.123	38.109	14
15	3.108	3.483	4.601	5.229	6.262	7.261	8.547	22.307	24.996	27.488	30.578	32.801	37.697	39.719	15
16	3.536	3.942	5.142	5.812	6.908	7.962	9.312	23.542	26.296	28.845	32.000	34.267	39.252	41.308	16
17	3.980	4.416	5.697	6.408	7.564	8.672	10.085	24.769	27.587	30.191	33.409	35.718	40.790	42.879	17
18	4.439	4.905	6.265	7.015	8.231	9.390	10.865	25.989	28.869	31.526	34.805	37.156	42.312	44.434	18
19	4.912	5.407	6.844	7.633	8.907	10.117	11.651	27.204	30.144	32.852	36.191	38.582	43.820	45.973	19
20	5.398	5.921	7.434	8.260	9.591	10.851	12.443	28.412	31.410	34.170	37.566	39.997	45.315	47.498	20

(Continued)

Table 9.4. (Continued).

ν \ α	0.9995	0.999	0.995	0.99	0.975	0.95	0.90	0.10	0.05	0.025	0.01	0.005	0.001	0.0005	α \ ν
21	5.896	6.447	8.034	8.897	10.283	11.591	13.240	29.615	32.671	35.479	38.932	41.401	46.797	49.010	21
22	6.404	6.983	8.643	9.542	10.982	12.338	14.041	30.813	33.924	36.781	40.289	42.796	48.268	50.511	22
23	6.924	7.529	9.260	10.196	11.688	13.091	14.848	32.007	35.172	38.086	41.638	44.181	49.728	52.000	23
24	7.453	8.085	9.886	10.856	12.401	13.848	15.659	33.196	36.415	39.364	42.980	45.558	51.179	53.479	24
25	7.991	8.649	10.520	11.524	13.120	14.611	16.473	34.382	37.652	40.646	44.314	46.928	52.618	54.947	25
26	8.538	9.222	11.160	12.198	13.844	15.379	17.292	35.563	38.885	41.923	45.642	48.290	54.052	56.407	26
27	9.093	9.803	11.808	12.879	14.573	16.151	18.114	36.741	40.113	43.194	46.963	49.645	55.476	57.858	27
28	9.656	10.391	12.461	13.565	15.308	16.928	18.939	37.916	41.337	44.461	48.278	50.993	56.892	59.300	28
29	10.227	10.986	13.121	14.256	16.047	17.708	19.768	39.087	42.557	45.722	49.588	52.336	58.301	60.735	29
30	10.804	11.588	13.787	14.953	16.791	18.493	20.599	40.256	43.773	46.979	50.892	53.672	59.703	62.162	30
40	16.906	17.916	20.707	22.164	24.433	26.509	29.051	51.805	55.758	59.342	63.691	66.766	73.402	76.095	40
50	23.461	24.674	27.991	29.707	32.357	34.764	37.689	63.167	67.505	71.420	76.154	79.490	86.661	89.561	50
60	30.340	31.738	35.535	37.485	40.482	43.188	46.459	74.397	79.082	83.298	88.379	91.952	99.607	102.695	60
70	37.467	39.036	43.275	45.442	48.758	51.739	55.329	85.527	90.531	95.023	100.425	104.215	112.317	115.578	70
80	44.791	46.520	51.172	53.540	57.153	60.391	64.278	96.578	101.879	106.629	112.329	116.321	124.839	128.261	80
90	52.276	54.155	59.196	61.754	65.647	69.126	73.291	107.565	113.145	118.136	124.116	128.299	137.208	140.782	90
100	59.896	61.918	67.328	70.075	74.222	77.929	82.358	118.498	124.342	129.561	135.807	140.169	149.449	153.167	100

Table 9.5. Critical values of the Snedecor's *F* distribution for $\alpha = 0.10$.

v_2 \ v_1	1	2	3	4	5	6	7	8	9	10	12	15	20	24	30	40	60	120	∞
1	39.86	49.50	53.59	55.83	57.24	58.20	58.91	59.44	59.86	60.19	60.71	61.22	61.74	62.00	62.26	62.53	62.79	63.06	63.33
2	8.53	9.00	9.16	9.24	9.29	9.33	9.35	9.37	9.38	9.39	9.41	9.42	9.44	9.45	9.46	9.47	9.47	9.48	9.49
3	5.54	5.46	5.39	5.34	5.31	5.28	5.27	5.25	5.24	5.23	5.22	5.20	5.18	5.18	5.17	5.16	5.15	5.14	5.13
4	4.54	4.32	4.19	4.11	4.05	4.01	3.98	3.95	3.94	3.92	3.90	3.87	3.84	3.83	3.82	3.80	3.79	3.78	3.76
5	4.06	3.78	3.62	3.52	3.45	3.40	3.37	3.34	3.32	3.30	3.27	3.24	3.21	3.19	3.17	3.16	3.14	3.12	3.10
6	3.78	3.46	3.29	3.18	3.11	3.05	3.01	2.98	2.96	2.94	2.90	2.87	2.84	2.82	2.80	2.78	2.76	2.74	2.72
7	3.59	3.26	3.07	2.96	2.88	2.83	2.78	2.75	2.72	2.70	2.67	2.63	2.59	2.58	2.56	2.54	2.51	2.49	2.47
8	3.46	3.11	2.92	2.81	2.73	2.67	2.62	2.59	2.56	2.54	2.50	2.46	2.42	2.40	2.38	2.36	2.34	2.32	2.29
9	3.36	3.01	2.81	2.69	2.61	2.55	2.51	2.47	2.44	2.42	2.38	2.34	2.30	2.28	2.25	2.23	2.21	2.18	2.16
10	3.29	2.92	2.73	2.61	2.52	2.46	2.41	2.38	2.35	2.32	2.28	2.24	2.20	2.18	2.16	2.13	2.11	2.08	2.06
11	3.23	2.86	2.66	2.54	2.45	2.39	2.34	2.30	2.27	2.25	2.21	2.17	2.12	2.10	2.08	2.05	2.03	2.00	1.97
12	3.18	2.81	2.61	2.48	2.39	2.33	2.28	2.24	2.21	2.19	2.15	2.10	2.06	2.04	2.01	1.99	1.96	1.93	1.90
13	3.14	2.76	2.56	2.43	2.35	2.28	2.23	2.20	2.16	2.14	2.10	2.05	2.01	1.98	1.96	1.93	1.90	1.88	1.85
14	3.10	2.73	2.52	2.39	2.31	2.24	2.19	2.15	2.12	2.10	2.05	2.01	1.96	1.94	1.91	1.89	1.86	1.83	1.80
15	3.07	2.70	2.49	2.36	2.27	2.21	2.16	2.12	2.09	2.06	2.02	1.97	1.92	1.90	1.87	1.85	1.82	1.79	1.76
16	3.05	2.67	2.46	2.33	2.24	2.18	2.13	2.09	2.06	2.03	1.99	1.94	1.89	1.87	1.84	1.81	1.78	1.75	1.72
17	3.03	2.64	2.44	2.31	2.22	2.15	2.10	2.06	2.03	2.00	1.96	1.91	1.86	1.84	1.81	1.78	1.75	1.72	1.69
18	3.01	2.62	2.42	2.29	2.20	2.13	2.08	2.04	2.00	1.98	1.93	1.89	1.84	1.81	1.78	1.75	1.72	1.69	1.66
19	2.99	2.61	2.40	2.27	2.18	2.11	2.06	2.02	1.98	1.96	1.91	1.86	1.81	1.79	1.76	1.73	1.70	1.67	1.63
20	2.97	2.59	2.38	2.25	2.16	2.09	2.04	2.00	1.96	1.94	1.89	1.84	1.79	1.77	1.74	1.71	1.68	1.64	1.61

(Continued)

Table 9.5. (Continued).

v_2 \ v_1	1	2	3	4	5	6	7	8	9	10	12	15	20	24	30	40	60	120	∞
21	2.96	2.57	2.36	2.23	2.14	2.08	2.02	1.98	1.95	1.92	1.87	1.83	1.78	1.75	1.72	1.69	1.66	1.62	1.59
22	2.95	2.56	2.35	2.22	2.13	2.06	2.01	1.97	1.93	1.90	1.86	1.81	1.76	1.73	1.70	1.67	1.64	1.60	1.57
23	2.94	2.55	2.34	2.21	2.11	2.05	1.99	1.95	1.92	1.89	1.84	1.80	1.74	1.72	1.69	1.66	1.62	1.59	1.55
24	2.93	2.54	2.33	2.19	2.10	2.04	1.98	1.94	1.91	1.88	1.83	1.78	1.73	1.70	1.67	1.64	1.61	1.57	1.53
25	2.92	2.53	2.32	2.18	2.09	2.02	1.97	1.93	1.89	1.87	1.82	1.77	1.72	1.69	1.66	1.63	1.59	1.56	1.52
26	2.91	2.52	2.31	2.17	2.08	2.01	1.96	1.92	1.88	1.86	1.81	1.76	1.71	1.68	1.65	1.61	1.58	1.54	1.50
27	2.90	2.51	2.30	2.17	2.07	2.00	1.95	1.91	1.87	1.85	1.80	1.75	1.70	1.67	1.64	1.60	1.57	1.53	1.49
28	2.89	2.50	2.29	2.16	2.06	2.00	1.94	1.90	1.87	1.84	1.79	1.74	1.69	1.66	1.63	1.59	1.56	1.52	1.48
29	2.89	2.50	2.28	2.15	2.06	1.99	1.93	1.89	1.86	1.83	1.78	1.73	1.68	1.65	1.62	1.58	1.55	1.51	1.47
30	2.88	2.49	2.28	2.14	2.05	1.98	1.93	1.88	1.85	1.82	1.77	1.72	1.67	1.64	1.61	1.57	1.54	1.50	1.46
40	2.84	2.44	2.23	2.09	2.00	1.93	1.87	1.83	1.79	1.76	1.71	1.66	1.61	1.57	1.54	1.51	1.47	1.42	1.38
60	2.79	2.39	2.18	2.04	1.95	1.87	1.82	1.77	1.74	1.71	1.66	1.60	1.54	1.51	1.48	1.44	1.40	1.35	1.29
120	2.75	2.35	2.13	1.99	1.90	1.82	1.77	1.72	1.68	1.65	1.60	1.55	1.48	1.45	1.41	1.37	1.32	1.26	1.19
∞	2.71	2.30	2.08	1.94	1.85	1.77	1.72	1.67	1.63	1.60	1.55	1.49	1.42	1.38	1.34	1.30	1.24	1.17	1.00

Table 9.6.   Critical values of the Snedecor's F distribution for $\alpha = 0.05$

v_2 \ v_1	1	2	3	4	5	6	7	8	9	10	12	15	20	24	30	40	60	120	∞	v_2
1	161.4	199.5	215.7	224.6	230.2	234.0	236.8	238.9	240.5	241.9	243.9	245.9	248.0	249.1	250.1	251.1	252.2	253.3	254.3	1
2	18.51	19.00	19.16	19.25	19.30	19.33	19.35	19.37	19.38	19.39	19.41	19.43	19.45	19.45	19.46	19.47	19.48	19.49	19.50	2
3	10.13	9.55	9.28	9.12	9.01	8.94	8.89	8.85	8.81	8.79	8.74	8.70	8.66	8.64	8.62	8.59	8.57	8.55	8.53	3
4	7.71	6.94	6.59	6.39	6.26	6.16	6.09	6.04	6.00	5.96	5.91	5.86	5.80	5.77	5.75	5.72	5.69	5.66	5.63	4
5	6.61	5.79	5.41	5.19	5.05	4.95	4.88	4.82	4.77	4.74	4.68	4.62	4.56	4.53	4.50	4.46	4.43	4.40	4.36	5
6	5.99	5.14	4.76	4.53	4.39	4.28	4.21	4.15	4.10	4.06	4.00	3.94	3.87	3.84	3.81	3.77	3.74	3.70	3.67	6
7	5.59	4.74	4.35	4.12	3.97	3.87	3.79	3.73	3.68	3.64	3.57	3.51	3.44	3.41	3.38	3.34	3.30	3.27	3.23	7
8	5.32	4.46	4.07	3.84	3.69	3.58	3.50	3.44	3.39	3.35	3.28	3.22	3.15	3.12	3.08	3.04	3.01	2.97	2.93	8
9	5.12	4.26	3.86	3.63	3.48	3.37	3.29	3.23	3.18	3.14	3.07	3.01	2.94	2.90	2.86	2.83	2.79	2.75	2.71	9
10	4.96	4.10	3.71	3.48	3.33	3.22	3.14	3.07	3.02	2.98	2.91	2.85	2.77	2.74	2.70	2.66	2.62	2.58	2.54	10
11	4.84	3.98	3.59	3.36	3.20	3.09	3.01	2.95	2.90	2.85	2.79	2.72	2.65	2.61	2.57	2.53	2.49	2.45	2.40	11
12	4.75	3.89	3.49	3.26	3.11	3.00	2.91	2.85	2.80	2.75	2.69	2.62	2.54	2.51	2.47	2.43	2.38	2.34	2.30	12
13	4.67	3.81	3.41	3.18	3.03	2.92	2.83	2.77	2.71	2.67	2.60	2.53	2.46	2.42	2.38	2.34	2.30	2.25	2.21	13
14	4.60	3.74	3.34	3.11	2.96	2.85	2.76	2.70	2.65	2.60	2.53	2.46	2.39	2.35	2.31	2.27	2.22	2.18	2.13	14
15	4.54	3.68	3.29	3.06	2.90	2.79	2.71	2.64	2.59	2.54	2.48	2.40	2.33	2.29	2.25	2.20	2.16	2.11	2.07	15
16	4.49	3.63	3.24	3.01	2.85	2.74	2.66	2.59	2.54	2.49	2.42	2.35	2.28	2.24	2.19	2.15	2.11	2.06	2.01	16
17	4.45	3.59	3.20	2.96	2.81	2.70	2.61	2.55	2.49	2.45	2.38	2.31	2.23	2.19	2.15	2.10	2.06	2.01	1.96	17
18	4.41	3.55	3.16	2.93	2.77	2.66	2.58	2.51	2.46	2.41	2.34	2.27	2.19	2.15	2.11	2.06	2.02	1.97	1.92	18
19	4.38	3.52	3.13	2.90	2.74	2.63	2.54	2.48	2.42	2.38	2.31	2.23	2.16	2.11	2.07	2.03	1.98	1.93	1.88	19
20	4.35	3.49	3.10	2.87	2.71	2.60	2.51	2.45	2.39	2.35	2.28	2.20	2.12	2.08	2.04	1.99	1.95	1.90	1.84	20

(Continued)

Table 9.6. (Continued).

v_2 \ v_1	1	2	3	4	5	6	7	8	9	10	12	15	20	24	30	40	60	120	∞
21	4.32	3.47	3.07	2.84	2.68	2.57	2.49	2.42	2.37	2.32	2.25	2.18	2.10	2.05	2.01	1.96	1.92	1.87	1.81
22	4.30	3.44	3.05	2.82	2.66	2.55	2.46	2.40	2.34	2.30	2.23	2.15	2.07	2.03	1.98	1.94	1.89	1.84	1.78
23	4.28	3.42	3.03	2.80	2.64	2.53	2.44	2.37	2.32	2.27	2.20	2.13	2.05	2.01	1.96	1.91	1.86	1.81	1.76
24	4.26	3.40	3.01	2.78	2.62	2.51	2.42	2.36	2.30	2.25	2.18	2.11	2.03	1.98	1.94	1.89	1.84	1.79	1.73
25	4.24	3.39	2.99	2.76	2.60	2.49	2.40	2.34	2.28	2.24	2.16	2.09	2.01	1.96	1.92	1.87	1.82	1.77	1.71
26	4.23	3.37	2.98	2.74	2.59	2.47	2.39	2.32	2.27	2.22	2.15	2.07	1.99	1.95	1.90	1.85	1.80	1.75	1.69
27	4.21	3.35	2.96	2.73	2.57	2.46	2.37	2.31	2.25	2.20	2.13	2.06	1.97	1.93	1.88	1.84	1.79	1.73	1.67
28	4.20	3.34	2.95	2.71	2.56	2.45	2.36	2.29	2.24	2.19	2.12	2.04	1.96	1.91	1.87	1.82	1.77	1.71	1.65
29	4.18	3.33	2.93	2.70	2.55	2.43	2.35	2.28	2.22	2.18	2.10	2.03	1.94	1.90	1.85	1.81	1.75	1.70	1.64
30	4.17	3.32	2.92	2.69	2.53	2.42	2.33	2.27	2.21	2.16	2.09	2.01	1.93	1.89	1.84	1.79	1.74	1.68	1.62
40	4.08	3.23	2.84	2.61	2.45	2.34	2.25	2.18	2.12	2.08	2.00	1.92	1.84	1.79	1.74	1.69	1.64	1.58	1.51
60	4.00	3.15	2.76	2.53	2.37	2.25	2.17	2.10	2.04	1.99	1.92	1.84	1.75	1.70	1.65	1.59	1.53	1.47	1.39
120	3.92	3.07	2.68	2.45	2.29	2.17	2.09	2.02	1.96	1.91	1.83	1.75	1.66	1.61	1.55	1.50	1.43	1.35	1.25
∞	3.84	3.00	2.60	2.37	2.21	2.10	2.01	1.94	1.88	1.83	1.75	1.67	1.57	1.52	1.46	1.39	1.32	1.22	1.00

Table 9.7. Critical values of the Snedecor's F distribution for $\alpha = 0.025$.

v_2 \ v_1	1	2	3	4	5	6	7	8	9	10	12	15	20	24	30	40	60	120	∞
1	647.8	799.5	864.2	899.6	921.8	937.1	948.2	956.7	963.3	968.6	976.7	984.9	993.1	997.2	1001	1006	1010	1014	1018
2	38.51	39.00	39.17	39.25	39.30	39.33	39.35	39.37	39.39	39.40	39.41	39.43	39.45	39.46	39.46	39.47	39.48	39.49	39.50
3	17.44	16.04	15.44	15.10	14.88	14.73	14.62	14.54	14.47	14.42	14.34	14.25	14.17	14.12	14.08	14.04	13.99	13.95	13.90
4	12.22	10.65	9.98	9.60	9.36	9.20	9.07	8.98	8.90	8.84	8.75	8.66	8.56	8.51	8.46	8.41	8.36	8.31	8.26
5	10.01	8.43	7.76	7.39	7.15	6.98	6.85	6.76	6.68	6.62	6.52	6.43	6.33	6.28	6.23	6.18	6.12	6.07	6.02
6	8.81	7.26	6.60	6.23	5.99	5.82	5.70	5.60	5.52	5.46	5.37	5.27	5.17	5.12	5.07	5.01	4.96	4.90	4.85
7	8.07	6.54	5.89	5.52	5.29	5.12	4.99	4.90	4.82	4.76	4.67	4.57	4.47	4.42	4.36	4.31	4.25	4.20	4.14
8	7.57	6.06	5.42	5.05	4.82	4.65	4.53	4.43	4.36	4.30	4.20	4.10	4.00	3.95	3.89	3.84	3.78	3.73	3.67
9	7.21	5.71	5.08	4.72	4.48	4.32	4.20	4.10	4.03	3.96	3.87	3.77	3.67	3.61	3.56	3.51	3.45	3.39	3.33
10	6.94	5.46	4.83	4.47	4.24	4.07	3.95	3.85	3.78	3.72	3.62	3.52	3.42	3.37	3.31	3.26	3.20	3.14	3.08
11	6.72	5.26	4.63	4.28	4.04	3.88	3.76	3.66	3.59	3.53	3.43	3.33	3.23	3.17	3.12	3.06	3.00	2.94	2.88
12	6.55	5.10	4.47	4.12	3.89	3.73	3.61	3.51	3.44	3.37	3.28	3.18	3.07	3.02	2.96	2.91	2.85	2.79	2.72
13	6.41	4.97	4.35	4.00	3.77	3.60	3.48	3.39	3.31	3.25	3.15	3.05	2.95	2.89	2.84	2.78	2.72	2.66	2.60
14	6.30	4.86	4.24	3.89	3.66	3.50	3.38	3.29	3.21	3.15	3.05	2.95	2.84	2.79	2.73	2.67	2.61	2.55	2.49
15	6.20	4.77	4.15	3.80	3.58	3.41	3.29	3.20	3.12	3.06	2.96	2.86	2.76	2.70	2.64	2.59	2.52	2.46	2.40
16	6.12	4.69	4.08	3.73	3.50	3.34	3.22	3.12	3.05	2.99	2.89	2.79	2.68	2.63	2.57	2.51	2.45	2.38	2.32
17	6.04	4.62	4.01	3.66	3.44	3.28	3.16	3.06	2.98	2.92	2.82	2.72	2.62	2.56	2.50	2.44	2.38	2.32	2.25
18	5.98	4.56	3.95	3.61	3.38	3.22	3.10	3.01	2.93	2.87	2.77	2.67	2.56	2.50	2.44	2.38	2.32	2.26	2.19
19	5.92	4.51	3.90	3.56	3.33	3.17	3.05	2.96	2.88	2.82	2.72	2.62	2.51	2.45	2.39	2.33	2.27	2.20	2.13
20	5.87	4.46	3.86	3.51	3.29	3.13	3.01	2.91	2.84	2.77	2.68	2.57	2.46	2.41	2.35	2.29	2.22	2.16	2.09

(Continued)

Table 9.7. (Continued).

v_2 \ v_1	1	2	3	4	5	6	7	8	9	10	12	15	20	24	30	40	60	120	∞	v_1 \ v_2
21	5.83	4.42	3.82	3.48	3.25	3.09	2.97	2.87	2.80	2.73	2.64	2.53	2.42	2.37	2.31	2.25	2.18	2.11	2.04	21
22	5.79	4.38	3.78	3.44	3.22	3.05	2.93	2.84	2.76	2.70	2.60	2.50	2.39	2.33	2.27	2.21	2.14	2.08	2.00	22
23	5.75	4.35	3.75	3.41	3.18	3.02	2.90	2.81	2.73	2.67	2.57	2.47	2.36	2.30	2.24	2.18	2.11	2.04	1.97	23
24	5.72	4.32	3.72	3.38	3.15	2.99	2.87	2.78	2.70	2.64	2.54	2.44	2.33	2.27	2.21	2.15	2.08	2.01	1.94	24
25	5.69	4.29	3.69	3.35	3.13	2.97	2.85	2.75	2.68	2.61	2.51	2.41	2.30	2.24	2.18	2.12	2.05	1.98	1.91	25
26	5.66	4.27	3.67	3.33	3.10	2.94	2.82	2.73	2.65	2.59	2.49	2.39	2.28	2.22	2.16	2.09	2.03	1.95	1.88	26
27	5.63	4.24	3.65	3.31	3.08	2.92	2.80	2.71	2.63	2.57	2.47	2.36	2.25	2.19	2.13	2.07	2.00	1.93	1.85	27
28	5.61	4.22	3.63	3.29	3.06	2.90	2.78	2.69	2.61	2.55	2.45	2.34	2.23	2.17	2.11	2.05	1.98	1.91	1.83	28
29	5.59	4.20	3.61	3.27	3.04	2.88	2.76	2.67	2.59	2.53	2.43	2.32	2.21	2.15	2.09	2.03	1.96	1.89	1.81	29
30	5.57	4.18	3.59	3.25	3.03	2.87	2.75	2.65	2.57	2.51	2.41	2.31	2.20	2.14	2.07	2.01	1.94	1.87	1.79	30
40	4.42	4.05	3.46	3.13	2.90	2.74	2.62	2.53	2.45	2.39	2.29	2.18	2.07	2.01	1.94	1.88	1.80	1.72	1.64	40
60	5.29	3.93	3.34	3.01	2.79	2.63	2.51	2.41	2.33	2.27	2.17	2.06	1.94	1.88	1.82	1.74	1.67	1.58	1.48	60
120	5.15	3.80	3.23	2.89	2.67	2.52	2.39	2.30	2.22	2.16	2.05	1.94	1.82	1.76	1.69	1.61	1.53	1.43	1.31	120
∞	5.02	3.69	3.12	2.79	2.57	2.41	2.29	2.19	2.11	2.05	1.94	1.83	1.71	1.64	1.57	1.48	1.39	1.27	1.00	∞

Table 9.8.   Critical values of the series distribution.

α = 0.05

n₂ \ n₁	2	3	4	5	6	7	8	9	10	11	12	13	14	15	16	17	18	19	20
2																			
3																			
4			2																
5		2	2	3															
6		2	3	3															
7		2	3	3	4	4													
8	2	2	3	3	4	4	5												
9	2	2	3	4	4	5	5	6											
10	2	3	3	4	5	5	6	6	6										
11	2	3	3	4	5	5	6	6	7	7									
12	2	3	4	4	5	6	6	7	7	8	8								
13	2	3	4	4	5	6	6	7	8	8	9	9							
14	2	3	4	5	5	6	7	7	8	8	9	9	10						
15	2	3	4	5	6	6	7	7	8	9	9	10	10	11					
16	2	3	4	5	6	6	7	8	8	9	10	10	11	11	11				
17	2	3	4	5	6	7	7	8	9	9	10	10	11	11	12	12			
18	2	3	4	5	6	7	8	8	9	10	10	11	11	12	12	13	13		
19	2	3	4	5	6	7	8	8	9	10	10	11	12	12	13	13	14	14	
20	2	3	4	5	6	7	8	9	9	10	11	11	12	12	13	13	14	14	15

α = 0.95

n₂ \ n₁	2	3	4	5	6	7	8	9	10	11	12	13	14	15	16	17	18	19	20
2	4																		
3	5	6																	
4	5	6	7																
5	5	7	8	8															
6	5	7	8	9	10														
7	5	7	8	9	10	11													
8	5	7	9	10	11	12	12												
9	5	7	9	10	11	12	13	13											
10	5	7	9	10	11	12	13	14	15										
11	5	7	9	11	12	13	14	14	15	16									
12	5	7	9	11	12	13	14	15	16	16	17								
13	5	7	9	11	12	13	14	15	16	16	17	18							
14	5	7	9	11	12	13	15	15	16	17	17	18	19						
15	5	7	9	11	12	14	15	16	16	17	18	18	19	19					
16	5	7	9	11	13	14	15	16	17	18	18	19	19	20	20				
17	5	7	9	11	13	14	15	16	17	18	19	19	20	20	21	22			
18	5	7	9	11	13	14	15	17	18	18	19	20	20	21	21	22	23		
19	5	7	9	11	13	14	15	17	18	19	19	20	21	21	22	22	23	24	
20	5	7	9	11	13	14	16	17	18	19	20	20	21	22	22	23	24	25	26

(Continued)

Table 9.8. (Continued).

α = 0.025

n_2＼n_1	2	3	4	5	6	7	8	9	10	11	12	13	14	15	16	17	18	19	20
2																			
3																			
4																			
5			2	2															
6			2	2	3														
7			2	2	3	3													
8			2	3	3	3	4												
9			2	3	3	4	5	5											
10			2	3	3	4	5	5	6										
11			2	3	4	4	5	6	6	7									
12				2	2	3	4	5	6	7	7								
13				2	2	3	4	5	6	7	7	8							
14				2	2	3	4	6	6	7	8	8	9						
15				2	3	3	4	6	7	7	8	9	9	10					
16				2	3	4	5	6	7	8	8	9	10	10	11				
17				2	3	4	5	6	7	8	9	10	10	11	11	11			
18				2	3	4	5	6	7	8	9	10	10	11	11	12	12		
19				2	3	4	5	6	7	8	9	10	11	11	12	12	13	13	
20				2	3	4	5	6	7	8	9	10	10	12	12	13	13	13	14

α = 0.975

n_1＼n_2	2	3	4	5	6	7	8	9	10	11	12	13	14	15	16	17	18	19	20
2	4																		
3	5	6																	
4	5	7	8																
5	5	7	8	9															
6	5	7	9	10	11														
7	5	7	9	10	11	12													
8	5	7	9	10	11	12	13												
9	5	7	9	11	12	13	13	14											
10	5	7	9	11	12	13	14	15	15										
11	5	7	9	11	12	13	14	15	16	16									
12	5	7	9	11	12	13	15	15	16	17	18								
13	5	7	9	11	13	14	15	16	17	18	18	19							
14	5	7	9	11	13	14	15	16	17	18	19	19	20						
15	5	7	9	11	13	14	15	17	17	18	19	20	21	21					
16	5	7	9	11	13	15	16	17	18	19	20	20	21	22	22				
17	5	7	9	11	13	15	16	17	18	19	20	21	22	22	23	24			
18	5	7	9	11	13	15	16	17	18	19	20	21	22	23	24	24	25		
19	5	7	9	11	13	15	16	17	19	20	21	22	22	23	24	25	25	26	
20	5	7	9	11	13	15	16	17	19	20	21	22	23	24	24	25	26	26	27

Table 9.9. Critical values of the Cochran statistic for α = 0.05.

v_x / K	1	2	3	4	5	6	7	8	9	10	16	36	144	∞
2	0.9985	0.9750	0.9392	0.9057	0.8772	0.8534	0.8332	0.8159	0.8010	0.7880	0.7341	0.6602	0.5813	0.5000
3	0.9669	0.8709	0.7977	0.7457	0.7071	0.6771	0.6530	0.6333	0.6167	0.6025	0.5466	0.4748	0.4031	0.3333
4	0.9065	0.7679	0.6841	0.6287	0.5895	0.5598	0.5365	0.5175	0.5017	0.4884	0.4366	0.3720	0.3093	0.2500
5	0.8412	0.6838	0.5981	0.5441	0.5065	0.4783	0.4564	0.4387	0.4241	0.4118	0.3645	0.3066	0.2513	0.2000
6	0.7808	0.6161	0.5321	0.4803	0.4447	0.4184	0.3980	0.3817	0.3682	0.3568	0.3135	0.2612	0.2119	0.1667
7	0.7271	0.5612	0.4800	0.4307	0.3974	0.3726	0.3535	0.3384	0.3259	0.3154	0.2756	0.2278	0.1833	0.1429
8	0.6798	0.5157	0.4377	0.3910	0.3595	0.3362	0.3185	0.3043	0.2926	0.2829	0.2462	0.2022	0.1616	0.1250
9	0.6385	0.4775	0.4027	0.3584	0.3286	0.3067	0.2901	0.2768	0.2659	0.2568	0.2226	0.1820	0.1446	0.1111
10	0.6020	0.4450	0.3733	0.3311	0.3029	0.2823	0.2666	0.2541	0.2439	0.2353	0.2032	0.1655	0.1308	0.1000
12	0.5410	0.3924	0.3264	0.2880	0.2624	0.2439	0.2299	0.2187	0.2098	0.2020	0.1737	0.1403	0.1100	0.0833
15	0.4709	0.3346	0.2758	0.2419	0.2195	0.2034	0.1911	0.1815	0.1736	0.1671	0.1429	0.1144	0.0889	0.0667
20	0.3894	0.2705	0.2205	0.1921	0.1735	0.1602	0.1501	0.1422	0.1357	0.1303	0.1108	0.0879	0.0675	0.0500
24	0.3434	0.2354	0.1907	0.1656	0.1493	0.1374	0.1286	0.1216	0.1160	0.1113	0.0942	0.0743	0.0567	0.0417
30	0.2929	0.1980	0.1593	0.1377	0.1237	0.1137	0.1061	0.1002	0.0958	0.0921	0.0771	0.0604	0.0457	0.0333
40	0.2370	0.1576	0.1259	0.1082	0.0968	0.0887	0.0827	0.0780	0.0745	0.0713	0.0595	0.0462	0.0347	0.0250
60	0.1737	0.1131	0.0895	0.0765	0.0682	0.0623	0.0583	0.0552	0.0520	0.0497	0.0411	0.0316	0.0234	0.0167
120	0.0998	0.0632	0.0495	0.0419	0.0371	0.0337	0.0312	0.0292	0.0279	0.0266	0.0218	0.0165	0.0120	0.0083
∞	0	0	0	0	0	0	0	0	0	0	0	0	0	0

Table 9.10.   Critical values of the Hartley statistic for $\alpha = 0.05$.

$n-1$ \ K	2	3	4	5	6	7	8	9	10	11	12
2	39.0	87.5	142	202	266	333	403	475	550	626	704
3	15.4	27.8	39.2	50.7	62.0	72.9	83.5	93.9	104	114	124
4	9.60	15.5	20.6	25.2	29.5	33.6	37.5	41.1	44.6	48.0	51.4
5	7.15	10.8	13.7	16.3	18.7	20.8	22.9	24.7	26.5	28.2	29.9
6	5.82	8.38	10.4	12.1	13.7	15.0	16.3	17.5	18.6	19.7	20.7
7	4.99	6.94	8.44	9.70	10.8	11.8	12.7	13.5	14.3	15.1	15.8
8	4.43	6.00	7.18	8.12	9.03	9.78	10.5	11.1	11.7	12.2	12.7
9	4.03	5.34	6.31	7.11	7.80	8.41	8.95	9.45	9.91	10.3	10.7
10	3.72	4.85	5.67	6.34	6.92	7.42	7.87	8.28	8.66	9.01	9.34
12	3.28	4.16	4.79	5.30	5.72	6.09	6.42	6.72	7.00	7.25	7.48
15	2.86	3.54	4.01	4.37	4.68	4.95	5.19	5.40	5.59	5.77	5.93
20	2.46	2.95	3.29	3.54	3.76	3.94	4.10	4.24	4.37	4.49	4.59
30	2.07	2.40	2.61	2.78	2.91	3.02	3.12	3.21	3.29	3.36	3.39
60	1.67	1.85	1.96	2.04	2.11	2.17	2.22	2.26	2.30	2.33	2.36
∞	1.00	1.00	1.00	1.00	1.00	1.00	1.00	1.00	1.00	1.00	1.00

Table 9.11.   Quantiles of the Poisson distribution.

d \ α	0.99993	0.9999	0.9993	0.999	0.993	0.99	0.95
0	0.000070	0.000100	0.000700	0.00100	0.00702	0.01005	0.05129
1	0.01188	0.01421	0.03789	0.04540	0.12326	0.14855	0.35536
2	0.07633	0.08618	0.16824	0.19053	0.38209	0.43604	0.81769
3	0.21115	0.23180	0.38894	0.42855	0.74108	0.82325	1.36632
4	0.41162	0.44446	0.68204	0.73937	1.17032	1.27911	1.97013
5	0.66825	0.71375	1.03236	1.10710	1.65152	1.78528	2.61301
6	0.97222	1.03840	1.42874	1.52034	2.17293	2.33021	3.28532
7	1.31628	1.38697	1.86297	1.97041	2.72659	2.90611	3.98082
8	1.69465	1.77758	2.32894	2.45242	3.30682	3.50746	4.69523
9	2.10271	2.19758	2.82197	2.96052	3.90942	4.13020	5.42541
10	2.53672	2.64323	3.33840	3.49148	4.53113	4.77125	6.16901
11	2.99367	3.11150	3.87531	4.04244	5.16960	5.42818	6.92421
12	3.47103	3.59988	4.43033	4.61106	5.82265	6.09907	7.68958
13	3.96672	4.10632	5.00152	5.19544	6.48871	6.78235	8.46394
14	4.47896	4.62904	5.58725	5.79398	7.16642	7.47673	9.24633
15	5.00626	5.16657	6.18615	6.40533	7.85464	8.18111	10.03596
16	5.54732	5.71762	6.79705	7.02835	8.55241	8.89457	10.83214

(Continued)

Table 9.11. (Continued).

d \ α	0.93	0.90	0.80	0.70	0.60	0.50	0.40
0	0.07257	0.10536	0.22314	0.35667	0.51082	0.69315	0.91629
1	0.43081	0.53181	0.82439	1.09733	1.37642	1.67835	2.02231
2	0.94230	1.10206	1.53504	1.91379	2.28508	2.67406	3.10338
3	1.53414	1.74477	2.29679	2.76371	3.21132	3.67206	4.17526
4	2.17670	2.43259	3.08954	3.63361	4.14774	4.67091	5.23662
5	2.85488	3.15190	3.90366	4.51714	5.09098	5.67016	6.29192
6	3.53984	3.89477	4.73366	5.41074	6.03924	6.66966	7.34265
7	4.28584	4.65612	5.57606	6.31217	6.99137	7.66925	8.38977
8	5.02895	5.43247	6.42848	7.21993	7.94661	8.66895	9.43395
9	5.78633	6.22130	7.28922	8.13293	8.90441	9.66871	10.47568
10	6.55583	7.02075	8.15702	9.05036	9.86440	10.66852	11.51533
11	7.33581	7.82934	9.03090	9.97161	10.82624	11.66836	12.55317
12	8.12496	8.64594	9.91010	10.89620	11.78972	12.66823	13.58944
13	8.92222	9.46962	10.79398	11.82373	12.75462	13.66811	14.62431
14	9.72672	10.29962	11.68206	12.75388	13.72081	14.66802	15.65793
15	10.53773	11.13539	12.57389	13.68639	14.68814	15.66793	16.68043
16	11.35465	11.97613	13.46913	14.62103	15.65651	16.66785	17.72191

d \ α	0.30	0.20	0.10	0.05	0.025	0.01	0.005
0	1.20397	1.60944	2.30258	2.99573	3.68888	4.60517	5.29832
1	2.43922	2.99431	3.88972	4.74386	5.57164	6.63835	7.43013
2	3.61557	4.27903	5.32232	6.29579	7.22469	8.40595	9.27379
3	4.76223	5.51504	6.68078	7.75366	8.76727	10.04512	10.97748
4	5.89036	6.72098	7.99359	9.15352	10.24159	11.60462	12.59409
5	7.00555	7.90599	9.27467	10.51303	11.66833	13.10848	14.14976
6	8.11105	9.07538	10.53207	11.84240	13.05947	14.57062	15.65968
7	9.20895	10.23254	11.77091	13.14811	14.42268	15.99996	17.13359
8	10.30068	11.37977	12.99471	14.43465	15.76319	17.40265	18.57822
9	11.38727	12.51875	14.20599	15.70522	17.08480	18.78312	19.99842
10	12.46951	13.65073	15.40664	16.96222	18.39036	20.14468	21.39783
11	13.54798	14.77666	16.59812	18.20751	19.68204	21.48991	22.77926
12	14.62316	15.89731	17.78158	19.44257	20.96158	22.82084	24.14494
13	15.69544	17.01328	18.95796	20.66857	22.23040	24.13912	25.49669
14	16.78512	18.12509	20.12801	21.88648	23.48962	25.44609	26.83398
15	17.83246	19.23316	21.29237	23.09713	24.74022	26.74289	28.16406
16	18.89769	20.93782	22.45158	24.30118	25.98300	28.03045	29.48198

Table 9.12.   Critical values in Kolmogorov test of goodness-of-fit.

n \ α	0.20	0.10	0.05	0.02	0.01
1	0.900 00	0.950 00	0.975 00	0.990 00	0.995 00
2	0.683 77	0.776 39	0.841 89	0.900 00	0.929 29
3	0.564 81	0.636 04	0.707 60	0.784 56	0.829 00
4	0.492 65	0.563 22	0.623 94	0.688 87	0.734 24
5	0.446 98	0.509 45	0.563 28	0.627 18	0.668 53
6	0.410 37	0.467 99	0.519 26	0.577 41	0.616 61
7	0.381 48	0.436 07	0.483 42	0.538 44	0.575 81
8	0.358 31	0.409 62	0.454 27	0.506 54	0.541 79
9	0.339 10	0.387 46	0.430 01	0.479 60	0.513 32
10	0.322 60	0.368 66	0.409 25	0.456 62	0.488 93
11	0.308 29	0.352 42	0.391 22	0.436 70	0.467 70
12	0.295 77	0.338 15	0.375 43	0.419 18	0.449 05
13	0.284 70	0.325 49	0.361 43	0.403 62	0.432 47
14	0.274 81	0.314 17	0.348 90	0.389 70	0.417 62
15	0.265 88	0.303 97	0.337 60	0.377 13	0.404 20
16	0.257 78	0.294 72	0.327 33	0.365 71	0.392 01
17	0.250 39	0.286 27	0.317 96	0.355 28	0.380 86
18	0.243 60	0.278 51	0.309 36	0.345 69	0.370 62
19	0.237 35	0.271 36	0.301 43	0.336 85	0.361 17
20	0.231 56	0.264 73	0.294 08	0.328 66	0.352 41
21	0.226 17	0.258 58	0.287 24	0.321 04	0.344 27
22	0.221 15	0.252 83	0.280 87	0.313 94	0.336 66
23	0.216 45	0.247 46	0.274 90	0.307 28	0.329 54
24	0.212 05	0.242 42	0.269 31	0.301 04	0.322 86
25	0.207 90	0.237 68	0.264 04	0.295 16	0.316 57

n \ α	0.20	0.10	0.05	0.02	0.01
51	0.146 97	0.167 96	0.186 59	0.208 64	0.223 86
52	0.145 58	0.166 37	0.184 82	0.206 67	0.221 74
53	0.144 23	0.164 83	0.183 11	0.204 75	0.219 68
54	0.142 92	0.163 32	0.181 44	0.202 89	0.217 68
55	0.141 64	0.161 86	0.179 81	0.201 07	0.215 74
56	0.140 40	0.160 44	0.178 23	0.199 30	0.213 84
57	0.139 19	0.159 06	0.176 69	0.197 58	0.211 99
58	0.138 01	0.157 71	0.175 19	0.195 90	0.210 19
59	0.136 86	0.156 39	0.173 73	0.194 27	0.208 44
60	0.135 73	0.155 11	0.172 31	0.192 67	0.206 73
61	0.134 64	0.153 85	0.170 91	0.191 12	0.205 06
62	0.133 57	0.152 63	0.169 56	0.189 60	0.203 43
63	0.132 53	0.151 44	0.168 23	0.188 12	0.201 84
64	0.131 51	0.150 27	0.166 93	0.186 67	0.200 29
65	0.130 52	0.149 13	0.165 67	0.185 25	0.198 77
66	0.129 54	0.148 02	0.164 43	0.183 87	0.197 29
67	0.128 59	0.146 93	0.163 22	0.182 52	0.195 84
68	0.127 66	0.145 87	0.162 04	0.181 19	0.194 42
69	0.126 75	0.144 83	0.160 88	0.179 90	0.193 03
70	0.125 86	0.143 81	0.159 75	0.178 63	0.191 67
71	0.124 99	0.142 81	0.158 64	0.177 39	0.190 34
72	0.124 13	0.141 83	0.157 53	0.176 18	0.189 03
73	0.123 29	0.140 87	0.156 49	0.174 98	0.187 76
74	0.122 47	0.139 93	0.155 44	0.173 82	0.186 50
75	0.121 67	0.139 01	0.154 42	0.172 68	0.185 28

n					
26	0.203 99	0.233 20	0.259 07	0.289 62	0.310 64
27	0.200 30	0.228 98	0.254 38	0.284 38	0.305 02
28	0.196 80	0.224 97	0.249 93	0.279 42	0.299 71
29	0.193 48	0.221 17	0.245 71	0.274 71	0.294 66
30	0.190 32	0.217 56	0.241 70	0.270 23	0.289 87
31	0.187 32	0.214 12	0.237 88	0.265 96	0.285 30
32	0.184 45	0.210 85	0.234 24	0.261 89	0.280 94
33	0.181 71	0.207 71	0.230 76	0.258 01	0.276 77
34	0.179 09	0.204 72	0.227 43	0.254 29	0.272 79
35	0.176 59	0.201 85	0.224 25	0.250 73	0.268 97
36	0.174 18	0.199 10	0.221 19	0.247 32	0.265 32
37	0.171 88	0.196 46	0.218 26	0.244 04	0.261 80
38	0.169 66	0.193 92	0.215 44	0.240 89	0.258 43
39	0.167 53	0.191 48	0.212 73	0.237 86	0.255 18
40	0.165 47	0.189 13	0.210 12	0.234 94	0.232 03
41	0.163 49	0.186 87	0.207 60	0.232 13	0.249 04
42	0.161 58	0.184 68	0.205 17	0.229 41	0.246 13
43	0.159 74	0.182 57	0.202 83	0.226 79	0.243 32
44	0.157 96	0.180 53	0.200 56	0.224 26	0.240 60
45	0.156 23	0.178 56	0.198 57	0.221 81	0.237 98
46	0.154 57	0.176 65	0.196 25	0.219 44	0.235 44
47	0.152 95	0.174 81	0.194 20	0.217 15	0.232 98
48	0.151 39	0.173 02	0.192 21	0.214 93	0.230 59
49	0.149 87	0.171 28	0.190 28	0.212 77	0.228 28
50	0.148 40	0.169 59	0.188 41	0.210 68	0.226 04

n					
76	0.120 88	0.138 11	0.153 42	0.171 55	0.184 08
77	0.120 11	0.137 23	0.152 44	0.170 45	0.182 90
78	0.119 35	0.136 36	0.151 47	0.169 38	0.181 74
79	0.118 60	0.135 51	0.150 52	0.168 32	0.180 60
80	0.117 87	0.134 67	0.149 60	0.167 28	0.179 49
81	0.117 16	0.133 85	0.148 68	0.166 26	0.178 40
82	0.116 45	0.133 05	0.147 79	0.165 26	0.177 32
83	0.115 76	0.132 26	0.146 91	0.164 28	0.176 27
84	0.115 08	0.131 48	0.146 05	0.163 31	0.175 23
85	0.114 42	0.130 72	0.145 20	0.162 36	0.174 21
86	0.113 76	0.129 97	0.144 37	0.161 43	0.173 21
87	0.113 11	0.129 23	0.143 55	0.160 51	0.172 23
88	0.112 48	0.128 50	0.142 74	0.159 61	0.171 26
89	0.111 86	0.127 79	0.141 95	0.158 73	0.170 31
90	0.111 25	0.127 09	0.141 17	0.157 86	0.169 38
91	0.110 64	0.126 40	0.140 40	0.157 00	0.168 46
92	0.110 05	0.125 72	0.139 65	0.156 16	0.167 55
93	0.109 47	0.125 06	0.138 91	0.155 33	0.166 66
94	0.108 89	0.124 40	0.138 18	0.154 51	0.165 79
95	0.108 33	0.123 75	0.137 46	0.153 71	0.164 93
96	0.107 77	0.123 12	0.136 75	0.152 91	0.164 08
97	0.107 22	0.122 49	0.136 06	0.152 14	0.163 24
89	0.106 68	0.121 87	0.135 37	0.151 37	0.162 42
99	0.106 15	0.121 26	0.134 69	0.150 61	0.161 61
100	0.105 63	0.120 67	0.134 03	0.149 87	0.160 81

Table 9.13. Critical values of a linear correlation coefficient and a partial correlation coefficient.

N \ α	0.1	0.05	0.02	0.01	0.001
1	0.98769	0.99692	0.999507	0.999877	0.9999988
2	0.90000	0.95000	0.980000	0.990000	0.99900
3	0.8054	0.8783	0.93433	0.95873	0.99116
4	0.7293	0.8114	0.8822	0.91720	0.97406
5	0.6694	0.7545	0.8329	0.8745	0.95074
6	0.6215	0.7067	0.7887	0.8343	0.92493
7	0.5822	0.6664	0.7498	0.7977	0.8982
8	0.5494	0.6319	0.7155	0.7646	0.8721
9	0.5214	0.6021	0.6851	0.7348	0.8471
10	0.4973	0.5760	0.6581	0.7079	0.8233
11	0.4762	0.5529	0.6339	0.6835	0.8010
12	0.4575	0.5324	0.6120	0.6614	0.7800
13	0.4409	0.5139	0.5923	0.6411	0.7603
14	0.4259	0.4973	0.5742	0.6226	0.7420
15	0.4124	0.4821	0.5577	0.6055	0.7246
16	0.4000	0.4683	0.5425	0.5897	0.7084
17	0.3887	0.4555	0.5285	0.5751	0.6932
18	0.3783	0.4438	0.5155	0.5614	0.6787
19	0.3687	0.4329	0.5034	0.5487	0.6652
20	0.3598	0.4227	0.4921	0.5368	0.6524
25	0.3233	0.3809	0.4451	0.4869	0.5974
30	0.2960	0.3494	0.4093	0.4487	0.5541
35	0.2746	0.3246	0.3810	0.4182	0.5189
40	0.2573	0.3044	0.3578	0.3932	0.4896
45	0.2428	0.2475	0.3384	0.3721	0.4648
50	0.2306	0.2732	0.3218	0.3541	0.4433
60	0.2108	0.2500	0.2948	0.3248	0.4078
70	0.1954	0.2319	0.2737	0.3017	0.3799
80	0.1829	0.2172	0.2565	0.2830	0.3568
90	0.1726	0.2050	0.2422	0.2673	0.3375
100	0.1638	0.1946	0.2301	0.2540	0.3211

Table 9.14. Critical values of the Spearman's rank correlation coefficient.

	α	
n	0.05	0.01
4	1.000	–
5	0.900	1.000
6	0.829	0.943
7	0.714	0.893
8	0.643	0.833
9	0.600	0.783
10	0.564	0.746
12	0.506	0.712
14	0.456	0.645
16	0.425	0.601
18	0.399	0.564
20	0.377	0.534
22	0.359	0.508
24	0.343	0.485
26	0.329	0.465
28	0.317	0.448
30	0.306	0.432

Table 9.15. Critical values of a multiple correlation coefficient.

N \ k	3	4	5	6
1	0.999	0.999	0.999	0.999
	1.000	1.000	1.000	1.000
2	0.975	0.983	0.987	0.990
	0.995	0.997	0.997	0.998
3	0.930	0.950	0.961	0.968
	0.977	0.983	0.987	0.990
4	0.881	0.912	0.930	0.942
	0.949	0.962	0.970	0.975
5	0.836	0.874	0.898	0.914
	0.917	0.937	0.949	0.957
6	0.795	0.839	0.867	0.886
	0.886	0.911	0.927	0.938
7	0.758	0.807	0.838	0.860
	0.855	0.885	0.904	0.918
8	0.726	0.777	0.811	0.835
	0.827	0.860	0.882	0.898
9	0.697	0.750	0.786	0.812
	0.800	0.837	0.861	0.878
10	0.671	0.726	0.763	0.790
	0.776	0.814	0.840	0.859

(Continued)

Table 9.15. (Continued).

N \ k	3	4	5	6
11	0.648	0.703	0.741	0.770
	0.753	0.793	0.821	0.841
12	0.627	0.683	0.722	0.751
	0.732	0.773	0.802	0.824
13	0.608	0.664	0.703	0.733
	0.712	0.755	0.785	0.807
14	0.590	0.646	0.686	0.717
	0.694	0.737	0.768	0.791
15	0.574	0.630	0.670	0.701
	0.677	0.721	0.752	0.776
16	0.559	0.615	0.656	0.687
	0.662	0.706	0.738	0.762
17	0.545	0.601	0.641	0.673
	0.647	0.691	0.724	0.749
18	0.532	0.587	0.628	0.660
	0.633	0.678	0.710	0.736
19	0.520	0.575	0.615	0.647
	0.620	0.665	0.697	0.723
20	0.509	0.563	0.604	0.636
	0.607	0.652	0.685	0.712
21	0.498	0.552	0.593	0.624
	0.596	0.641	0.674	0.700
22	0.488	0.542	0.582	0.614
	0.585	0.630	0.663	0.690
23	0.479	0.532	0.572	0.604
	0.574	0.619	0.653	0.679
24	0.470	0.523	0.562	0.594
	0.565	0.609	0.643	0.669
25	0.462	0.514	0.553	0.585
	0.555	0.600	0.633	0.660
26	0.454	0.506	0.543	0.576
	0.546	0.590	0.624	0.651
27	0.446	0.498	0.536	0.568
	0.538	0.582	0.615	0.642
28	0.439	0.490	0.529	0.560
	0.529	0.573	0.607	0.633
29	0.432	0.483	0.521	0.552
	0.522	0.565	0.598	0.625
30	0.425	0.476	0.514	0.545
	0.514	0.557	0.591	0.618
40	0.373	0.419	0.455	0.484
	0.454	0.494	0.526	0.552
60	0.308	0.348	0.380	0.406
	0.377	0.414	0.442	0.467
120	0.221	0.251	0.275	0.295
	0.272	0.300	0.322	0.342

The upper row is for $\alpha = 0.05$; the lower row for $\alpha = 0.01$.

Table 9.16a. Critical values $D_{n,m}(\alpha)$ in the Smirnov test of goodness-of-fit for two empirical distributions.

n	m	α 0.10		0.05		0.02		0.01		k
3	3	3	0.100		—		—		—	3
4	4	4	0.029	4	0.029		—		—	4
	3	12	0.057		—		—		—	12
5	5	4	0.079	5	0.008	5	0.008	5	0.008	5
	4	16	0.079	20	0.016	20	0.016		—	20
	3	15	0.036	15	0.036		—			15
	2	10	0.095		—		—		—	10
6	6	5	0.026	5	0.0026	6	0.002	6	0.002	6
	5	24	0.048	24	0.048	30	0.004	30	0.004	30
	4	9	0.095	10	0.048	12	0.010	12	0.010	12
	3	5	0.095	6	0.024		—		—	6
	2	6	0.071		—		—		—	6
7	7	5	0.053	6	0.008	6	0.008	6	0.008	7
	6	28	0.091	30	0.038	35	0.015	36	0.009	42
	5	25	0.066	29	0.030	30	0.015	35	0.003	35
	4	21	0.067	24	0.030	28	0.006	28	0.006	28
	3	18	0.067	21	0.017	21	0.017		—	21
	2	14	0.056		—		—		—	14
8	8	5	0.087	6	0.019	6	0.019	7	0.002	8
	7	34	0.087	40	0.033	42	0.013	48	0.005	56
	6	15	0.093	17	0.043	20	0.009	20	0.009	24
	5	27	0.079	29	0.042	35	0.009	35	0.009	40
	4	6	0.085	7	0.020	8	0.004	8	0.004	8
	3	21	0.048	21	0.048	24	0.012		—	24
	2	8	0.044	8	0.044		—		—	8
9	9	6	0.034	6	0.034	7	0.006	7	0.006	9
	8	40	0.079	46	0.047	54	0.011	55	0.008	72
	7	36	0.098	42	0.034	47	0.015	49	0.008	63
	6	11	0.095	13	0.028	14	0.014	15	0.006	18
	5	30	0.086	35	0.028	36	0.014	40	0.006	45
	4	27	0.062	28	0.042	32	0.014	36	0.003	36
	3	7	0.091	8	0.036	9	0.009	9	0.009	9
	2	18	0.036	18	0.036		—		—	18
10	10	6	0.052	7	0.012	7	0.012	8	0.002	10
	9	50	0.084	53	0.045	61	0.018	63	0.007	90
	8	22	0.095	24	0.050	28	0.012	30	0.007	40
	7	40	0.087	46	0.036	50	0.016	56	0.006	70
	6	18	0.092	20	0.042	22	0.019	24	0.009	30
	5	7	0.061	8	0.019	8	0.019	9	0.004	10
	4	14	0.084	15	0.046	18	0.010	18	0.010	20
	3	24	0.070	27	0.028	30	0.007	30	0.007	30
	2	9	0.091	10	0.030		—		—	10
11	11	6	0.075	7	0.021	8	0.004	8	0.004	11
	10	57	0.092	60	0.043	69	0.017	77	0.008	110
	9	52	0.089	59	0.039	63	0.019	70	0.007	99
	8	48	0.081	53	0.047	61	0.013	64	0.007	88

(Continued)

Table 9.16a.   (Continued).

n	m	α 0.10		0.05		0.02		0.01		k
	7	44	0.083	48	0.049	55	0.014	59	0.006	77
	6	38	0.092	43	0.048	49	0.013	54	0.006	66
	5	35	0.074	39	0.044	44	0.014	45	0.010	55
	4	29	0.098	33	0.035	40	0.007	40	0.007	44
	3	27	0.055	30	0.022	33	0.005	33	0.005	33
	2	20	0.077	22	0.026		—		—	22
12	12	6	0.100	7	0.031	8	0.008	8	0.008	12
	11	64	0.091	72	0.050	77	0.017	86	0.009	132
	10	30	0.093	33	0.049	37	0.020	40	0.007	60
	9	19	0.078	21	0.041	23	0.018	25	0.007	36
	8	13	0.091	15	0.032	16	0.018	17	0.009	24
	7	46	0.098	53	0.034	58	0.017	60	0.010	84
	6	8	0.046	8	0.046	9	0.015	10	0.004	12
	5	36	0.096	43	0.033	48	0.010	50	0.007	60
	4	9	0.048	9	0.048	10	0.016	11	0.005	12
	3	9	0.088	10	0.044	11	0.018	12	0.004	12
	2	11	0.066	12	0.022		—		—	12
13	13	7	0.044	7	0.044	8	0.013	9	0.003	13
	12	71	0.091	81	0.049	92	0.019	95	0.009	156
	11	67	0.100	75	0.048	86	0.015	91	0.009	143
	10	64	0.094	70	0.049	78	0.018	84	0.010	130
	9	59	0.098	65	0.042	73	0.018	78	0.008	117
	8	54	0.099	62	0.039	67	0.019	72	0.009	104
	7	50	0.094	56	0.046	63	0.017	65	0.007	91
	6	46	0.086	52	0.034	54	0.019	59	0.007	78
	5	40	0.087	45	0.040	50	0.015	52	0.007	65
	4	35	0.089	39	0.038	44	0.013	48	0.004	52
	3	30	0.071	33	0.036	36	0.014	39	0.004	39
	2	24	0.057	26	0.019	26	0.019		—	26
14	14	7	0.059	8	0.019	8	0.019	9	0.005	14
	13	78	0.097	89	0.049	102	0.017	104	0.010	182
	12	39	0.087	43	0.044	47	0.020	52	0.008	84
	11	73	0.090	82	0.041	90	0.017	96	0.009	154
	10	34	0.091	37	0.049	42	0.016	45	0.008	70
	9	63	0.082	70	0.046	80	0.015	84	0.008	126
	8	29	0.091	32	0.046	36	0.018	38	0.009	56
	7	8	0.083	9	0.033	10	0.012	11	0.003	14
	6	24	0.086	27	0.037	30	0.014	32	0.008	42
	5	42	0.079	46	0.047	51	0.019	56	0.006	70
	4	19	0.072	21	0.030	24	0.010	24	0.010	28
	3	33	0.050	36	0.029	39	0.012	42	0.003	42
	2	12	0.100	13	0.050	14	0.017		—	14
15	15	7	0.075	8	0.026	9	0.008	9	0.008	15
	14	92	0.100	98	0.044	111	0.017	123	0.009	210
	13	87	0.088	96	0.047	107	0.019	115	0.008	195
	12	28	0.078	31	0.040	34	0.017	36	0.010	60
	11	76	0.099	84	0.048	95	0.018	102	0.009	165
	10	15	0.077	16	0.050	18	0.018	20	0.006	30
	9	38	0.073	25	0.042	28	0.015	30	0.007	45

(Continued)

Table 9.16a. (Continued).

n	m	α								k
		0.10		0.05		0.02		0.01		
	8	60	0.086	67	0.042	75	0.014	81	0.010	120
	7	56	0.079	62	0.047	70	0.014	75	0.009	105
	6	17	0.087	19	0.040	21	0.016	23	0.006	30
	5	10	0.052	11	0.023	12	0.009	12	0.009	15
	4	40	0.086	44	0.042	48	0.018	52	0.008	60
	3	11	0.086	12	0.049	14	0.010	14	0.010	15
	2	26	0.088	28	0.044	30	0.015	—		30
16	16	7	0.093	8	0.035	9	0.011	10	0.003	16
	15	101	0.093	114	0.048	120	0.019	133	0.009	240
	14	48	0.093	53	0.048	60	0.017	63	0.009	112
	13	91	0.089	101	0.047	112	0.018	121	0.009	208
	12	22	0.084	24	0.047	27	0.017	29	0.008	48
	11	80	0.086	89	0.049	100	0.019	106	0.009	176
	10	38	0.088	42	0.044	47	0.017	50	0.009	80
	9	69	0.100	78	0.043	87	0.016	94	0.007	144
	8	9	0.058	10	0.024	11	0.009	11	0.009	16
	7	59	0.094	64	0.048	73	0.018	77	0.010	112
	6	27	0.086	30	0.042	33	0.019	36	0.008	48
	5	48	0.088	54	0.041	59	0.018	64	0.007	80
	4	11	0.070	12	0.034	13	0.014	14	0.006	16
	3	36	0.072	39	0.041	45	0.008	45	0.008	48
	2	14	0.078	15	0.039	16	0.013	—		16
17	17	8	0.045	8	0.045	9	0.016	10	0.005	17
	16	109	0.096	124	0.045	139	0.020	143	0.009	272
	15	105	0.094	116	0.049	131	0.018	142	0.009	255
	14	100	0.096	111	0.048	125	0.018	134	0.009	238
	13	96	0.091	105	0.050	118	0.019	127	0.008	221
	12	90	0.093	100	0.046	112	0.017	119	0.009	204
	11	85	0.092	93	0.046	104	0.018	110	0.010	187
	10	79	0.097	89	0.044	99	0.017	106	0.008	170
	9	74	0.091	82	0.049	92	0.016	99	0.009	153
	8	68	0.097	77	0.044	85	0.018	88	0.009	136
	7	61	0.099	68	0.046	77	0.019	84	0.007	119
	6	56	0.084	62	0.040	68	0.016	73	0.010	102
	5	50	0.094	55	0.048	63	0.015	68	0.005	85
	4	44	0.085	48	0.046	56	0.012	60	0.005	68
	3	36	0.098	42	0.035	45	0.018	48	0.007	51
	2	30	0.070	32	0.035	34	0.012	—		34
18	18	8	0.056	9	0.021	10	0.007	10	0.007	18
	17	118	0.091	133	0.047	150	0.018	164	0.009	306
	16	58	0.093	64	0.048	71	0.019	77	0.009	144
	15	37	0.096	41	0.046	46	0.017	49	0.010	90
	14	52	0.100	58	0.048	65	0.018	70	0.008	126
	13	99	0.098	110	0.050	123	0.020	131	0.010	234
	12	16	0.094	18	0.042	20	0.016	21	0.010	36
	11	88	0.097	97	0.048	108	0.019	118	0.009	198
	10	41	0.099	46	0.047	52	0.016	54	0.010	90
	9	9	0.088	10	0.041	11	0.017	12	0.007	18

(Continued)

Table 9.16a. (Continued).

n	m	α 0.10		0.05		0.02		0.01		k
	8	36	0.088	40	0.040	44	0.017	47	0.010	72
	7	65	0.095	72	0.046	83	0.014	87	0.008	126
	6	11	0.053	12	0.025	13	0.011	14	0.004	18
	5	52	0.099	60	0.038	65	0.019	70	0.008	90
	4	23	0.090	25	0.049	28	0.019	30	0.010	36
	3	13	0.084	15	0.030	16	0.015	17	0.006	18
	2	16	0.063	17	0.032	18	0.011		—	18
19	19	8	0.068	9	0.027	10	0.009	10	0.009	19
	18	133	0.097	142	0.049	160	0.019	176	0.009	342
	17	126	0.095	141	0.046	158	0.018	166	0.010	323
	16	120	0.096	133	0.048	151	0.018	160	0.009	304
	15	114	0.100	127	0.048	142	0.020	152	0.010	285
	14	110	0.095	121	0.047	135	0.019	148	0.008	266
	13	104	0.097	114	0.046	130	0.018	138	0.009	247
	12	99	0.090	108	0.050	121	0.019	130	0.010	228
	11	92	0.094	102	0.049	114	0.018	122	0.009	209
	10	85	0.089	94	0.047	104	0.019	113	0.009	190
	9	80	0.092	89	0.046	99	0.016	107	0.009	171
	8	74	0.097	82	0.049	93	0.018	98	0.010	152
	7	69	0.088	76	0.044	86	0.016	91	0.009	133
	6	64	0.082	70	0.043	77	0.019	83	0.009	114
	5	56	0.082	61	0.043	70	0.015	71	0.009	95
	4	49	0.077	53	0.041	57	0.019	64	0.008	76
	3	42	0.073	45	0.045	51	0.013	54	0.005	57
	2	32	0.095	36	0.029	38	0.010	38	0.010	38
20	20	8	0.081	9	0.034	10	0.012	11	0.004	20
	19	144	0.098	160	0.049	171	0.019	187	0.010	380
	18	68	0.098	76	0.048	85	0.019	91	0.009	180
	17	130	0.099	146	0.048	163	0.019	175	0.009	340
	16	32	0.089	35	0.049	39	0.020	42	0.009	80
	15	25	0.079	27	0.046	30	0.019	32	0.010	60
	14	57	0.095	63	0.049	71	0.019	76	0.009	140
	13	108	0.099	120	0.049	135	0.018	143	0.009	260
	12	26	0.091	29	0.043	32	0.018	35	0.007	60
	11	96	0.096	107	0.046	118	0.020	127	0.010	220
	10	10	0.062	11	0.029	12	0.012	12	0.005	20
	9	84	0.095	93	0.049	104	0.020	111	0.010	180
	8	20	0.087	22	0.44	25	0.014	26	0.009	40
	7	72	0.085	79	0.043	91	0.016	93	0.008	140
	6	33	0.082	36	0.035	40	0.016	44	0.007	60
	5	12	0.085	13	0.047	15	0.012	16	0.005	20
	4	13	0.087	15	0.027	16	0.013	17	0.007	20
	3	42	0.095	48	0.040	54	0.011	57	0.005	60
	2	17	0.087	19	0.026	20	0.009	20	0.009	20

Table 9.16b. Distribution of the Smirnov statistic $D_{n,m}$ $P\{D_{n,m} \le k/n\}$.

k \ n	1	2	3	4	5	6	7
1	1.000 000	0.666 666	0.400 000	0.228 571	0.126 984	0.069 264	0.037 296
2		1.000 000	0.900 000	0.771 428	0.642 857	0.525 974	0.424 825
3			1.000 000	0.971 428	0.920 634	0.857 142	0.787 878
4				1.000 000	0.992 063	0.974 025	0.946 969
5					1.000 000	0.997 835	0.991 841
6						1.000 00	0.999 417
7							1.000 000

k \ n	8	9	10	11	12	13	14
1	0.019 891	0.010 530	0.005 542	0.002 903	0.001 514	0.000 787	0.000 408
2	0.339 860	0.269 888	0.213 070	0.167 412	0.131 018	0.102 194	0.079 484
3	0.717 327	0.648 292	0.582 476	0.520 849	0.463 902	0.411 803	0.364 515
4	0.912 975	0.874 125	0.832 178	0.788 523	0.744 224	0.700 079	0.656 679
5	0.981 351	0.966 433	0.947 552	0.925 339	0.900 453	0.873 512	0.845 065
6	0.997 513	0.993 706	0.987 659	0.979 260	0.968 563	0.955 727	0.940 970
7	0.999 844	0.999 259	0.997 943	0.995 633	0.992 140	0.987 350	0.981 217
8	1.000 000	0.999 958	0.999 783	0.999 345	0.998 503	0.997 125	0.995 100
9		1.000 000	0.999 989	0.999 937	0.999 795	0.999 500	0.998 979
10			1.000 000	0.999 997	0.999 982	0.999 937	0.999 836
11				1.000 000	0.999 999	0.999 995	0.999 981
12					1.000 000	0.999 999	0.999 998
13						1.000 000	1.000 000

k \ n	15	16	17	18	19	20	21
1	0.000 211	0.000 109	0.000 056	0.000 028	0.000 014	0.000 007	0.000 003
2	0.061 668	0.047 743	0.036 892	0.028 460	0.021 922	0.016 863	0.012 955
3	0.321 861	0.283 588	0.249 392	0.218 952	0.191 938	0.168 030	0.146 921
4	0.614 453	0.573 706	0.534 647	0.497 409	0.462 071	0.428 664	0.397 187
5	0.815 583	0.785 465	0.755 040	0.724 581	0.694 310	0.664 409	0.635 020
6	0.924 535	0.906 673	0.887 622	0.867 606	0.846 826	0.825 466	0.803 687
7	0.973 751	0.965 002	0.955 047	0.943 981	0.931 910	0.918 942	0.905 183
8	0.992 344	0.988 800	0.984 439	0.979 252	0.973 250	0.966 458	0.958 911
9	0.998 162	0.996 984	0.995 389	0.993 331	0.990 776	0.987 701	0.984 094
10	0.999 646	0.999 329	0.998 847	0.998 160	0.997 232	0.996 032	0.994 532
11	0.999 947	0.999 880	0.999 761	0.999 570	0.999 285	0.998 884	0.998 343
12	0.999 994	0.999 983	0.999 960	0.999 916	0.999 843	0.999 729	0.999 561
13	0.999 999	0.999 998	0.999 994	0.999 987	0.999 971	0.999 944	0.999 899
14	1.000 000	0.999 999	0.999 999	0.999 998	0.999 995	0.999 990	0.999 980
15		1.000 000	1.000 000	0.999 999	0.999 999	0.999 998	0.999 996
16				1.000 000	0.999 999	0.999 999	0.999 999
17					1.000 000	1.000 000	1.000 000

(Continued)

Table 9.16b. (Continued).

k \ n	22	23	24	25	26	27	28
1	0.000 001	0.000 001	0.000 000	0.000 000	0.000 000	0.000 000	0.000 000
2	0.009 942	0.007 622	0.005 838	0.004 468	0.003 417	0.002 611	0.001 993
3	0.128 321	0.111 963	0.097 599	0.085 006	0.073 980	0.064 337	0.055 914
4	0.367 613	0.339 899	0.313 982	0.289 796	0.267 262	0.246 302	0.226 833
5	0.606 260	0.578 218	0.550 963	0.524 546	0.499 004	0.474 362	0.450 633
6	0.781 631	0.759 421	0.737 166	0.714 957	0.692 876	0.670 992	0.649 361
7	0.890 738	0.875 705	0.860 177	0.844 239	0.827 971	0.811 443	0.794 721
8	0.950 653	0.941 731	0.932 196	0.922 101	0.911 498	0.900 437	0.888 969
9	0.979 952	0.975 279	0.970 086	0.964 388	0.958 206	0.951 561	0.944 480
10	0.992 710	0.990 548	0.988 034	0.985 162	0.981 927	0.978 330	0.974 375
11	0.997 641	0.996 759	0.995 679	0.994 385	0.992 865	0.991 109	0.989 109
12	0.999 326	0.999 009	0.998 598	0.998 079	0.997 439	0.996 666	0.995 750
13	0.999 831	0.999 732	0.999 594	0.999 409	0.999 167	0.998 861	0.998 482
14	0.999 963	0.999 936	0.999 895	0.999 837	0.999 756	0.999 647	0.999 505
15	0.999 993	0.999 986	0.999 976	0.999 960	0.999 936	0.999 901	0.999 853
16	0.999 998	0.999 997	0.999 995	0.999 991	0.999 985	0.999 975	0.999 961
17	0.999 999	0.999 999	0.999 999	0.999 998	0.999 996	0.999 994	0.999 990
18	1.000 000	1.000 000	0.999 999	0.999 999	0.999 999	0.999 998	0.999 998
19			1.000 000	0.999 999	0.999 999	0.999 999	0.999 999
20				1.000 000	1.000 000	0.999 999	0.999 999

k \ n	29	30	31	32	33	34	35
1	0.000 000	0.000 000	0.000 000	0.000 000	0.000 000	0.000 000	0.000 000
2	0.001 521	0.001 160	0.000 884	0.000 674	0.000 513	0.000 390	0.000 297
3	0.048 563	0.042 153	0.036 570	0.031 710	0.027 482	0.023 808	0.020 615
4	0.208 772	0.192 036	0.176 546	0.162 222	0.148 989	0.136 773	0.125 505
5	0.427 822	0.405 929	0.384 946	0.364 860	0.345 656	0.327 315	0.309 815
6	0.628 305	0.607 054	0.584 454	0.566 263	0.546 505	0.527 197	0.508 355
7	0.777 865	0.760 926	0.743 954	0.726 991	0.710 076	0.693 241	0.676 518
8	0.877 140	0.864 996	0.852 579	0.839 930	0.827 085	0.814 080	0.800 946
9	0.936 988	0.929 112	0.920 879	0.912 317	0.903 453	0.894 313	0.884 922
10	0.970 069	0.965 419	0.960 438	0.955 137	0.949 530	0.943 629	0.937 451
11	0.986 859	0.984 356	0.981 599	0.978 588	0.975 325	0.971 814	0.968 060
12	0.994 681	0.993 451	0.992 054	0.990 483	0.988 735	0.986 806	0.984 695
13	0.998 020	0.997 469	0.996 821	0.996 069	0.995 206	0.994 228	0.993 128
14	0.999 325	0.999 100	0.998 825	0.998 494	0.998 102	0.997 644	0.997 113
15	0.999 790	0.999 706	0.999 600	0.999 466	0.999 302	0.999 104	0.998 868
16	0.999 940	0.999 912	0.999 875	0.999 825	0.999 762	0.999 683	0.999 586
17	0.999 984	0.999 976	0.999 964	0.999 947	0.999 925	0.999 896	0.999 859
18	0.999 996	0.999 994	0.999 990	0.999 985	0.999 978	0.999 968	0.999 955
19	0.999 999	0.999 998	0.999 997	0.999 996	0.999 994	0.999 991	0.999 987
20	0.999 999	0.999 999	0.999 999	0.999 999	0.999 998	0.999 997	0.999 996

(Continued)

Table 9.16b.   (Continued).

k \ n	36	37	38	39	40
1	0.000 000	0.000 000	0.000 000	0.000 000	0.000 000
2	0.000 226	0.000 171	0.000 130	0.000 099	0.000 075
3	0.017 844	0.015 440	0.013 354	0.011 546	0.009 980
4	0.115 119	0.105 553	0.096 746	0.088 644	0.081 194
5	0.293 133	0.277 243	0.262 120	0.247 737	0.234 068
6	0.489 989	0.472 106	0.454 713	0.437 810	0.421 399
7	0.659 934	0.643 511	0.627 272	0.611 234	0.595 412
8	0.787 713	0.774 409	0.761 059	0.747 686	0.734 312
9	0.875 305	0.865 485	0.855 485	0.845 325	0.835 027
10	0.931 011	0.924 322	0.917 402	0.910 264	0.902 925
11	0.964 067	0.959 843	0.955 395	0.950 731	0.945 858
12	0.982 400	0.979 921	0.977 260	0.974 418	0.971 396
13	0.991 904	0.990 551	0.989 067	0.987 450	0.985 698
14	0.996 507	0.995 820	0.995 049	0.994 189	0.993 239
15	0.998 589	0.998 265	0.997 891	0.997 464	0.996 981
16	0.999 467	0.999 325	0.999 156	0.998 958	0.998 729
17	0.999 812	0.999 754	0.999 683	0.999 598	0.999 496
18	0.999 938	0.999 916	0.999 888	0.999 854	0.999 812
19	0.999 981	0.999 973	0.999 963	0.999 950	0.999 934
20	0.999 994	0.999 992	0.999 988	0.999 984	0.999 978

Table 9.17. Critical values $\vartheta_\alpha(2n, 2m)$ of the ϑ distribution.

$\alpha = 0.99$

n \ m	1	2	3	4	5	6	7	8	9	10	15	20	∞
1	539.35	188.89	146.30	130.89	123.02	118.26	115.07	112.79	111.07	109.74	105.9	104.1	99
2	13.91	46.36	31.16	25.94	23.37	21.84	20.84	20.12	19.59	19.18	18.0	17.4	16
3	10.62	7.86	19.60	15.71	13.76	12.69	11.96	11.44	11.07	10.78	9.9	9.5	8.5
4	9.32	6.77	5.75	12.32	10.66	9.69	9.04	8.57	9.24	7.98	7.2	6.9	6.0
5	8.62	6.18	5.21	4.70	9.14	8.25	7.56	7.23	6.92	6.86	6.0	5.7	4.8
6	8.19	5.78	4.88	4.37	4.05	7.38	6.82	6.42	6.13	5.91	5.2	4.9	4.2
7	7.90	5.54	4.65	4.14	3.84	3.63	6.28	5.89	5.61	5.39	4.8	4.5	3.7
8	7.69	5.37	4.48	3.98	3.68	3.47	3.32	5.52	5.24	5.03	4.4	4.1	3.4
9	7.54	5.52	4.35	3.86	3.56	3.36	3.20	3.09	4.97	4.77	4.2	3.9	3.1
10	7.41	5.13	4.24	3.76	3.46	3.26	3.11	3.00	2.91	4.56	4.0	3.7	2.9
15	7.06	4.82	3.94	3.48	3.18	2.98	2.83	2.72	2.63	2.36	3.4	3.1	2.4
20	6.9	4.7	3.8	3.4	3.1	2.9	2.7	2.6	2.5	2.2	2.1	2.8	2.1
∞	6.39	4.28	3.44	2.98	2.69	2.49	2.34	2.222	2.1	1.84	1.69	1.0	1.0

(leftmost column values: 66.12, 32.76, 26.76, 24.37, 23.10, 22.31, 21.77, 21.39, 21.10, 20.87, 20.21, 19.9, 19.0)

$\alpha = 0.95$

n \ m	1	2	3	4	5	6	7	8	9	10	15	20	∞

References

Adcock, R.J. 1878. A problem in least squares. The Analyst (Annals of Mathematics) 5 (2), pp. 53–54.

Amoroso, L. 1925. Ricerche intorno alla curva dei redditi. Annali di Matematica Pura ed Applicata. 421, pp. 123–159.

Amstrong, J.S. 2001. Principles of forecasting. A handbook for researchers and practitioners. Kluwer Academic Publishers, Norwell, Mass.

Andersson, O. 2012. Experiment!: planning, implementing and interpreting. Wiley.

Андрейев, С.Е., Товаров, В.В., Петров, В.А. 1959. Закономерности измелчения и усчисление характеристик гранулометрического состава. Недра, Москва.

Antoniak, J., Brodziński, S., Czaplicki, J. 1976–1980. Reliability investigation of hoisting installations. Research work. Mining Mechanization Institute, Silesian University of Technology, Gliwice, (unpublished, in Polish).

Barbaro, R.W., Rosenshine, M. 1986. Evaluating the productivity of shovel-truck materials haulage system using a cyclic queuing model. SME Fall Meeting, St. Luis, Missouri, Sept. 7–10.

Barlow, R.E., Proschan, F. 1975. Statistical theory of reliability and life testing. Probability models. Holt, Rinehart & Winston. New York.

Benjamin, J.R., Cornell, C.A. 1970. Probability, statistics and decision for civil engineers. Mc Graw-Hill, Inc.

Bennet, B.M., Nakamura, E. 1963. Tables for testing significance in 2×3 contingency table. Technometrics 5, pp. 501–511.

Bhaudury, B., Basu, S.K. 2003. Terotechnology: Reliability engineering and maintenance management. Asian Books.

Bielińska, E. 2007. Forecasting of time sequences. Silesian University of Technology, Gliwice,

Birnbaum, Z.W., Hall, R.A. 1960. Small sample distributions for multi-sample statistics of the Smirnov type. Annals of Mathematical Statistics 31, pp. 710–720.

Болъшев, Л.Н., Смирнов, Н.В. 1965. Таблицы математической статистики. Наука, Москва.

Bortkiewicz, W. 1898. Das Gesetz der kleinen Zahlen. Leipzing, Germany: B.G. Teubner.

Bousfiha, A., Delaporte, B., Limnios, N. 1996. Evaluation numérique de la fiabilité des systèmes semi-markoviens. Journal Européen des Systémes Automatisés. 30, 4, (In French), pp. 557–571.

Bousfiha, A., Limnios, N. 1996. Ph-distribution method for reliability evaluation of semi-Markov systems. Proceedings of ESREL-97. Lisbon, June: 2149–2154. 1997.

Box, G.E.P., Jenkins, G.M. 1976. Time series analysis: forecasting and control. San Francisco Holden-Day, 2nd ed.

Box, G.E.P., Jenkins, G.M., Reinsel, G.C. 2008. Time series analysis: forecasting and control. Wiley.

Bowerman, B., O'Connell, R., Koehler, A. 2005. Forecasting, time series, and regression. Thomson Books Cole.

Bretscher, O. 1995. Linear algebra with applications. 3rd ed. Upper Saddle River NJ: Prentice Hall.

Breusch, T.S. 1979. Testing for Autocorrelation in Dynamic Linear Models. Australian Economic Papers, 17, pp. 334–355.

Броди, С.М., Погосян, И.А. 1973. Вложенные стохастические процессы в теории массого обслуживания. Кийев.

Broś, J., Czaplicki, J., Ścieszka, S. 1976. Investigations of abrasive-wear characteristics of braking materials in model conditions. Scientific Problems of Machines Operation and Maintenance, 4 (28), pp. 413–434 (In Polish).

Brown, O.F., Frimpong, S. 2012. Nonlinear finite element analysis of blade-formation interactions in excavation. Mining Engineering, SME, November, pp. 60–67.

Brown, R.G. 1959. Statistical forecasting for inventory control. Mc Graw-Hill Book, New York.

Butkiewicz, J., Hys. 1977. Multi-dimensional generalization of Weibull distribution and exponential Gumbel distribution. Ossolineum. (In Polish).

Butkiewicz, J. 1997. A review of multidimensional probability distributions. In: Statistical methods in quality control of production. Ossolineum. (In Polish).

Carbogno, A., Sajdok, H., Herok, R. 2001. Durability of elements of hoist guide system in shafts of Pniówek Colliery. IX International Conference TEMAG'2001, Gliwice-Ustroń 24–26 Oct., pp. 57–64 (In Polish).

Cardou, M., Clerici, C., Morandini, A., Ocella, E. 1993. An experimental research on the comminution law and work index in jaw. XVII IMPC, Sydney.

Carroll, R.J., Ruppert, D., Stefański, L.A., Creiniceanu, C. 2006. Measurement Error in Nonlinear Models: A Modern Perspective. CRC.

Ceylanoğlu, A., Görgülü, K. 1997. The performance measurement results of stone cutting machines and their relations with some material properties. Conference Mine Planning and Equipment Selection'97, Ostrava, 3–6 Oct., pp. 397.

Chatfield, C. 2001. Time-series forecasting. Chapman & Hall/CRC.

Church, H.K. 1981. Excavation Handbook. McGraw-Hill Book Co.

Ciechanowicz, K. 1972. Generalized gamma distribution and power distribution as models for component durability. Archiwum Elektrotechniki, XXI, 3, pp. 489–512. (In Polish).

Cieślak, M. (ed). 2001. Economic forecasting. Methods and applications. PWN, Warsaw. (In Polish).

Çinclar, E. 1969. On semi-Markov processes on arbitrary spaces. Proc. Cambridge Philos. Soc., 66, pp. 381–392.

Clements, M., Hendry, D. 1998. Forecasting economic time series. Cambridge Univ. Press.

Corder, G.W., Foreman, D.I. 2009. Nonparametric Statistics for Non-Statisticians: A Step-by-Step Approach. Wiley.

Counningham, L.A., Lwin, T., Singh, N. 1973. On prediction problems in reliability. Microelectronics & Reliability, 12.

Cramer, H. 1999. Mathematical methods of statistics. Princeton Univ. Press.

Czaplicki, J.M. 1974. On cyclic component in the process of changes of states. Scientific Problems of Machines Operation and Maintenance, 4 (20), pp. 545–555.

Czaplicki, J.M. 1975. Prediction of reliability and exploitation indices and characteristics for mine continuous transport systems based on theoretical investigations. Ph.D. dissertation, Mining Faculty of Silesian Technical University, Gliwice. (In Polish).

Czaplicki, J.M. 1975. Significant non-uniformity of the process of change of conditions of an element or system during working cycle. Scientific Problems of Machines Operation and Maintenance, 4 (24), pp. 525–531.

Czaplicki, J.M. 1976. Certain model of a process of failures and the problem of its prediction. Scientific Problems of Machines Operation and Maintenance, 4, pp. 481–491. (In Polish).

Czaplicki, J.M. 1977a. Complex indices prediction model of technical object. Scientific Problems of Machines Operation and Maintenance, 2, pp. 273–289.

Czaplicki, J.M. 1977b. Interval estimation and prediction of some reliability indices of renewable technical objects and their systems. Statistical Review, R.XXIV, 4, pp. 507–520. (In Polish).

Czaplicki, J.M. 1978. Estimation (with information a priori) of the density function of work time of object. Scientific Problems of Machines Operation and Maintenance, 3, 33, pp. 189–205. (In Polish).

Czaplicki, J.M. 1980. Some problems of prediction of time of durability investigation. (Part I). Scientific Problems of Machines Operation and Maintenance, 4, 44, pp. 511–521. (In Polish).

Czaplicki, J.M. 1981. On a certain two-dimensional population. IV Conference "Reliability and Quality Control". Jabłonna, pp. 7–8. (In Polish).

Czaplicki, J.M. 1981a. Some problems of prediction of time of durability investigation. (Part II). Scientific Problems of Machines Operation and Maintenance, 1, pp. 85–93. (In Polish).

Czaplicki, J.M., Lutyński, A. 1982. Vertical transport. Reliability problems. Silesian Univ. of Tech., 1052, Gliwice (In Polish).

Czaplicki, J.M., Lutyński, A. 1987. Horizontal transport. Reliability problems. Silesian Univ. of Tech., 1330, Gliwice (In Polish).

Czaplicki, J.M. 1999. Random component autocorrelation in the process of cumulating wire breaks in hoisting ropes. Conference: Mining 2000, Gliwice, pp. 71–77.

Czaplicki, J.M. 2000. Investigation of the random component autocorrelation in the process of cumulating wire breaks in hoisting ropes. Conference "Energy-saving and Reliable Mine Machinery", Mining of equilibration development. Silesian Univ. of Tech., 246, Gliwice, pp. 123–130.

Czaplicki, J.M. 2004. Elements of theory and practice of cyclic systems in mining and earthmoving engineering. Silesian Univ. of Tech., Gliwice (In Polish).

Czaplicki, J.M. 2006. Description of accumulation function for increasing number of wire cracks of hoist head ropes by means of modified autoregression function. X International Conference "Quality, Reliability and Safety of Ropes and Rope Transportation Devices". Cracow 25–26 Sept., Papers AGH, 37, pp. 69–84 (In Polish).

Czaplicki, J.M. 2009. Shovel-truck systems. Modelling, analysis and calculation. Francis & Taylor CRC/Balkema.

Czaplicki, J.M. 2009. Reliability of a system: loader—crusher—belt conveying for surface mining. Surface Mining, 4–5, pp. 35–39.

Czaplicki, J.M. 2010a. Mining equipment and systems. Theory and practice of exploitation and reliability. Francis & Taylor CRC/Balkema.

Czaplicki, J.M. 2010b. On a certain family of exploitation processes of series system in mining engineering. Mining Review, 11–12, pp. 26–30. (In Polish).

Czaplicki, J.M. 2012. Reliability in mine mechanization and earthmoving engineering. Silesian University of Technology, Gliwice. (In Polish).

Czaplicki, J.M., Kulczycka A.M. 2011. Steady-state availability of a multi-element symmetric pair. Scientific Problems of Machines Operation and Maintenance, 3 (167), vol. 46, pp. 15–26.

Czaplicki, J.M., Ścieszka, S. 1974. On certain conditional method of analytical determination of operational characteristics of equipment. 3, (19), pp. 407–418. (In Polish).

Czekała, M. 2004. Groups of untypical observations—a multidimensional test. In "Progress in Econometrics". Barczak ed. Katowice, pp. 15–21. (In Polish).

Deming, W.E. 1943. Statistical adjustment of data. Wiley, NY (Dover Publications edition, 1985).

Dhillon, B.S. 2008. Mining equipment reliability, maintainability and safety. Springer.

Dovitch, R.A. 1990. Relaibility statistics. ASQ Quality Press, Milwaukee, Wisconsin.

Downarowicz, O. 1997. Operational system. Management of technical store. Exploitation Institute, Gdańsk-Radom. (In Polish).

Draper, N.R., Smith, H. 1998. Applied regression analysis. Wiley.

Durbin, J. 1953. A note on regression when there is extraneous information about one of the coefficients. Journal of the American Statistical Association, 48, December, pp. 799–808.

Durbin, J., Watson, G.S. 1950. Testing for serial correlation in least squares regression. Biometrica, 37, December, pp. 409–428.

Durbin, J., Watson, G.S. 1950. Testing for serial correlation in least squares regression II. Biometrica, 38, June, pp. 159–178.

Dworczyńska, M., Gładysiewicz, L., Kawalec, W.A model of stream of rock excavated for main belt conveyors design. The School of Surface Mining, 27–28 March 2012, Cracow. (In Polish).

Dziembała, L. 1967. Analysis of cyclic fluctuates of number of fires in month. Statistical Review, R. XIV, 2, pp. 203–208 (In Polish).

Dziembała, L. 1972. On some parameters' estimators of stochastic structure in conditional (single) least square method. Statistical Review, R. XIX, 4, (In Polish).

Dziembała, L., Zadora, K. 1971. Application of spectral analysis for investigation of cyclic fluctuations. Statistical Review, R. XVIII, 3–4, (In Polish).

Epstein, B. 1948. Logaritmico-normal distributions in breakage of solids. Ind. Eng. Chem., 40, pp. 2289–2291.

Everitt, B.S. 2002. Cambridge dictionary of statistics. CUP.

Feller, W. 1957. An introduction to probability theory and its applications. J. Wiley and Sons, Inc.

Firkowicz, S. 1969. On the power distribution. Archiwum Elektrotechniki, XVIII, 1, pp. 29–40.

Firkowicz, S. 1970. Statistical testing of products. WNT, Warsaw. (In Polish).

Fisher, R.A. 1912. On an absolute criterion for fitting frequency curves. Mess. of Math. 41. P. 155.

Fisher, R.A.: Test of significance in harmonic analysis. Proc. Roy. Soc. A 125, 1929, pp. 54–59.

Fréchet, M. 1927. Sur la loi de probabilité de l'écart maximum. Annales de la Société Polonaise de Mathematique, Cracovie, 6, pp. 93–116.

Fuller, W.A. 2006. Measurement error models. Wiley.

Gelper, S., Fried, R., Croux, C. 2008. Robust forecasting with exponential and Holt-Winters smoothing. (http://www.econ.kuleuven.be/public/ndbae06/pdf-files/robust_holt_winters.pdf).

Герцбах, И.Б., Кордонский, Х.Б., 1966. Модели отказов. Изд. Советское Радио. МоскВа.

Gauss, C.F. 1880. Werke. Vol. 4, Göttingen. (Reprint).

Gawenda, T., 2004. An assessment of the influence of the physic-chemical properties of rock materials and technological parameters of jaw crushers on effects of crushing. Ph.D. dissertation, Mining and Geoengineering Faculty, AGH, Cracow (In Polish).

Gibbons, J.D., Chakraborti S., 2003. Nonparametric Statistical Inference. 4th Ed. CRC.

Гихман, И.И. and Скороход, А.В. 1965. Введение в теорию случайных просессов. Изд. Наука, Москва.

Gillard, J.W. 2006. An historical overview of linear regression with errors in both variables. Cardiff University.

Giorski, A.C. 1968. Beware of the Weibull euphoria. IEEE Transactions on Reliability, R-17, pp. 202–203.

Гнеденко, Б.В. 1941. Предельные теоремы дла максимального члена вариационного ряда. ДАН СССР, т. 32.

Gnyedenko, B.V., Belayev, Yu.K., and Solovyev, A.D. 1969. Mathematical methods of reliability theory. Academic Press. New York.

Гнеденко, Б.В., and Коваленко, И.Н. 1966. Введение в теорию массового обслуживания. Изд. Наука. МоскBа.

Гнеденко, Б.В., 1969. Резервирование с восставлением и суммирование случайново числа слагаемых. Colloquium on Reliability Theory. Supplement to preprint volume. pp. 1–9.

Godfrey, L.G. 1978. Testing Against General Autoregressive and Moving Average Error Models when the Regressors Include Lagged Dependent Variables, Econometrica, 46, pp. 1293–1302.

Godfrey, L.G.: Specification test in econometrics. Cambridge 1988.

Goldberger, A.S. 1966. Econometric Theory. J. Wiley & Sons.

Gopal, K. Kanji. 2006. 100 statistical tests. London: SAGE Publ. Ltd.

Grabski, F., Jaźwiński, J. 2001. Bayesian methods in reliability and diagnostics. WKŁ, Warsaw (In Polish).

Granger, C.W., Hatanaka, H. 1969. Analyse spectrale des séries temporelles en économie. Dunod, Paris.

Greń, J. 1978. Methodological aspects of economical prediction. Statistical Survey, 1. (In Polish).

Griffin, W. 1989. Perturbation theory applied to truck-shovel system simulation. Proceedings of the International Symposium on Off-Highway Haulage in Surface Mines, Edmonton 15–17 May, pp. 83–87.

Grubbs, F.E. 1969. Procedures for detecting outlying observations in samples. Technometrics 11, pp. 1–21.

Gumbel, E.J. Statistical theory of extremes and some practical applications. National Bureau of Standards, Applied Mathematics Series, 33.

Харламов Б. 2008. Репрерывныйе полумарковские просессы. Петербург.

Hartley, H.O. 1950. The maximum *F*-ratio as a short-cut test of heterogeneity of variance. Biometrika, 37, pp. 308–312.

Hellwig, Z. 1963. Statistical prognoses and their significance in forecasting of future courses of economic phenomena and processes. Papers of High School of Economics, 16, Wrocław. (In Polish).

Hogg, R.V., Craig, A.T. 1995. Introduction to mathematical statistics. 5th ed. Macmillan.

Holt, C.C. 1957. Forecasting seasonals and trends by exponentially weighted moving averages. ONR Research Memorandum 52.

http://en.wikipedia.org/wiki/Poisson_distribution#cite_note-4

Huber, P.J. 1981. Robust statistics. Wiley.

Hufford, G., Griffin, W.H., Sturgul, J.R. 1981. An application of queuing theory to equipment selection for a large open pit copper mine. SME-AIME Meeting, Denver, Colorado, 18–20 November.

Hustrulid, W, Kuchta, M. 2006. Open pit mine. Planning and design. Vol. 1. Fundamentals. Taylor & Francis. London.

Jóźwiak, J., Podgórski, J. 1997. Foundations of statistics. 5th ed. PWE. Warsaw (In Polish).

Jurdziak, L.: Application of extreme value theory for dimensioning of belt conveyors cooperating with BWE. Industrial Transport, 3, (25), 2006, pp. 38–43. (In Polish).

Kasiewicz, S. 2005. Constructing the value of a company in operational management. SGH, Warsaw.

Kaźmierczak, J. 2000. Exploitation of technical systems. Silesian Univ. of Tech., Gliwice (In Polish).

Kendall M.G., Babington Smith. 1938. Randomness and random sampling numbers. Journal of the Royal Statistical Society, 101, 1, pp. 147–166.

Kokosz, Z. 2010. Advantages of isolinear map construction by kriging. Surface Mining, 2, pp. 91–95 (In Polish).

Kolmogorov A.N. 1933. Grundbegriffe der Wahrscheinlichkeitsrechnung. Berlin: Julius Springer (In German).

Translation: Kolmogorov, A.N. 1956. Foundations of the theory of probability. New York, Chelsea.

Лщлмогоров, А.Н. 1941. Щ логарифмическг нормлном законе разпределеня размеров частиц при дроблений. АН СССР, 31, 99–101.

Koopmans, T.C. 1937. Linear regression analysis of economic time series. DeErven F. Bohn, Haarlem, Netherlands.

Kopociński, B. 1973. An outline of renewal and reliability theory. PWN, Warsaw. (In Polish).

Кщролюук В.С., Турбин А.Ф. 1976. Полумарковские процессы и их приложения. Наук. Думка, Киев.

Kowalczyk, J. 1957. The fatigue index for hoist head ropes. Central Mining Institute Bulletin 187. (In Polish).

Коваленко И.Н., Кузнецов Н.Ю., Шуренков В.М. 1983. Случайнуе процессы. Справочник. Наукова Думка.

Koyck, L. 1954. Distributed lags and investment analysis. North-Holland, Amsterdam.

Koźniewska, I., and Włodarczyk, M. 1978. Models of renewal, reliability and mass servicing. PWN, Warsaw. (In Polish).

Kummell, C.H. 1879. Reduction of observation equations which contain more than one observed quantity. The Analyst (Annals of Mathematics) 6 (4), pp. 97–105.

Lehmann, E.L., D'Abrera, H.J.M. 2006. Nonparametrics: statistical methods based on ranks. Springer.

Lévy, P. 1954. Processus semi-markoviens. Pro. Int. Cong. Math. Amsterdam, pp. 416–426.

Limnios, N., Oprişan, G. 2001. Semi-Markov processes and reliability. Statistics for Industry and Technology. Birkhäuser.

Lin Zaikang, Yang Jian & Nie Zhenggang. 1997.: Design capacity of belt and bunkers by computer simulators in Barakupuria Mine of Bangladesh. Conference Mine Planning and Equipment Selection'97, Ostrava, 3–6 Oct., p. 259.

Lwin, T., Singh, N. 197

Makridakis, S., Wheelwright, S., Hyndman, R. 1998. Forecasting, methods and applications. Wiley.

Maliński, M. 2004. Computer aided verification of statistical hypotheses. Silesian Univ. of Tech., Gliwice (In Polish).

Manowska, A. 2010. Forecasting of hard coal sale for a group of mines. Ph. D. dissertation. Mining Faculty of Silesian University of Technology, Gliwice, Poland. (In Polish).

Martin, J. 1972. Foundations of mathematics and statistics for biologists, physicians and chemists. PZWL, Warsaw. (In Polish).

Melchers, R.E. 1999. Structural reliability analysis and prediction. J. Wiley & Sons.

Meyer and Seaman, 2006. Expanded tables of critical values for the Kruskal-Wallis H statistic. Paper presented at the annual meeting of the American Educational Research Association, San Francisco.

Mianowski, A. 1988. The characterization of size distribution as a reciprocal hyperbolical function. Int. Journal of Mining Proc. 24, pp. 11–26.

Migdalski, J. (ed.) 1982. Reliability manual. Mathematical foundations. WEMA, Warsaw (In Polish).

Migdalski, J. (ed.) 1992. Reliability engineering. Manual. ATR Bydgoszcz, ZETOM Warsaw (In Polish).

Moore, D.S. McCabe, G.P. 1999. Introduction to the Practice of Statistics. New York: W.H. Freeman.

Morgenstern, D. 1956. Einfache Beispiele zweidimensionaler Verteilungen. Mittel-Amtsblatt für Math., Stat. 8, pp. 234–235.

Nagelkerke, N. 1991. A note on a general definition of the coefficient of determination. Biometrica, 78, 3, pp. 691–692.

Nerlove, M. 1958. Distributed lags and demand analysis for agricultural and other commodities. U.S. Department of Agriculture, Washington.

Neyman, J. 1937. Outline of a theory of statistical estimation based on the classical theory of probability. Philosophical Transactions of the royal Society of London A, 236, pp. 333–380.

Nipl, R. 1979. On application of generalized gamma distribution to approximation of grain composition curves. Proc. of XII Cracow Conference on Ore Dressing. Cracow.

O'Connor, P.D.T. 2005. Practical reliability engineering. John Wiley & Sons.

Papoulis, A. 1984. Probability, random variables and stochastic processes. McGraw-Hill.

Pawłowski, Z. 1973. Econometric prognoses. PWN, Warsaw (In Polish).

Pearson, K. 1894. Contributions to the mathematical theory of evolution. Philosophical Transactions of the Royal Society A: Mathematical, Physical and Engineering Sciences, 185, p. 71.

Pearson, K. 1895. Contributions to the Mathematical Theory of Evolution. II. Skew Variation in Homogeneous Material. Philosophical Transactions of the Royal Society A: Mathematical, Physical and Engineering Sciences, 186, pp. 343–326.

Pearson, K. 1898. Contribution etc. IV: On the probable errors of frequency constants and on the influence of random selection on variation and correlation. Philosophical Transactions of the Royal Society A: Mathematical, Physical and Engineering Sciences, 191, p. 229.

Pfeiffer, P.E. 1978. Concepts of probability theory. Courier Dover Publ. pp. 47–48.

PN-82/N-04001. Exploitation of Technical Objects. General Terminology. (In Polish).

Poisson, S.D. 1837. Probabilité des jugements en matière criminelle et en matière civile, précédées des règles générales du calcul des probabilitiés. Paris, France: Bachelier.

Pyke, R. 1961a. Markov renewal processes: definitions and preliminary properties. Ann. of Math. Statist. 32, pp. 1231–1242.

Pyke, R. 1961b. Markov renewal processes with finitely many states. Ann. of Math. Statist. 32, pp. 1243–1259.

Pyke, R., Schaufele, R., 1964. Limit theorems fir Markov renewal processes. Ann. of Math. Statist. 35, pp. 1746–1764.

Pyra, J., Sołtys, A., Winzer, J. 2009. Some remarks on estimation of reaction of shock wave in blasting. Mining Review, 11–12, pp. 94–102 (In Polish).

Ravenis, J.V. II. 1969. A potentially universal probability density function for scientists and engineers. Proc. Intern. Conf. on Quality Control. Tokyo, Sept. pp. 523–526.

Rojo, J. (ed.) 2012. Selected works of E.L. Lehmann. Springer.

Rosin, P., Rammler, E. 1933. The Laws Governing the Fineness of Powdered Coal, Journal of the Institute of Fuel, 7, pp. 29–36.

Ross, S.M. 1995. Stochastic processes. J. Wiley & Sons.

Rousseeuw, P.J., Leroy, A.M. 1987. Robust regression and outliers detection. Wiley.

Rybicki, W. 1976. An optimal forecasting, predictors F-points and statistical games. Statistical Review, 4. (In Polish).

Rumsey, D. 2006. Probability of dummies. Wiley, p. 58.

Sadowski, W. 1977. Decisions and prognoses. PWE, Warsaw. (In Polish).

Sajkiewicz, J.: Foundations of theory systems with continuous technological structures (applied in mining). Ossolineum 1982 (In Polish).

Saporito, V., Self, F.M. 2012. Taking the mining industry to a higher level of excellence. Mining Engineering, SME, October, pp. 20–24.

Saramak, D., Tumidajski, T. 2006. The role and meaning of the approximation particle size distribution curves for mineral resources. AGH Mining and Geoengineering, 3/1, pp. 301–314. (In Polish).

Senatorski, J., Pasteruk, T. 2001. Technological methods to increase durability of tribological nodes of vehicles and mine machinery. IX International Conference TEMAG'2001, Gliwice-Ustroń 24–26 Oct., pp. 239–246 (In Polish).

Serfling, R.J. 1980. Approximation theorems of mathematical statistics. J. Wiley & Sons.

Siegel, S., Castellan, N.J. Jr. 1988. Nonparametric statistics fort the behavioural sciences. Mc Graw-Hill. Inc.

Sivazlian, B.D., Wang, K.H. 1989. System characteristics and economic analysis of the G/G/R machine repair problem with warm standbys using diffusion approximation. Microelectronics & Reliability, 29, 5. pp. 829–848.

Smirnow, S. 1979. Nutzanwendungen für erweiterte Korngroessenverteilungs-funktionen. Aufber. Technik, 6.

Smith, D.J. 2007. Reliability, maintainability and risk. Practical methods for engineers including reliability centered maintenance and safety-related systems. Elsevier.

Smith, W.L. 1955. Regenerative stochastic processes. Proc. Roy. Soc. London. Ser. A, 232, pp. 6–31.

Sobczyk, M. 1996. Statistics. PWN, Warsaw (In Polish).

Sokoła-Szewioła, V. 2011. Research studies and modelling of the changing in time vertical displacements of the mining area in time at the period of tremor occurrence inducted by longwall mining. D.Sc. Dissertation. Silesian Univ. of Technology, Gliwice, Poland. (In Polish).

Spurrier, J.D. 2003. On the null distribution of the Kruskal-Wallis statistic. Journal of Nonparametric Statistics 15, (6), pp. 685–691.

Stacy, E.W. 1962. A generalization of the gamma distribution. Annals of Mathematical Statistics, 28, pp. 1187–1192.

Steele, C. 2008. Use of the lognormal distribution for the coefficients of friction and wear. Reliability Engineering & System Safety, **93** (10), pp. 1574–2013.

Szepke, R. 1967. Applied radiometry. WNT, Warsaw. (In Polish).

Ścieszka, S. 1971. The hoist brakes wear process investigation. Silesian Univ. of Tech. Research work, unpublished, Gliwice. (In Polish).

Ścieszka, S. 1972. Problems of dry friction in brakes in mine hoist. Ph.D. dissertation, Mining Faculty, Silesian Univ. of Tech., Gliwice. (In Polish).

Takács, L. 1954. Some investigations concerning recurrent stochastic processes of a certain type. Magyar Tud. Akad. Mat. Kutato Int. Kzl., 3, pp. 115–128.

Takács, L. 1955. On a sojourn time problem in the theory of stochastic processes. Trans. Amer. Math. Soc. 93, pp. 631–540.

Timbergen, J. 1951. Business cycles in the United Kingdom 1870–1914. Amsterdam.

Tumidajski, T. 1992. Selected problems of stochastic analysis of properties of granular materials and ore dressing processes. Papers of AGH, Mining 159, Cracow. (In Polish).

Tumidajski, T. 1993. Application of statistical methods in analysis of ore dressing processes. Śląskie Wyd. Techniczne, Katowice. (In Polish).

Tumidajski, T., Saramak, D. 2009. Methods and models of mathematical statistics in ore dressing. AGH, Cracow (In Polish).

Walpole, R.E., Myers, R.H. 1989. Probability and statistics for engineers and scientists. 4th ed. Macmillan, New York.

Warszyński, M. 1988. Reliability in construction computations. PWN, Warsaw. (In Polish).

Wasserman, L. 2007. All of nonparametric statistics. Springer.

Weibull, W. 1939. A statistical theory of strength of materials. Proc. Roy. Swed, Inst. Eng. V. 15.

Weibull, W. 1956. Scatter of fatigue life and fatigue strength in aircraft structural materials and parts. Fatigue in Aircraft Structures, 4.

Weisstein, Eric W. Pearson Mode Skewness. MathWorld.

Winters, P.R. 1960. Forecasting sales by exponentially weighted mooring averages. Management Science 6, 1960.

Wold, H. (ed.) 1964. Prediction by chain principle: essays on the casual chain approach, Amsterdam.

Wu C.F.J., Hamada M.S. 2009. Experiments: planning, analysis, and optimization. Wiley.

Yan-Gang Zhao, Zhao-Hui Lu. 2007. Fourth-moment standardization for structural reliability assessment. Journal of Structural Engineering, July, pp. 916–924.

Zadora, K. 1974. Modelling and prediction of stochastic fields. Statistical Review, XXI, 3, pp. 435–449. (In Polish).

Zeliaś, A. 1996. Methods of outliers tracing in economic research. Statistical News, 8. (In Polish).

Zieliński, R. 1972. Statistical tables. PWN, Warsaw. (In Polish).

Subject index